Eastern and Southern Afric

PEARSON
Education

We work with leading authors to develop the
strongest educational materials in geography,
bringing cutting-edge thinking and best
learning practice to a global market.

Under a range of well-known imprints, including
Prentice Hall, we craft high quality print and
electronic publications which help readers to understand
and apply their content, whether studying or at work.

To find out more about the complete range of our
publishing, please visit us on the World Wide Web at:
www.pearsoned.com

DEVELOPING AREAS RESEARCH GROUP
THE ROYAL GEOGRAPHICAL SOCIETY
(WITH THE INSTITUTE OF BRITISH GEOGRAPHERS)

DARG Regional Development Series No. 4
Series Editor: David Simon

Eastern and Southern Africa
Development challenges in a volatile region

edited by Deborah Potts and Tanya Bowyer-Bower

Harlow, England • London • New York • Boston • San Francisco • Toronto
Sydney • Tokyo • Singapore • Hong Kong • Seoul • Taipei • New Delhi
Cape Town • Madrid • Mexico City • Amsterdam • Munich • Paris • Milan

Pearson Education Limited
Edinburgh Gate
Harlow
Essex CM20 2JE
England

and Associated Companies throughout the world

Visit us on the World Wide Web at:
www.pearsoned.com

First published 2004

© Pearson Education Limited 2004

ISBN 0130 26468 7

British Library Cataloguing-in-Publication Data
A catalogue record for this book is available from the British Library

10 9 8 7 6 5 4 3 2 1
08   07   06   05   04

Typeset in 10/11pt Palatino by 35
Printed in Malaysia, CLP
The publisher's policy is to use paper manufactured from sustainable forests.

Contents

Figures

Tables

Boxes

Contributors

Tanya Bowyer-Bower is a Lecturer in Geography at King's College London. Her principal research interests are in physical processes in semi-arid environments and environment and development issues in dryland Africa. She has conducted research in Alberta, Canada and, within Africa, Kenya, Swaziland, South Africa, Nigeria, Zambia and Zimbabwe. Major projects have included ODA- and DFID-funded research into the policy implications of urban agriculture in Harare and Lusaka; work for the NRI Food for Africa project on urban agriculture; and an eight-country field study for a UNDCP project on drugs in Africa for which she was the co-ordinator. From 1997 to 1998 she was Chairman of the Centre of African Studies of the University of London. Recent publications include *Land Reform in Zimbabwe: Constraints and prospects* (Ashgate 2000) (edited with Colin Stoneman). An increasing future focus for her research is the political ecology and political economy of development activities in dryland Africa.

John Briggs is Professor of Geography at the University of Glasgow. He has also taught and researched at the Universities of Dar es Salaam (Tanzania), Khartoum (Sudan), South Valley, Aswan (Egypt) and North-West (South Africa), as well as the former University of Botswana, Lesotho and Swaziland. His current research interests are in local (indigenous) environmental knowledges in African rural communities, and the ways in which such knowledges might contribute to rural development strategies. He also has research interests in the peri-urban zones of African cities. Recent publications on such work include articles in *Applied Geography* (1999), *Scottish Geographical Journal* (1999), *Urban Studies* (2000) and *Area* (2001).

Richard Gibb is Reader in Human Geography at the University of Plymouth. His research interests focus on regional economic and political integration, with a particular focus on southern Africa and the European Union. Recent publications include articles in *Third World Quarterly*, *Journal of Modern African Studies*, *Journal of Southern African Studies* and *Area*. His latest research examines southern Africa's trading relationships with the EU under the Cotonou Agreement and the South Africa–EU Trade, Development and Co-operation Agreement. He is currently involved in a detailed project examining the impact of the EU's sugar regime on the countries of southern Africa.

Katherine Homewood is Professor of Anthropology at University College London. She has also taught and researched at the University of Dar es Salaam. From 1994 to 1997 she was Principal Investigator for research funded by the Department for International Development on conservation with development in East African rangelands, focusing on community use of natural resources around Mkomazi Game Reserve, Tanzania, and supervising PhD theses on reserve-adjacent pastoralist and farming communities. She also coordinated a European Union-funded international collaborative research programme on land-use, household viability and migration in the West African Sahel. Her recent research for the EU and DFID investigates the long-term outcomes for habitat, wildlife and people of different land-use policies in the savanna buffer zones around the Serengeti–Mara ecosystem in East Africa, with a particular focus on factors driving conversion of rangeland to cultivation. Her current work is developing a synthesis of changing land-use and livelihoods in Maasailand, including the poverty implications of development interventions on pastoralist production systems. Current publications include K. Homewood *et al.* (2001) Long-term changes in Serengeti-Mara wildebeest and land cover: pastoralism, population or policies? *Proc. Nat Acad. Sci* **98** (22): 12544–12549 and an edited volume, *Rural Resources and Local Livelihoods in Africa* (James Currey, in press, 2003).

Akim J. Mturi is Associate Professor of Demography, School of Development Studies, University of Natal in Durban, South Africa. Before moving to Durban, Akim was a Senior Lecturer in demography at the National University of Lesotho. He has been involved in various research projects in both East and southern Africa. His research interests include fertility transition in sub-Saharan Africa, sexual and reproductive health among young people, HIV/AIDS and its impact on demographic outcomes and child labour. He is currently a team leader of a UNFPA-funded research project entitled 'Understanding the changing family composition and structure in South Africa in the era of HIV/AIDS pandemic'. He has published a number of articles in scholarly journals and chapters in books and monographs. His publications include works in the *Journal of Biosocial Science, Health Policy and Planning, Population Studies, Journal of Southern African Studies, African Population Studies, African Journal of Reproductive Health* and *United Nations Population Bulletin.*

Anthony O'Connor is Reader in Geography at University College London, having taught earlier in his career at four African universities, including Makerere (Uganda) and Dar es Salaam (Tanzania). His interests remain focused on eastern Africa. He is currently working on replacements for his now out-of-print books on *The African City* (1983) and *Poverty in Africa* (1991), in each case retaining tropical Africa as its spatial scope. Much of his teaching is on the poorest one-third of the world's people, essentially in tropical Africa and South Asia, complementing most colleagues' teaching on the richest one-third. Recent publications include, Information needs for urban policy making in Africa, in I. Livingstone and D. Belshaw (eds) *Renewing Development in Sub-Saharan Africa* (Routledge 2001) and Lagos, in L. Beckel (ed.) *Megacities* (Geospace Verlag 2001).

Deborah Potts is a human geographer specialising on southern Africa and is Senior Lecturer in Geography at King's College London, having previously taught and researched in the Geography Department of the School of Oriental and African Studies for 22 years. Her research focuses on urbanisation and migration in sub-Saharan Africa, particularly southern Africa. She also works on land and environmental issues in the region in the context of political ecology. She is a leading specialist on Zimbabwe. She has published work on these themes in journals such as *Third World Planning Review, Development and Change, Geographical Journal, Environment and Urbanization* and *Review of African Political Economy*. Recent publications include co-editing three special issues of the *Journal of Southern African Studies* on *African Environments: Past and Present, Fertility in Southern Africa* and *Malawi*. She is currently co-editing a book entitled *African Urban Economies: Viability, vitality or vitiation of major cities in East and southern Africa?* with Deborah Bryceson and furthering her research on urban growth patterns and trends in circular migration in sub-Saharan Africa.

Marcus Power is a Lecturer in Human Geography at the School of Geographical Sciences, University of Bristol. His research interests include the study of space, 'development' and power in Africa and Latin America; the uses of media communications technologies in African societies; the politics of class, cultural and gender identity formation in the 'Global South' and the study of geopolitics and 'post-colonialism' in Portugal and the lusophone world. He has also recently written a teaching text for undergraduate students interested in development entitled *Rethinking Development Geographies*.

Colin Stoneman was trained originally as a scientist and holds a doctorate in physical chemistry as well as professional qualifications in economic and social statistics. He has worked in the Universities of Zimbabwe, Hull and York, where he was Lecturer in the Centre for Southern African Studies from 1986 to 1996. He is currently Visiting Research Fellow in the Centre for Development Studies in the University of Leeds. He has acted as consultant to 15 development NGOs or governments and was the author or co-author of five major reports; he has also written or edited four books and over 70 academic papers, mainly on southern Africa. He was the author of the Economist Intelligence Unit's Country Reports on Zimbabwe for 12 years. He is editorial co-ordinator of the *Journal of Southern African Studies* and has recently been appointed as editor-in-chief of the Agritrade website for the Technical Centre for Agricultural and Rural Co-operation (CTA) in Wageningen, the Netherlands.

Sian Sullivan is Research Fellow at the Centre for the Study of Globalisation and Regionalisation, University of Warwick. She came to Warwick from

King's College London where she was Lecturer in Environment and Development. Previously she held a British Academy Postdoctoral Research Fellowship in the Anthropology and Geography Departments of the School of Oriental and African Studies. Her work has focused on conflicts of interest in environmental policy, exploring disjunctions between conservation and degradation discourses held at local, national and transnational levels. She has extensive experience of these issues from fieldwork conducted over an eight-year period in Namibia. Publications arising from this work include *Political Ecology: Science, myth and power* (edited with Philip Stott) (Edward Arnold 2000) and articles in several anthropology and ecology journals including *Anthropos, Journal of Biogeography, Global Ecology and Biogeography, Africa é Mediterraneo Societa é Cultura*, and chapters in a number of edited volumes. Influenced by local protest against donor and NGO facilitation of conservation initiatives in north-west Namibia, and by recent anti-capitalism protests in 'the north', Sian is now working with theoretical, ethnographic and experiential material to elucidate areas of significance within an emerging and radical 'global anti-capitalism'. Her particular interests are the social and structural significance of non-hierarchical, open networks of organisation; and interpreting pro-justice and -autonomy politics as a politics of experience and subjectivity that challenges the 'normal' and economistic rationality assumed under modernity.

Richard Taylor is a Lecturer in the Department of Geography at University College London, and Adjunct Lecturer in the Department of Geology at Makerere University, Uganda. His research interests revolve around the development and use of groundwater for potable water supplies. A key focus of his research, funded by IDRC (Canada), DFID (UK) and a NSERC (Canada) Postdoctoral Fellowship, is deeply weathered crystalline rock aquifers that underlie much of sub-Saharan Africa. He has worked as a consultant to DANIDA and World Bank-funded water supply projects as well as a variety of international and indigenous non-governmental organisations. He has, since 1992, worked closely with staff from the Directorate of Water Development (Uganda) and Makerere University. His publications included works in scholarly periodicals (e.g. *Journal of Hydrology, Water Research*) and popular journals such as *Waterlines*. Recently, he has contributed book chapters to monographs for the World Health Organisation (*Protecting Groundwater for Health*) and UNESCO (*Urban Groundwater Pollution*). In 2003, he will lead an international, multidisciplinary study of the impact of climate change on the water resources and aquatic ecosystems of the snowcapped Rwenzori Mountains.

Series Preface

In the late 1980s, the Developing Areas Research Group (DARG) of the then Institute of British Geographers produced a series of three edited student texts under the general editorship of Prof. Denis Dwyer (University of Keele). These volumes, focusing on Latin America, Asia and Tropical Africa respectively, were published by Longman and achieved wide circulation, thereby also contributing to the financial security of DARG and enabling it to expand its range of activities. These are now out of print. The Latin American volume was revised and published in a second edition in 1996.

However, there have been dramatic changes in the global political economy, in the nature of development challenges facing individual developing countries and regions, and in debates on development theory over the last decade or so. In a review of the situation in 1997/8, the DARG Committee and Matthew Smith, the Senior Acquisitions Editor at Addison Wesley Longman (now Pearson Education) therefore felt that mere updating of the existing texts would not do these circumstances justice or catch the imagination of a new generation of students. Accordingly, we have launched an entirely new series.

This is both different in conception and larger, enabling us to address smaller, more coherent continental or subcontinental regions in greater depth. The organising principles of the current series are that the volumes should be thematic and issue-based rather than having a traditional sectoral focus, and that each volume should integrate perspectives on development theory and practice. The objective is to ensure topicality and clear coherence of the series, while permitting sufficient flexibility for the editors and contributors to each volume to highlight regional specificities and their own interests. Another important innovation is that the series was launched in January 1999 by a book devoted entirely to provocative contemporary analyses of *Development as Theory and Practice; current perspectives on development and development co-operation*. Edited by David Simon and Anders Närman, this provides a unifying foundation for the regionally focused texts and is designed for use in conjunction with one or more of the regional volumes.

The complete series is expected to include titles on Central America and the Caribbean, Southern and East Africa, West Africa, South Asia, Pacific

Asia, the transitional economies of central Asia, and Latin America. While the editors and many contributors are DARG members, other expertise – not least from within the respective regions – is being specifically included to provide more diverse perspectives and representativeness. Once again, DARG is benefiting substantially from the royalties. In addition, a generous number of copies of each volume will be supplied to impoverished higher education institutions in developing countries in exchange for their departmental publications, thereby contributing in a small way to overcoming one pernicious effect of the debt crisis, namely the dearth of new imported literature available to staff and students in those countries.

David Simon
Royal Holloway, University of London

Series Editor
(Chair of DARG 1996–8)

Acknowledgements

We are grateful to everyone who has helped to bring this volume to fruition. David Simon, the series editor, has been exceptionally supportive at all stages and we cannot thank him enough. We would also like to thank all the authors who have generally been both good humoured and patient about queries and delays, and have produced such excellent contributions. Roma Beaumont, the cartographer at the Geography department, King's College London, helped us with many of the diagrams. We have been through several changes of personnel at Pearson; in the final stages of the book Tim Parker has been particularly supportive.

Publisher's Acknowledgements

We are grateful to the following for permission to reproduce copyright material:

Tables 1.1, 2.2, 4.1, 4.3 and 4.4 include information from *Human Development Report 2002* by United Nations Development Programme, © 2002 by the United Nations Development Programme. Used by permission of Oxford University Press, Inc.; Table 5.1 adapted from 'Conservation with a human face: conflict and reconciliation in African land use planning', by R. Bell from *Conservation in Africa: People, Policies and Practice*, published and reprinted by permission of Cambridge University Press (Anderson and Grove (eds), 1987); Table 5.2 partially adapted from *An African Savannah: Synthesis of the Nylsvley Study* published and reprinted by permission of Cambridge University Press (Barrow, 1991); Figure 6.1 from *Household resilience, food security and recurrent exogenous shocks: a study from the semi-arid communal areas of Zimbabwe* (Alderson, 1999). Unpublished PhD thesis, Durham University; Tables 6.1, 6.2, 6.3, 6.4, 6.5 and 6.6 from *FAOSTAT Agricultural Data* at http://apps.fao.org (1999), reprinted by permission of FAO; Table 7.2 from *Discharge of selected rivers of Africa* published and reprinted by permission of UNESCO (1995); Table 7.3 from *Africa's floodplains: a hydrological overview* by J. Thompson in *Water Management and Wetlands in sub-Saharan Africa* (Acreman and Hollis (eds) 1996), reprinted by permission of IUCN, Gland; Table 7.6 from www.who.int/water_sanitation_health/Globassessment/GlobalTOC.htm, reprinted by permission of WHO (2000); Figure 5.1 from *The Vegetation of Africa. A descriptive memoir to accompany the UNESCO/AETFAT/UNSO vegetation map of Africa* (White 1983), reprinted by permission of UNESCO, Paris; Figures 8.1, 8.2 and Table 8.3 from *Land Degradation and Breakdown of Terrestrial Environments* published and reprinted by permission of Cambridge University Press (Barrow, 1991); Table 8.4 from *Desertification: financial support for the biosphere* published and reprinted by permission of Hodder Arnold (Ahmad and Kassas, 1987).

We are grateful to Elsevier for extracts from 'Land degradation is not a necessary outcome of communal pastoralism in arid Namibia' by D. Ward, B.T. Ngairorue, J. Kathen, R. Samuels and Y. Ofran published in *Journal of Arid Environments* 40 (1989) and 'Factors influencing erosion in Zimbabwe: a statistical analysis' by R. Whitlow and B. Campbell published in *Journal of Environmental Management* 29 (1989), and UNCCD for an extract from their Factsheet 11: *Action Plan to Combat Desertification in Africa* published at www.unccd.info.

Whilst every effort has been made to trace the owners of copyright material, in a few cases this has proved impossible and we take this opportunity to offer our apologies to any copyright holders whose rights we may have unwittingly infringed.

CHAPTER 1

Development challenges and debates in eastern and southern Africa

Deborah Potts

Introduction

The decision to produce a new series of regional volumes for the Developing Areas Research Group of the Royal Geographical Society/Institute of British Geographers was based on the view that the 'dramatic changes in the global political economy, in the nature of development challenges facing individual developing countries, and in debates on development theory' (Simon 1999a: xii) required a re-evaluation of major world regions from a geographical perspective. This volume covers the vast region of eastern and southern Africa – and there can be no doubt that the 1990s and the turn of the century have wrought major transformations in this region in terms of all three issues identified in the quotation above. A specific regional perspective on each of these issues is provided in this introductory chapter, thereby linking these themes from the global to the national and local level in the region. First, it is necessary to examine the interactions and factors which give this country grouping a degree of coherence, and then to discuss the developmental characteristics of the region.

Eastern and southern Africa: defining a diverse region

Defining regions within sub-Saharan Africa which lend themselves to critical analysis is no easy task. Different themes or disciplinary perspectives may logically choose somewhat varying regional boundaries. The geographical distribution of fundamental characteristics (be they environmental, historical, cultural, political or economic) rarely displays easy, definite patterns. Indeed, within sub-Saharan Africa, as Chapter 9 in this volume cogently argues, national boundaries themselves are exceedingly arbitrary (and are themselves a constraint on development). The region upon which this volume is based, eastern and southern Africa, covers 16 countries: South Africa; Botswana, Lesotho, Swaziland and Namibia; Angola and Mozambique; Zimbabwe, Zambia and Malawi; the Democratic Republic of the Congo (DRC), Rwanda and Burundi; and Tanzania, Uganda and Kenya (see Figure 1.1). While most of these countries share some common features in terms of, for example, a colonial past and economic structures typical of less developed countries, this does not set them apart from other African countries or regions as considered, for example, in a forthcoming text in this DARG series: *West African Worlds*. Within this group there are, however, some clear subgroupings which have long-standing historical, political and geographical coherence which will be briefly considered before turning to some overarching themes of interaction which help to define this region.

South Africa has very specific status within this region as the most economically developed country. Its mineral wealth combined with its

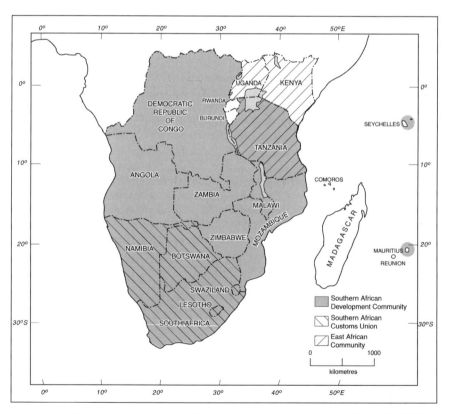

Figure 1.1 Eastern and southern African countries and membership of SADC, SACU and EAC

Note: The region as defined in the text includes mainland countries only

capacity for economic and political self-determination from 1910 (when independence from Great Britain was granted) gave South Africa the opportunity to develop as a minerals–energy economy with a significant degree of industrialisation (Fine and Rustomjee 1995) and, in regional terms, incomparable infrastructure. However as a white settler state which chose to pursue a political economic model based on racial inequality, the *human* development outcomes of apartheid South Africa's policies for most of its people were manifestly disastrous. South Africa's early escape from British colonial rule gave it particular local regional influence. Swaziland, Botswana and Lesotho (then Bechuanaland and Basutoland) spent much of their colonial period under threat of absorption into the nearby racist regime, which was a defining characteristic of their colonial experience and consequent neglect of their development (Spence 1965, 1968; Stevens 1967, 1972). They were also incorporated into a regional customs union dominated by South Africa: the Southern African Customs Union (SACU – see Gibb, Chapter 10 for further details). Namibia (the former South West Africa) suffered an even greater degree of dominance by South Africa which, during the First World War, occupied what was then a German colony in 1917 and was subsequently rewarded with a League of Nations mandate to administer the territory and essentially, if illegally, proceeded to incorporate it into South Africa as a fifth province. It did not attain independence until 1990, when it was Africa's last major colony. Namibia is also a member of SACU, so these five countries form a particularly coherent subgrouping within which South Africa's dominance is exceptional.

The next subgrouping indicated in the opening paragraph to this section is Angola and Mozambique. The essential binding link here is their experience of the specificities of Portuguese colonialism which both retarded their economic development throughout most of the colonial period and delayed their independence until 1975 after long wars of liberation against the colonists, both factors which facilitated their independent governments' subsequent 'choice' of Marxism-Leninism as the framework for their political and economic policies (Abshire and Samuels 1969; Hammond 1966; Smith 1974; Bender 1978; Newitt 1995; Vail and White 1980; Munslow 1982). This was a defining moment for southern African regional interactions because it brought them both into devastating conflict with apartheid South Africa, which perceived the loss, to Marxist-Leninist governments, of two of its 'buffer states' against the tide of black majority rule as an unacceptable threat. Mozambique, particularly its southern region, had long been within South Africa's economic ambit due to the logic of local transport geography, which made Lourenço Marques/Maputo the Transvaal's natural port (Henshaw 1998), and its key role in South Africa's migrant labour force. Angola was now also drawn firmly into the apartheid regime's regional ambit played out through its so-called 'total strategy' through which it was endeavouring, by every means possible, to undermine the potential success of regional black majority rule and to prevent regional support for South African and Namibian liberation movements (Hanlon 1986, 1989; Johnson and Martin 1989; Davies and O'Meara 1985; Potts 1992). In Angola and Mozambique this included South African backing for anti-government forces which led to devastating destabilisation and the reversal of any positive developmental trends. Both countries also became key players in the anti-apartheid regional grouping of the Front Line States (FLS) which included Botswana, Zambia and Tanzania and was another important political linkage within the region.

Zimbabwe, Zambia and Malawi form another coherent subgrouping on a number of grounds. They were all British colonies; they are all landlocked; they all formed part of an earlier, institutionalised regional grouping, the Federation of Rhodesia and Nyasaland (1953–63) which was dominated by Southern Rhodesia (i.e. colonial Zimbabwe). Both Zambia and Malawi gained their independence in 1964. Zimbabwe, on the other hand, had to wait until 1980 and its African people, as in Angola and Mozambique, had to wage a long and bloody liberation war against white settlers. It then joined the FLS, providing yet another regional linkage.

The southern African regional grouping, the Southern African Development Coordination Conference (SADCC), was also formed in 1980, its essential binding economic and political aim being to reduce dependence on apartheid South Africa. The original membership included all 10 countries so far considered (except, of course, South Africa) plus Tanzania (largely because of it FLS status). The Namibian liberation movement, SWAPO, had observer status. The political linkages and status forged through this grouping, and the liberation struggles against white minority rule more generally, have proved long-lasting (Sidaway 1998) and remain essential factors in explaining contemporary regional politics. South Africa's first democratic elections in 1994 paved the way for its shift from SADCC's ultimate enemy to becoming the potential hegemon of a new southern African group, the Southern African Development Community (SADC) (Ahwireng-Obeng and McGowan 1998; Simon and Johnston 1999; Simon 2001).

SADC now includes the DRC, thus pushing much further north the logical boundaries of regional study. The inclusion of this huge country, by far the largest in the region (see Table 1.1), into the ambit of any coherent study of southern Africa, if SADC membership be deemed to be a defining characteristic, has however tipped the balance of regional coverage in a significant way. To be brief, as the issues are covered in detail in Chapter 9, development issues in the DRC cannot be coherently considered without coverage of Rwanda and Uganda, both of which have been key players in its 'liberation' from Mobutu and then its civil war, which also brought in Angola, Namibia and Zimbabwe (providing further political linkages within the region defined in this book, although also leading to tensions within SADC) (*Financial Times* 2001; Regional Roundup 2002). Since South Africa, Zambia and Botswana have all played important diplomatic roles in trying to end this conflict – a regional example of the concept of 'African solutions for African problems' – the *political* geography rationale for the region as covered in this book is enhanced. As the DRC's economic resources were a key reason for its SADC membership, with South Africa a key player in that decision, and Zimbabwe, Namibia and Angola are now also economically involved in the DRC, the *economic* geography rationale is also growing.

Table 1.1 Eastern and southern Africa: socio-economic indices 2000/1 (ranked by HDI value)

	Surface area (1,000 km²)	Population		Human development			Economy		
		Total 2000 (millions)	Density (per km²)	HDI[a] 2000 (rank)	Life expectancy at birth, 2000	Adult literacy,[b] 2000 (%)	GNI[c] per capita, 2001 (US$)	PPP GNI[d] per capita, 2001	Debt service as % exports, 2000
Burundi	28	6.4	229	0.313 (171)	41	48	100	591	37
Mozambique	802	18.3	23	0.322 (170)	39	44	210	854	11
Malawi	118	11.3	96	0.400 (163)	40	60	170	615	12
Angola	1,247	13.1	10	0.403 (161)	45	42[e]	500	2,187	15
Rwanda	26	7.6	292	0.403 (162)	40	67	220	943	25
Congo (DRC)	2,345	50.9	22	0.431 (155)	51	61	110[f]	765[g]	1[f]
Zambia	753	10.4	14	0.433 (153)	41	78	320	780	24
Tanzania	884	35.1	40	0.440 (151)	51	75	270	523	16
Uganda	241	23.3	97	0.444 (150)	44	67	280	1,208	16
Kenya	580	30.7	53	0.513 (134)	51	82	340	1,022	17
Lesotho	30	2.0	67	0.535 (132)	46	83	550	2,031	12
Zimbabwe	391	12.6	32	0.551 (128)	43	89	480	2,635	19
Botswana	582	1.5	3	0.572 (126)	40	77	3,630	7,184	2
Swaziland	17	0.9	53	0.577 (125)	44	80	1,300	4,492	2
Namibia	824	1.8	2	0.610 (122)	45	82	1,960	6,431	–
South Africa	1,221	43.3	35	0.695 (107)	52	85	2,900	9,401	10

a. Human Development Index and world rank in 2000 in brackets.
b. Literacy rates are for the 15+ population.
c. Gross national income, the term now used in World Bank tables, is the same as gross national product.
d. Purchasing power parity dollars.
e. UNICEF estimate for 2001.
f. World Bank (2000) estimate for 1998.
g. 1998 estimate.
Sources: GNI and PPP per head from World Bank (2002); other figures from UNDP (2002) unless otherwise noted.

The inclusion of Kenya and Burundi becomes inevitable once the region includes Rwanda, Uganda and Tanzania. Rwanda and Burundi share a host of environmental, historical, political and economic features and are in any case encapsulated by the region as defined. Kenya, Uganda and Tanzania form their own economic grouping, the East African Community, and are frequently considered as a regional unit. Otherwise Kenya is the one country in the region covered in this volume for which there is arguably no strong political or economic rationale for studying it alongside southern Africa. However, like so many of the countries to the south, Kenya was a white settler colony, which very firmly links it to the south in terms of its developmental heritage from colonialism (Mosley 1983; Kennedy 1987).

There are also some important environmental features which help to define the region covered in this volume, although there are exceptions in every case. As Figure 7.1 indicates, the boundary between 'high' and 'low' Africa provides, to some extent, a natural boundary to the west of the region, although it divides the DRC in two. Also, as illustrated by Chapter 8, significant proportions of many of the countries in the region are designated as drylands. Yet again, though, the DRC is exceptional here, as are Rwanda and Burundi. Inevitably, any region contains discontinuities, and arguments can be made for or against the inclusion of its geographically outlying members and are, to some degree, subjective. For example, the coverage of eastern and southern Africa here is confined to the mainland, and excludes Madagascar, Mauritius and the Seychelles. The latter two are, nevertheless, SADC members, but it has been deemed that the specificities of their small island economies and their lack of involvement in intra-regional political affairs make them too exceptional to be included logically in a study of development trends and challenges (although it makes sense to mention them in Chapter 10 on trade as they are affected in similar ways to other countries in the region by recent trade regulation changes). Similarly it is arguable that if Kenya is covered, then why not Somalia, Djibouti and Ethiopia? However, this would stretch the intra-regional links south to the tip of the continent beyond any meaningful rationale.[1]

Overall, the essential justification for the coverage of the 16 countries included here as 'eastern and southern Africa' is based on their level of interactions. It is worth noting, in terms of the developmental prospects and problems of the region, that these have been characterised as much, if not more, by conflict than co-operation. Furthermore, the countries of the region are highly diverse in terms of their natural resource base, and levels of economic and human development (see Table 1.1). It includes Burundi, Mozambique, Malawi, Rwanda and the DRC which, in terms of both crude per capita income levels and the UNDP's Human Development index (which is a better measure of 'development' as it combines income, education and life expectancy data), are among the least developed

---

[1.] The regional trade grouping, the Common Market for Eastern and Southern Africa (COMESA) – although its name is suggestive – is also far too disparate in its membership (see Figure 10.3) to be a useful linking regional theme.

countries in the world. Angola, as can be seen, has much higher per capita income levels mainly because of its vast oil revenues, but these have been of little relevance to most of its population whose welfare and livelihoods have been ravaged by decades of war. Nearly all the countries of the region are classified by the World Bank as 'less developed' or low-income countries, which is another binding characteristic of central significance to the themes of this volume. Botswana (which has the highest nominal per capita income, having now surpassed South Africa in this respect), Swaziland, Namibia and South Africa are, however, 'middle-income' countries and, as discussed by Gibb in Chapter 10, South Africa is deemed to have 'transitional' development status in trade negotiations. Tony O'Connor deconstructs regional statistics related to poverty (and their reliability) in great detail in Chapter 4, to which the reader is directed for further analysis. It is worth noting here the following points. First, although great inequalities in income and welfare between people is an important characteristic of all the countries of eastern and southern Africa, South Africa's and Namibia's indices need particular care in interpretation as race is still such an important correlate with privilege in these two countries (as, indeed, it still is in Zimbabwe, although there the recent economic crisis and the very small share of the national population accounted for by whites mean this country is slowly slipping down such 'development' tables). Second, regional experience shows that minerals can be key elements in a country's *economic* development, as evidenced by South Africa, Namibia, Botswana and Zimbabwe, but that conflict and poor governance can all too easily negate this relationship, as in the DRC and Angola. Finally it is worth reiterating the extreme variation in size and natural resource endowment within the region, from densely populated Rwanda and Burundi to the DRC and Angola (although the table shows how little there is to choose between them currently in terms of human welfare). Third, it is important to note that, in the absence of the impact of AIDS on life expectancy, the gap between the countries with the worst and best HDI rankings would be greater (cf. the purchasing power parity (PPP) per capita statistics) since life expectancy in many southern African countries has been savagely reduced by the epidemic.

This chapter now returns to the three sets of changes identified in the opening paragraph which have shaped this regional volume, along with others in the series – changes in the global political economy, in development challenges and in development theory – and reflects on each from a regional perspective.

Global political economy changes: regional perspectives

The ending of the Cold War unquestionably brought about, or contributed to, extraordinary shifts in regional political formations in the southern and central parts of the region. In the first four years of the 1990s the list of major political changes was truly startling: the release of Mandela and other political prisoners in apartheid South Africa and the beginning of the negotiations to bring an end to white minority rule; the independence of Namibia from South Africa's illegal occupation; a ceasefire and elections in Angola (sadly abrogated by Savimbi when the outcome was not to his liking); a ceasefire in Mozambique; and, in 1994 alone, there were democratic elections not only in South Africa, bringing an end to its terrible history of racist rule, but also in Mozambique and even in Malawi. In Malawi's case, once the 'Cold War' blinkers dropped from the eyes of Western donors, it became clear that this country was not, as represented in old World Bank literature, a well-run capitalist enclave in a sea of African socialism, but an exceedingly repressive dictatorship where Banda's personal business empire and parastatals held almost complete sway. Once recognised, donor conditionality did the rest. In essence, then, the early 1990s saw the end of South African-backed destabilisation in southern Africa – which had cost this region some US$60 billion during the 1980s (Smith 1990) – and a shift away from largely rhetorical support by the West (driven by the need to balance the appeasement of voters against perceived strategic interests in southern Africa) for regional destabilisation to end, and for the demise of white minority rule, to the imposition of widespread political conditionalities attached to aid and loans, adding to the economic conditions which had become ubiquitous in the 1980s. The new global political economy also soon saw Zaire, previously one of the main beneficiaries of the West's Cold War strategic support in the region (despite its horrendous political and economic record, with Mobutu's name virtually a synonym for corruption), teetering on the edge of political change. The positive political doors opened in southern Africa in 1994, however, were counterbalanced in eastern Africa by the decade's most terrible event – the genocide in Rwanda. The ripples from this tragedy in a tiny country have spread throughout the region covered in this book – to Namibia, Angola, Zimbabwe, Zambia, South Africa, Burundi, Uganda and Tanzania – and had the most profound outcomes in Zaire (soon to be renamed the Democratic Republic of the Congo (DRC)). These are detailed by Power in Chapter 9. Suffice it to say that, two wars later, a festering stalemate in the DRC was still not fully resolved at the time of writing in late 2002, but the violence is estimated to have resulted in over 3.5 million deaths (Pottier, 2003).

The new global political economy has also wrought an economic policy straitjacket for most of the countries in southern and eastern Africa. The now familiar policies of trade liberalisation, devaluation of national currencies, reduction of government budgets and a general 'rolling back' of the state, privatisation of services and parastatals and the general imposition of market-based allocation of resources had already become the norm in much of the region by the end of the 1980s. Pockets of resistance were mopped up as the 1990s proceeded and the last Cold War cards were played. At the beginning of the twenty-first century the utter pervasiveness of this new economic order can be exemplified by three changes of symbolic significance, given the region's political and economic history. First, Tanzania, whose Arusha Declaration in 1967 was one of the great markers of 'African socialism' and the drive for a modernising, African developmental state, and which was a key player in the FLS' active opposition to apartheid South Africa, is now viewed as a preferable investment opportunity to Kenya: by South African investors, no less (Simon 2001)! Second, in 2001 after years of opposition and endless negotiations, Zambia's copper mines were reprivatised (largely by South Africa's Anglo-American, one of the original owners). Third, in 2001 Malawi sold off its maize stocks at the behest of the international financial institutions (IFIs) on the grounds that such large-scale storage was costly and inefficient – along the lines that butter mountains in the EU used to be inefficient. These maize stocks, much of them in massive silos in the capital Lilongwe, were a matter of national pride in the past when President Banda was obsessed with national food security and self-sufficiency and 'growing more maize in the fields' (a frequent exhortation in Malawian newspapers). But maize is not butter; it is the staple food. And southern Africa is not Europe – it has a disequilibrial environment, driven by unreliable and unpredictable rainfall. And Malawi remains one of the world's most rural countries, with most people deeply reliant on those rains, and one, moreover, in which most households are food purchasers at the end of the year. That is why it had built up the stocks, and why the Ministry of Agriculture had for years tried to persuade the IFIs of the specificities of the country's agricultural situation (Harrigan 1991). It is also deeply affected by AIDS, making many rural households more than usually vulnerable to any shock. Furthermore it is landlocked. So, when the 2001–2 rainy season proved disastrous for the crops, there was a famine. Efforts to alleviate this crisis gathered pace in the early summer of 2002 as various agencies pleaded for donations and government aid to import food into Malawi (and other countries in the region affected by the poor rains). Had the maize stocks still been there and had they been used wisely[2] they could have alleviated part of the suffering that had been caused. No better, or rather worse, example of the triumph of neo-liberal ideology over the consideration that countries have different needs, strengths and weaknesses – indeed one might say different geographies – can be found. It is almost incredible that in 1990 the IFIs had managed to pull off a similar blunder in Zimbabwe, which also sold its significant maize stocks and was immediately faced with the region's worst drought of the century. From this one might deduce that the IFIs have both execrable timing, and an inability to learn from past mistakes.

The grip of the IFIs in the region of eastern and southern Africa is almost complete because nearly all the countries there are so heavily indebted and politically weak. This is an important differentiating factor when comparing this region to others, such as South Asia, the Pacific Rim or Latin America. Yet even post-apartheid South Africa has adopted the general economic regime – with predictable outcomes for the value of the rand (which, having drifted downwards, collapsed in 2002 as the ripples from the Argentinean economic crisis circled the globe) and, so far, no discernible benefits in terms of economic growth and employment creation. South Africa does, however, have the benefit of some room for

---

[2.] Much has been made of the fact that some of the revenues from the maize stocks 'disappeared' due, presumably, to corrupt dealings. Evidently, however, this was not the *cause* of the maize shortage.

manoeuvre as no formal structural adjustment policies (SAPs), policed by the IFIs, are in effect. In this there are parallels with, for example, the situation in South-East Asia as analysed by Chris Dixon in the first volume of this series (Dixon 1999). As Dixon shows, there is a spectrum of conformity to the 'ideals' of the neo-liberalist agenda and, although private entrepreneurialism is generally now encouraged, the state has not always been forced to roll so far back that the 'welfare baby' has been thrown out with the 'anti-private sector' bathwater. Sadly in much of southern and eastern Africa the stagnation or deterioration of health, education and poverty indices (see O'Connor, Chapter 4) demonstrates only too well how crude and undevelopmental (in the normative sense) unmediated SAPs can be. South Africa, on the other hand, had managed to allocate cash grants for a million houses by the turn of the century, to provide nutrition supplements to millions of schoolchildren, to connect millions of people to electricity and water,[3] and bring in radical water laws which guarantee, free, a minimal amount of potable water per person and essentially nationalise water rights (see Taylor, Chapter 7). None of these things has been as successful or fully implemented as was hoped; but none of them could even have been tried under a full-blown SAPs regime. This analysis has so far left aside the pertinent fact that the NICs of Pacific Asia had already achieved their 'economic miracles' under heavily state-interventionist policies – the antithesis of the neo-liberal agenda as practised on vulnerable African economies – and only subsequently liberalised their economies (Dixon 1999; see also Stoneman, Chapter 3). India also attained growth rates around three times those of most poor countries in the 1980s, and accomplished what many commentators judged 'to be a period of economic success and even "take-off" under a similarly illiberal and interventionist regime, before moving towards a more liberal economy in the 1990s' (Corbridge and Harriss 2002: 106–7).

Development challenges in the region

In terms of changes in the major development challenges facing eastern and southern Africa today, there is one clear problem which overwhelms all others and that is HIV/AIDS. Eastern and southern Africa are by far the worst affected regions in the world, with many countries in the southern part of the continent experiencing adult HIV infection rates of 25–30 per cent. The eastern African area, which was the original 'core' region for the African epidemic, has been overtaken by countries such as Malawi, Zimbabwe and Botswana in terms of the extent of infection. The impacts of this terrible disease on development in Africa are now well known and extensively covered in a burgeoning literature (e.g. Cross and Whiteside

---

[3.] Unfortunately millions of these connections have since been disconnected due to inability to pay the suppliers (see also Chapter 11). In countries where the majority of people are poor or very poor (as is still the case in South Africa despite its relatively strong economic position in the region) this is the, perhaps inevitable, downside of relying mainly on the private, or commercialised public, sector for service delivery.

1993; Caldwell 1997; Caldwell and Caldwell 1995; Banatar 2001; Barnett and Blaikie 1992; Barnett *et al.* 1995; Browne and Barrett 1995; Cuddington and Hancock 1995; Daniel 2000; Essex *et al.* 1994; Gregson *et al.* 1994; Gregson 1994; Grieser *et al.* 2001; Lamboray *et al.* 1992; Tibaijuka 1997; Wallman 1996; Webb 1997; *African Urban Quarterly* 1991, 1992). These include the loss of significant numbers of able-bodied working age adults since it is this cohort which is most susceptible to AIDS deaths and, therefore, an increasing dependency ratio. In agriculture this leads to shifts in production systems as shortages of able-bodied labour become acute. In urban areas the loss of key skilled workers in formal enterprises has national implications, and the death of both formal and informal sector breadwinners has devastating implications for their households. Vast numbers of households have been affected and new types of households are increasingly appearing, as orphaned children or grandparents are forced to take charge. Added to this is the dreadful burden of grief and trauma that so many families now bear and which must have some impact on national psyches. Mothers with HIV are faced with the dilemma of whether to breastfeed their babies and risk passing on HIV, or bottle feed and increase manifold the babies' chances of dying of malnutrition and diarrhoeal diseases (Potts and Marks 2001). In the worst case scenarios, a dreadful fatalism can creep in where people tend to feel that HIV is unavoidable, that all babies born to HIV mothers are HIV+, and that all pregnancies lead to a swift onset of AIDS in a HIV+ mother. Such fatalism prevents the very changes in sexual behaviour and health care that are, for the moment, the only realistic ways of reducing the spread of the epidemic. Men's efforts to avoid infection by shifting their attention to younger and younger cohorts of women and girls have had the terrible effect of greatly increasing the rate of infection in these female cohorts, compared to their male counterparts.

The epidemic is a huge *political* developmental challenge too. The health and demographic professionals call for strong leadership, urging better treatment of women within sexual relationships and for condoms to be used whenever possible: strategies that could save millions of lives. There seems to be evidence from Uganda that this can reduce HIV rates. But all the political leaders in eastern and southern Africa are men, and so far very few seem prepared to take the strong line on these issues that health professionals require. This is not only because of fears of a right-wing backlash (as experienced in many countries, including the US) of the type that argues that such issues should not be aired in the political arena and that reproductive health education encourages promiscuity, but also because these issues go to the very heart of male identity and the gender relations on which this is based. The frustration of health professionals and AIDS activists is palpable throughout the region – which is why Thabo Mbeki, the South African President, was pilloried for bringing into the political discussion debates about the correlation between the spread of HIV and poverty, and the toxicity of some of the drugs which can slow the onset of AIDS. The first point was in reality an orthodoxy among AIDS workers (Mbeki's real problem being his ambiguous stance on whether he believed that the corrrelation implied causation – that it does not is a statistical nicety that escapes many politicians), and the second was similarly unremarkable if viewed dispassionately. But such is the crisis that nothing *can* be viewed dispassionately, and such statements are bound to be reduced to their crudest form in the public mind, if not wilfully misconstrued. The lesson is that politicians in eastern and southern Africa need to be even more careful than normal when pronouncing on the AIDS crisis, and that no politician can afford to divert attention away from the essential messages about changing sexual behaviour and using condoms. Added to this has been the subsequent furore over the use of anti-retroviral drugs in South Africa. One of the most important successes of the new coalitions against globalisation was the campaign to make such drugs available much more cheaply than the drug companies would allow. But the South African government is resisting their widespread use. Again there is a relevant argument that in African countries, including South Africa, the health infrastructure and personnel are insufficient to ensure that the complex cocktails of drugs are taken properly so that drug-resistant strains of HIV are not widely released. South African health professionals, however, counter that Brazil has seemingly made a real impact on its AIDS death rate by using generic drugs. Furthermore, the use of a one-off drug to counter most mother-to-child transmission at birth seems clearly affordable and possible, but it remains a matter of political controversy in South Africa.

To the extent that national politics can be reduced to the struggle between groups to raise and allocate resources to realise specific outcomes, then AIDS in South Africa is deeply politicised. In other countries in the region, with the exception perhaps of Botswana and Namibia, health professionals are just as frustrated but, until a vaccine is found, even free drugs might not help and changes in people's behaviour are the best hope for reducing HIV spread. In this volume there is no specific chapter on AIDS – a decision that was made because the literature on AIDS in the region is already so rich and detailed that there seemed little point in repetition. Instead each themed chapter makes reference to AIDS where pertinent; it is also covered in its demographic impacts in Chapter 2.

The developmental challenges in the economic sphere faced by the region today are, from one perspective, rather less geographically variable than before inasmuch as in so many of the countries the realistic, or perceived, policy options are much more uniform than in the past. At a macro-level these are how best to play your cards to attract some global crumbs of foreign direct investment (FDI) in *new* enterprises (i.e. rather than the privatising of existing ones) – which is not to say that there is no FDI occurring but, without doubt, comparatively speaking, this region is extremely marginal in the global FDI stakes. Certainly some countries are learning the rules of this game, and promising acronyms abound: SDIs (spatial development initiatives); LED (local economic development); EPZs (export processing zones) (Nel 1996; Jauch 2002; Good and Hughes 2002; Sachikonye 1998; Simone 1998; Maharaj and Ramballi 1998). On the whole, however, these have not been successful in attracting (or keeping) FDI of real significance. Export crops and minerals can be expanded, and often have, in line with IFI exhortations, but price fluctuations (particularly for crops in increasingly saturated markets) remain a nightmare for realistic forward budgeting and planning. This was predictable and, indeed, widely predicted. Another option is to seek categorisation as an HIPC (highly indebted poor country), which allows some debt forgiveness, but if this 'card' can be played, then evidently the country in question is unlikely to have the skills and infrastructure which will encourage private investment in a competitive world (with the possible exception of the mineral and tourism sectors).

A notable exception to this conformity is Zimbabwe, where extreme and sudden divergence from the neo-liberal, capitalist, market-oriented package in the late 1990s has ushered in a period where property rights, which many had come to assume were sacrosanct under the rule of the IFIs throughout Africa, have been utterly transgressed. Many other keystones of structural adjustment there have also been unpicked, from liberalised currencies to market prices for key commodities. The extent of these changes has confounded many experienced analysts of Zimbabwe and has resulted, unsurprisingly, in pariah status for aid and IFI financial support, leading to virtual economic implosion. The conventional explanation is that this has occurred as a result of Mugabe and ZANU (PF)'s totally undemocratic reaction to massive internal political opposition which would probably have toppled them, had the parliamentary elections in 2000, and the presidential elections in 2002, been free and fair (see Power, Chapter 9). Such analysis is correct in so far as it goes; but it begs the question of what was driving that opposition? This is a question that will be returned to in the concluding chapter, along with the implications of the Zimbabwe crisis for the region as a whole.

Development theory and development debates

In terms of development debates, the countries of southern and eastern Africa have experienced a wide range of new approaches to 'development' as donor country and agency paradigms have shifted. As most of the countries are highly aid-dependent (some, like Malawi, almost wholly so for their non-recurrent expenditure), they have little choice, although some of the 'new' approaches are theoretically positive and the principles behind them, well-meaning. Participatory approaches in almost all spheres of development, from natural resource management (see Sullivan and Homewood, Chapter 5) to urban low-income housing (see Potts, Chapter 11), are commonplace today. The linked idea of decentralising the management and delivery of services (e.g. water in South Africa and Namibia) and governance more generally, is equally widespread. NGOs are favoured over government bodies (even if representative) for donor funding very often, as they are believed to be nearer to the 'grassroots' and less corruptible (although in fact they are not necessarily more accountable or effective). Programme, rather than project, aid is more frequent and capacity and institution building are important donor priorities in the belief that this will make other development objectives more sustained. There is no scope in this introduction to discuss fully these new approaches and their impacts in eastern and southern Africa (see Simon 1999b in the first volume of this DARG series for an analysis of these shifts). Suffice it to say that the gap between theory and practice is sometimes so large that the principles behind these approaches become virtually invisible; and that on occasion it appears that practitioners have become so fixated on the means (i.e. the approach) that they are reluctant to reconsider the policy even if it becomes clear that the developmental 'end' is not, in reality, being served. In decentralisation programmes, unfunded mandates abound – perhaps because there is not enough money, but also often because a weak (and maybe corrupt) central state correctly perceives decentralisation as highly threatening – so goes through the motions under pressure from donors but does its best not to 'put its money where its mouth is'.

Developmental paradigms in the IFIs have also shown some evidence of a significant shift – to the so-called post-Washington consensus (Stiglitz, 1998) which, in theory, allows for a little more nuance in policy packages to allow for individual circumstances. That the state can (and should) play an important (although not central) role in development is also allowed, as global and local economic problems stemming from a lack of regulation have become very evident. A more flexible approach to inflation within economies has also been seen as compatible with, or even contributory to, economic development – an idea which could have saved war-torn Mozambique a great deal of suffering (Hanlon 1997). On the other hand, the underlying principles and ideologies have not changed and neo-liberalism remains the order of the day (Fine *et al.* 2001); furthermore this paradigmatic shift has often not trickled down to IFI programmes at the country level.

One truly fundamental economic issue for the neo-liberal agenda remains highly contentious and contested in the region, namely land tenure. Much land (and the natural resources thereon) in the countries of eastern and southern Africa is still managed according to indigenous tenurial practices, even if, *de jure*, the land is often theoretically state-owned (see Sullivan and Homewood, Chapter 5). The situation in different areas is highly variable as practices have adapted to local circumstances (e.g. increasing population densities) and the whole subject is intensely complex but, broadly speaking, there is still a strong tradition of tenure being communal, rather than market-based. To use Sen's concepts, men are commonly regarded as having an 'entitlement' to land in the area of their birth or where they have kinship links – it is an essential right to the means of production. This link to the land is also important for the economic security of many rural–urban migrants (Potts and Mutambirwa 1990; Potts 1997, 2000a). Where population densities are high, such 'rights' often cannot be realised or become deeply contested (e.g. see Peters 2002, on the situation in southern Malawi), and land may be bought and sold – although this process may still be culturally mediated rather than being an essentially capitalist process. Indeed, attempts to impose freehold, titled land tenure in some parts of eastern and southern Africa have often been confounded by the subsequent reassertion of elements of communal, family-based tenure approaches and the obstruction of the development of a capitalist land market (e.g. on the Lilongwe Land Development Programme in Malawi, and in parts of Kenya) (Ng'ong'ola 1986; Platteau 1996; Palmer 1997). Furthermore, the essentially geographical fact that arable land is still quite abundant in relation to rural populations in some countries, or parts thereof (e.g. Angola, Zambia, Tanzania, Mozambique, the DRC), means that land markets for smallholder land are unlikely to 'work' largely because land has insufficient scarcity value. In other areas, however, strains on indigenous systems have rendered them increasingly dysfunctional, particularly in South Africa, where they have been substantially undermined by massive land alienation and centuries of attrition of the rural community-based traditions and authority needed for the potentially positive aspects of indigenous tenure to function (Adams *et al.* 1999).

For the World Bank the existence of indigenous, non-market forms of tenure is a somewhat confounding factor for its mission of economic liberalisation in Africa. How can the factors of production be allocated according to the laws of supply and (monetary) demand, when a key factor of production in so many countries remains allocated according to deeply entrenched birthrights for most of the (male) population? The Bank has been very persuasive in putting the argument that freehold land title for smallholders will facilitate the provision of vital agricultural credit lines to increase productivity by providing a source of collateral. The developmentally negative outcomes of the process are scarcely mentioned, yet poorer families inevitably end up losing their land – a process which in dryland eastern and southern Africa, where there is so little irrigation and such unpredictable rainfall, could be very swift. It has also been proven that, in most African situations, the cost of land surveying for *smallholder* land registration renders the process economically inefficient, and the lack of institutional capacity to implement and register land titles means it is often also infeasible (Platteau 1996, 2000). Many countries in the region are nevertheless involved in, or discussing, land reform processes which could transform their rural socio-economic and political landscapes. Mozambique, for example, toyed with the idea of introducing comprehensive freehold tenure in the early 1990s, but FRELIMO's remaining 'egalitarian' principles eventually decided against it. In the early twenty-first century it went through the process again, still under heavy IFI pressure (Hanlon 2002).

Despite the pressures, however, there are countervailing developmental forces around land issues. National reviews of land policy in southern and eastern African countries have broadly not transformed allocative practices for smallholders on customary land – in some cases these have even been slightly democratised (Okoth-Ogendo 2000; Adams *et al.* 1999). The more immediate threat comes from government decisions to privatise and sell off 'other' land to large-scale farmers and ranchers, in a bid to raise some government revenue and increase agricultural production. The problem here is that too often this 'other' land is not necessarily unoccupied, and may be perceived by smallholders as part of their 'customary' estate. Smallholders, pastoralists or hunter-gatherers may thus find that their land has been 'privatised' by somebody else, as has happened, for example in Mozambique (Myers 1994; Hanlon 2002), Tanzania (see Sullivan and Homewood, Chapter 5) and Botswana (Sporton *et al.* 1999); frequently their subsequent treatment has been brutal.

These external pressures for land privatisation in the 1990s accord poorly with the new development paradigms which revolve so strongly round 'community-based development', 'poverty alleviation' and 'livelihood security'. The British government's Department for International Development's (DfID) livelihood approach to research and development in Africa virtually requires adherence to its preferred methodology of assessing household 'capitals' – a pentagon of natural capital, social capital, human capital, financial capital and physical capital (Carney 1999; Ashley and Carney 1999; Rakodi with Lloyd-Jones 2002). Where communal tenure is working reasonably well this implies that there is an important element of local community-based 'social capital', the fostering of which is now often seen to be a key means to 'development' (but see Fine 2001). It also means that nearly all households have access to an essential resource, no matter the status of their 'financial capital'. Furthermore, functioning customary land rights provide an extremely widespread element of 'livelihood security' for rural and urban people (Potts 2000b). If privatisation of smallholder land in eastern and southern Africa were to be achieved in tune with the general thrust of neo-liberalism, however, the following scenario could be envisaged: well-meaning research programmes in the region would soon be recommending that the poorest rural households need to have their 'financial capital' strengthened in order to gain access to the 'natural capital' of land, and lamenting the loss of community-based 'social capital' for natural resource management. In other words, while the left hand strives for rural livelihood security, the right hand will have fundamentally undermined it. This is, in fact, explicitly recognised within DfID as exemplified by its sponsorship of the edited volume *Evolving Land Rights, Policy and Tenure in Africa* (Toulmin and Quan 2000), where it is argued that 'a new donor consensus reached by the main OECD member states is facilitating the development of a more human-centred approach to land rights' (Toulmin and Quan 2000: 3). An important message running through that book is that land policy which helps the poor should recognise that 'imported western notions of property rights are not the only principles which may be appropriate in Africa' (Toulmin and Quan 2000: 3). It is to be hoped that this will strengthen African governments' hands in inevitable future contestations with the World Bank over land tenure.

Another sphere of development debate which is of intense significance for eastern and southern Africa is the re-evaluation of environmental *changes* that are occurring, and their impact on people's livelihoods. The word 'change' is used here intentionally. After decades of greatly simplified analysis which tended to assume that almost any change in vegetation cover and the status of the soil were both 'bad' and very much the 'fault' of poor people's agricultural practices (usually African farmers and pastoralists) and population growth, there has been a critique of the environmental 'narratives' behind these widespread approaches (Leach and Mearns 1996; Binns 1995; Potts 2000c). A wide range of empirical research has also been conducted which challenges whether the environmental changes, which are undoubtedly occurring throughout the region as populations grow, are necessarily 'degradation' – an overused term in much of the environmentalist literature. These debates articulate both with the increasingly accepted view that African dryland environments (which dominate much of this region) should be analysed as disequilibrial, with highly temporally variable rainfall as the key influence, and that the nature of changes in the environment should be analysed in terms of the usefulness and value of outputs to communities and nations and trends in productivity and pollution. Evidently Western nations have wrought profound changes in their physical environments in order to produce different, more highly 'valued' goods, and this is bound to occur also in many parts of Africa. But southern and eastern Africa are undoubtedly special in global environmental terms because of the sheer, unparalleled size of their 'conservation estate' – the proportion and amount of land set aside for wildlife – and the awe-inspiring numbers of animals and types of megafauna (e.g. elephants, lions) which are there. The recent community-based conservation which endeavours to ensure that local communities benefit materially from this 'conservation estate' is a step towards addressing the question of why southern and eastern Africans should not convert their environments wholesale, as has happened elsewhere in the world. However, as indicated by Sullivan and Homewood (Chapter 5), this is probably insufficient, and the global significance of the region's environmental resources may also need some global political and economic solutions, including resource transfers from Western nations and agencies to compensate countries in this region for maintaining such areas.

One worrying agro-ecological trend in much of the region is the decrease in use of fertilisers. Eastern and southern African farmers generally used a fraction of the inorganic fertilisers typical of Asian, let alone European or North American, agricultural systems. Fallowing systems of various kinds often compensate, but these have to adapt as populations grow, and in areas such as Malawi, Lesotho, Rwanda, Burundi, the communal areas of Zimbabwe, and many other densely populated parts of other countries, fertilisers are increasingly necessary. Their price and distribution were often subsidised in various ways, but structural adjustment policies have largely ended these practices. As a result, prices have soared and farmers throughout the region have often found that the cost is now too high in relation to the benefits, and reduced their use dramatically (see, for example, Scoones 1997a, 1997b, on Zimbabwe; Whiteside 1998, Lele 1990 on Malawi; Oygard 1997 on Tanzania; Holden 1997, Copestake 1998 on Zambia). Both environments and livelihoods have thereby been put at much greater risk (Cleaver 1997) and this must surely be one of the most pressing policy issues in the region.

*****

This volume comprises 11 chapters, followed by a short, concluding essay. The themes covered provide coherent coverage of key contemporary development geography issues. Each fairly standard geographical theme (i.e. demographic change; natural resources; urbanisation; agriculture; water resources; political geography; trade) is developed therefore with reference to the guiding principles of the series: the role and nature of globalising trends in the region's development; the characteristics and trajectories of contemporary development challenges related to those themes; and how recent development approaches and theories have influenced patterns of development and change. In addition there are two chapters which very specifically address development in the region: one on structural adjustment policies and the other on poverty.

Chapters 2–4 provide the human and policy context of the region within which subsequent themes need to be placed. Akim Mturi, a Tanzanian demographer who has also worked in Lesotho and is now at the University of Natal, Durban, discusses the important changes that have occurred in regional demography over the past 20 years or so, including the fall in fertility now well established in the southern part of the region. It is this chapter that also provides the most direct discussion of the AIDS crisis which is the region's most severe developmental and humanitarian problem. Colin Stoneman provides a robust and damning critique of structural adjustment policies in the region from a development economist's perspective. One of the rebuttals to concerns about the impact of structural adjustment made by those who promote such policies, and favour an explanatory framework which lays the main blame for Africa's economic predicaments at the end of the 1970s on internal mismanagement, is that it cannot be proven that things would not have been worse without them, because evidently some change was necessary. Stoneman does not deny that hard economic decisions and policies needed to be followed in order to stabilise regional economies at this time, whatever the causes of the problems. For him, though, the IMF medicine was the essential rectification needed. Although the IMF is often viewed as the hardest taskmaster, he argues that it is the World Bank which has caused the most damage with its truly *structural* transformations which have opened up African economies to globalisation in such an unmediated and blindly ideological way, and in terms of such unequal power relations. Tony O'Connor's chapter assesses trends in poverty indices in the region and queries whether, in the light of the evidence, 'development' has been occurring in many countries in recent decades. In this he engages empirically with issues raised by Simon (1999b) about the meaning of development and the geographical variability of 'success', with most of Africa faring poorly compared to Southeast Asia and parts of Latin America. His analysis of factors implicated in the poverty of so many of the people of eastern and southern Africa broadens the discussion beyond the policy issues of Stoneman's chapter, and provides a typically geographical perspective in which the roles of both the physical and human environment, and their interaction, are embraced.

The next four chapters examine the nature of the region's physical environment, and those human livelihoods which depend upon it. Sian Sullivan and Katherine Homewood examine the significance of gathered and hunted resources in rural people's livelihoods and address the highly topical developmental debates about their management and use which have been briefly mentioned above. This chapter also discusses in detail the varied nature of, and common misunderstandings about, land tenure in the region and emphasises the importance of securing tenurial rights in order to secure rural livelihoods. Their perspectives are strongly informed by an anthropological, as well as geographical, perspective. In the next chapter, John Briggs analyses FAO data on agricultural production in the countries of eastern and southern Africa, and the implications of the downward trend in per capita production that is apparent in many. This is followed by an examination of water resource issues by Richard Taylor, key themes being the tremendous variability in rainfall and riverflow regimes experienced by different countries and the fundamental influence of this on developmental prospects and problems, and the difficulties associated with the usual models for assessing national water resources and their adequacy for human needs for this region. Tanya Bowyer-Bower's chapter looks at the debates about environmental change and the concept of desertification with reference to conflicting evidence from within the region on the desertification process and the impact such debates have on international policy.

Central themes in the region's economic and political geography are covered in the last three chapters. Marcus Power discusses the extra-ordinary and sweeping political changes which have occurred within the region since the 1980s, and the linkages with global political change. The existence of open, military conflict within countries has been a fundamental determinant of development prospects and outcomes in many countries, but the geography of this had changed very significantly by the turn of the century as the central locus of conflict shifted north to the DRC. His wide-ranging chapter, which is informed by a 'critical geopolitics' perspective, also examines the developmental problems stemming from the disjuncture between inherited colonial political boundaries and 'cultural spaces', and the struggle that states experience to assert their territorial sovereignty. Trade is often held up as the 'best' path to development for countries in the region (cf. the adage, 'trade not aid') and Chapter 10, by Richard Gibb, examines in detail the nature of international and regional trade. Struc-tural adjustment has made the significance of trade protocols and agree-ments ever more critical for economic development in individual African countries but, as Gibb shows, the nature of eastern and southern Africa's comparative (dis)advantage and its weakness in trade-related negotiations are such that there appears little prospect in the medium term for sig-nificant developmental impulses to emanate from a trade liberalisation which has usually taken place when economies were very weak. In the final themed chapter, Potts reviews the often not fully recognised changes that are taking place in urban growth patterns, and then examines the rela-tionship between urban livelihoods and urban housing and the shifting development policy environment.

*****

The geographical themes in this volume on eastern and southern Africa are all examined in relation to the 'development' context. Individual authors have interpreted that context and its implications for people's livelihoods and national economic and political progress in different ways, but all have engaged with key contemporary debates about the measure-ment and assessment of development and development problems, and/or the influence of development policy trends. All have endeavoured to provide regional coverage of their topics to allow intra-regional and international comparisons to be made: this in itself provides an essential pointer to the (continued) value of regional geography and regional specialisation (cf. Parnwell 1999), for the variability of the 'development geographies' of individual nations (and thus their development policy requirements) is thereby made clear. The role of contemporary regional geography in a globalising world will be returned to in the concluding chapter.

References

Abshire D.M. and **Samuels, M.A.** (eds) (1969) *Portuguese Africa: A handbook*. Pall Mall, London.

Adams, M., Sibanda, S. and **Turner, S.** (1999) Land tenure reform and rural livelihoods in Southern Africa, *Natural Resource Perspectives* **39**.

African Urban Quarterly (1991) Special issue on *AIDS, STDs and Urbanization in Africa* **6**(1/2).

African Urban Quarterly (1992) Special issue on *AIDS in Eastern and Southern Africa* **7**(3/4).

Ahwireng-Obeng, F. and **McGowan, P.** (1998) Partner or hegemon? Part I. South Africa in Africa, *Journal of Contemporary African Studies* **16**(1): 5–38.

Ashley, C. and **Carney, D.** (1999) *Sustainable Livelihoods: Lessons from early experience*. Department for International Development, London.

Banatar, S. (2001) South Africa's transition in a globalizing world: HIV/AIDS as a window and a mirror, *International Affairs* **77**(2): 347–75.

Barnett, I. *et al.* (1995) Field report: the social and economic impact of HIV/AIDS on farming systems and livelihoods in rural Africa: some experience and lessons from Uganda, Tanzania and Zambia, *Journal of International Development* **7**(1): 117–34.

Barnett, T. and **Blaikie, P.** (1992) *AIDS in Africa: Its present and future impact*. Pinter, London.

Bender, G.J. (1978) *Angola under the Portuguese: Myth and reality*. Heinemann Educational, London.

Binns, T. (ed.) (1995) *People and Environment in Africa*. John Wiley, Chichester.

Browne, A. and **Barrett, H.** (1995) *Children and AIDS in Africa*. School of International Studies and Law, Occasional Papers Series 2, Coventry University, Coventry.

Caldwell, J. (1997) The impact of the African AIDS epidemic, *Health Transition Review*, Supplement 2 to **7**: 169–88.

Caldwell, J.C. and **Caldwell, P.** (1995) The nature and limits of the sub-Saharan African AIDS epidemic: evidence from geographic and other patterns, in Orubuloye, I.O. Caldwell, J.C., Caldwell, P. and Santow, G. (eds) *Sexual Networking and AIDS in Sub-Saharan Africa: Behavioural research and the social context.* Australian National University, Canberra.

Carney, D. (1999) *Approaches to Sustainable Livelihoods for the Rural Poor.* Overseas Development Institute, London.

Cleaver, K. (1997) *Rural Development Strategies for Poverty Reduction and Environmental Protection in Sub-Saharan Africa.* World Bank, Washington, DC.

Copestake, J. (1998) Coping with change in Zambia: how farmers fare after structural adjustment and other shockwaves [nt1.ids.ac.uk/id21/static/1cJC1.htm].

Corbridge, S. and **Harriss, J.** (2002) The shock of reform: the political economy of liberalisation in contemporary India, in Bradnock, R. and Williams, G. (eds) *South Asia in a Globalising World: A reconstructed regional geography.* Pearson Education, Harlow.

Cross, S. and **Whiteside, A.** (eds) (1993) *Facing up to AIDS: The socio-economic impact in Southern Africa.* St. Martin's Press, New York; Macmillan, London.

Cuddington, J.T. and **Hancock, J.D.** (1995) The macro-economic impact of AIDS in Malawi: a dualistic, labour surplus economy, *Journal of African Economies* **4**(1): 1–28.

Daniel, M. (2000) The demographic impact of HIV/AIDS in sub-Saharan Africa, *Geography* **85**(1): 46–55.

Davies, R. and **O'Meara, D.** (1985) Total strategy in Southern Africa: an analysis of SA regional policy since 1978, *Journal of Southern African Studies* **11**(2): 183–211.

Dixon, D. (1999) The Pacific Asian challenge to neoliberalism, in Simon, D. and Närman, A. (eds) *Development as Theory and Practice: Current perspectives on development and development co-operation.* DARG Regional Development Series No. 1, Longman, Harlow: 205–29.

Essex, M., Mboup, S., Kanki, P. and **Kalengayi, M.** (ed.) (1994) *AIDS in Africa.* Raven, New York.

Financial Times (2001) Congo facilitator wary about history repeating itself at power-sharing talks, 12 September.

Fine, B. (2001) *Social Capital versus Social Theory: Political economy and social science at the turn of the millennium.* Routledge, London and New York.

Fine, B., Lapavitsas, C. and **Pincus, J.** (eds) (2001) *Development Policy in the Twenty-first Century: Beyond the post-Washington consensus.* Routledge, London and New York.

Fine, B. and **Rustomjee, Z.** (1995) *The Political Economy of South Africa: From minerals–energy complex to industrialisation?* Hurst, London.

Good, K. and **Hughes, S.** (2002) Globalization and diversification: two cases in southern Africa, *African Affairs* **101**(402): 39–60.

Gregson, S. (1994) Will HIV become a major determinant of fertility in sub-Saharan Africa? *Journal of Development Studies* **30**(3): 650–79.

Gregson, S., Garnett, G. and **Anderson, R.** (1994) Assessing the potential impact of the HIV-1 epidemic on the orphaned and the demographic structure of populations in sub-Saharan Africa, *Population Studies* **48**(3): 435–58.

Grieser, M., Gittelsohn, J., Shankar, A., Koppenhaver, T., Legrand, T., Marindo, R., Mavhu, W. and **Kenneth Hill, K.** (2001) Reproductive decision-making and the HIV/AIDS epidemic in Zimbabwe, *Journal of Southern African Studies*, special issue on Fertility in Southern Africa **27**(2): 225–43.

Hammond, R.J. (1966) *Portugal and Africa 1815–1910: A study in uneconomic imperialism.* Stanford University Press, Stanford.

Hanlon, J. (1986) *Beggar Your Neighbours: Apartheid power in Southern Africa.* James Currey, London; Indiana University Press, Bloomington.

Hanlon, J. (1989) *SADCC in the 1990s: Development in the front line.* Special report 1158, EIU, London.

Hanlon, J. (1997) *Mozambique: Peace without profit.* James Currey, Oxford.

Hanlon, J. (2002) The land debate in Mozambique: will foreign investors, the urban elite, advanced peasants or family farmers drive rural development? Research paper commissioned by Oxfam GB – Regional Management Center for Southern Africa [www.oxfam.org.uk/landrights/debatMoz.doc].

Harrigan, J. (1991) Malawi, in Mosley, P., Harrigan, J. and Toye, J. (eds) *Aid and Power: The World Bank and policy-based lending*, vol. 2. Routledge, London: 201–69.

Henshaw, P. (1998) The 'key to South Africa': Delagoa Bay and the origins of the South Africa war, 1890–1899, *Journal of Southern African Studies* **24**(3): 527–43.

Holden, S. (1997) Adjustment policies, peasant household resource allocation and deforestation in Northern Zambia: an overview and some policy considerations, *Forum for Development Studies* **1**: 117–34.

Jauch, H. (2002) Export processing zones and the quest for sustainable development: a southern African perspective, *Environment and Urbanization* [Globalization and Cities], **14**(1): 3–12.

Johnson, P. and **Martin, D.** (1989) *Apartheid Terrorism: The destabilization report.* Commonwealth Secretariat in association with James Currey, London; Indiana University Press, Bloomington.

Kennedy, D. (1987) *Islands of White: Settler society and culture in Kenya and Southern Rhodesia 1890–1939.* Duke Centre for International Studies Publications, Duke University Press, Durham, NC.

Lamboray, J., Elmendorf, L. and **Edward A.** (1992). *Combatting AIDS and Other Sexually Transmitted Diseases in Africa.* Discussion Paper No. 181, World Bank, Washington.

Leach, M. and **Mearns, R.** (eds) (1996) *The Lie of the Land.* James Currey, Oxford.

Lele, U. (1990) Structural adjustment, agricultural development and the poor: some lessons from the Malawian experience, *World Development* **18**(9): 1207–19.

Maharaj, B. and **Ramballi, K.** (1998) Local economic development strategies in an emerging democracy: the case of Durban in South Africa, *Urban Studies* **35**(1): 131–42.

Mosley, P. (1983) *The Settler Economies. Studies in the economic history of Kenya and Southern Rhodesia 1900–63.* Cambridge University Press, London, New York.

Munslow, B. (1982) *Mozambique: The revolution and its origins.* Longman, London.

Myers, G. (1994) Competitive rights, competitive claims: land access in post-war Mozambique, *Journal of Southern African Studies* **20**(4): 603–32.

Nel, E. (1996) Export processing zones in Zimbabwe: establishment considerations, *Geographical Journal of Zimbabwe* **27**: 1–10.

Newitt, M. (1995) *A History of Mozambique.* Hurst, London.

Ng'ong'ola, C. (1986) Rural developement and reorganization of customary land: lessons from the Lilongwe Land Development Programme, *Journal of Social Science [Malawi]*, special issue on Agriculture and related development in Malawi, **13**.

Okoth-Ogendo, H. (2000) Legislative approaches to customary tenure and tenure reform in East Africa, in Toulmin, C. and Quan, J. (eds) *Evolving Land Rights, Policy and Land Tenure in Africa.* DfID/IIED/NRI, London: 123–34.

Oygard, R. (1997) Structural adjustment policies and land degradation in Tanzania, *Forum for Development Studies* **1**: 75–93.

Palmer, R. (1997) *Contested Lands in Southern and Eastern Africa: A literature survey.* Oxfam Working Paper, Oxfam, Oxford.

Parnwell, M. (1999) Between theory and reality: the area specialist and the study of development, in Simon, D and Närman, A. (eds) *Development as Theory and Practice: Current perspectives on development and development co-operation.* DARG Regional Development Series no. 1, Longman, Harlow: 76–94.

Peters, P. (2002) Bewitching land: the role of land disputes in converting kin to strangers and in class formation in Malawi, *Journal of Southern African Studies*, special issue on Malawi **28**(1): 155–78.

Platteau, J.P. (1996) The evolutionary theory of land rights as applied to sub-Saharan Africa: a critical assessment, *Development and Change* **27**(1): 29–85.

Platteau, J.P. (2000) Does Africa need land reform? in Toulmin, C. and Quan, J. (eds) *Evolving Land Rights, Policy and Land Tenure in Africa*. DFID/IIED/NRI, London: 51–74.

Pottier, J. (2003) Pers.comm.

Potts, D. (1992) The changing geography of southern Africa, in Chapman, G. and Baker, K. (eds) *The Changing Geography of African and the Middle East*. Routledge, London: 12–51.

Potts, D. (1997) Urban lives: adopting new strategies and adapting rural links, in Rakodi, C. (ed.) *The Urban Challenge in Africa: Growth and management of its large cities*. United Nations University Press, Tokyo: 447–9.

Potts, D. (2000a) Urban unemployment and migrants in Africa: evidence from Harare, 1985–94, *Development and Change* **31**(4): 879–910.

Potts, D. (2000b) Worker-peasants and farmer-housewives in Africa: the debate about 'committed' farmers, access to land and agricultural production, *Journal of Southern African Studies*, special issue on Southern African Environments **26**(4): 807–32.

Potts, D. (2000c) Environmental myths and narratives: case studies from Zimbabwe, in Stott, Philip and Sullivan, Sian (eds) *Political Ecology: Science, myth and power*. Edward Arnold, London, New York; Oxford University Press: 45–65.

Potts, D. and **Marks, S.** (2001) Fertility in Southern Africa: the silent revolution, *Journal of Southern African Studies*, special issue on Fertility in Southern Africa **27**(2): 189–206.

Potts, D. and **Mutambirwa, C.C.** (1990) Rural–urban linkages in contemporary Harare: why migrants need their land, *Journal of Southern African Studies* **16**(4): 676–96.

Rakodi, C. with **Lloyd-Jones, T.** (eds) (2002) *Urban Livelihoods: A people-centred approach to reducing poverty*. Earthscan, London; Sterling, Va.

Regional Roundup (2002) Information from the SADC Press: Democratic Republic of Congo, 29 October [citing www.iht.com/articles/75123]

Sachikonye, L. (ed.) (1998) *Labour and Migration in Southern Africa* [Zimbabwe]. SAPES, Harare.

Scoones, I. (1997a) Landscapes, fields and soils: understanding the history of soil fertility management in southern Zimbabwe, *Journal of Southern African Studies* **23**(4): 615–34.

Scoones, I. (1997b) The dynamics of soil fertility change: historical perspectives on environmental transformation from Zimbabwe, *Geographical Journal*. **163**(2): 161–9.

Sidaway, J. (1998) The (geo)politics of regional integration: the example of the Southern African Development Community. *Environment and Planning D: Society and Space* **16**: 549–76.

Simon, D. (1999a) Series preface, in Simon, D and Närman, A. (eds), *Development as Theory and Practice: Current perspectives on development and development co-operation*. DARG Regional Development Series no. 1, Longman, Harlow: xi–xii.

Simon, D. (1999b) Development revisited: thinking about, practising and teaching development after the Cold War, in Simon, D and Närman, A. (eds) *Development as Theory and Practice: Current perspectives on development and development co-operation*. DARG Regional Development Series no. 1, Longman, Harlow: 17–54.

Simon, D. (2001) Trading spaces: imagining and positioning the 'new' South Africa within the regional and global economies, *International Affairs* **77**(2): 377–405.

Simon, D. and **Johnston, A.** (1999) *The Southern African Development Community: regional integration in ferment*, RIIA, Southern African Study Group, Briefing Paper No. 8.

Simone, A. (1998) The prospects for local economic development in Winterveld, in *Case Studies on LED and Poverty*. Isandla, Johannesburg.

Smith, A.K. (1974) Antonio Salazar and the reversal of Portuguese colonial history, *Journal of African History* **4**.

Smith, S. (1990) *Front Line Africa: The right to a future: an Oxfam report on conflict and poverty in southern Africa*. Oxfam, Oxford.

Spence, J.E. (1965) British policy towards the HCTs, *Journal of Modern African Studies* **2**(2).

Spence, J.E. (1968) *Lesotho: The politics of dependence*, Oxford University Press for the Institute of Race Relations, London.

Sporton, D., Thomas, D. and **Morrison, J.** (1999) Outcomes of social and environmental change in the Kalahari of Botswana: the role of migration, *Journal of Southern African Studies* **25**(3): 441–59.

Stevens, R.P. (1967) *Botswana, Lesotho and Swaziland*. Pall Mall, London.

Stevens, R.P. (1972) History of Anglo-South African conflict over the proposed incorporation of the HCTs, in Dale, C. and Potholm, R. (eds) *Southern Africa in Perspective: Essays in regional politics*. Free Press, New York; Collier-Macmillan, London.

Stiglitz, J. (1998) More instruments and broader goals: moving toward the post-Washington consensus, 1998 WIDER Annual Lecture, Helsinki, Finland, 7 January.

Tibaijuka, A.K. (1997) AIDS and economic welfare in peasant agriculture: case studies from Kagabiro village, Kagera Region, Tanzania. *World Development* **25**(6): 963–75.

Toulmin, C. and **Quan, J.** (2000) Evolving land rights, policy and land tenure in Africa, in Toulmin, C. and Quan, J. (eds) *Evolving Land Rights, Policy and Land Tenure in Af*rica. DfID/IIED/NRI, London.

UNDP (United Nations Development Programme) (2002) *Human Development Report 2002*. Oxford University Press, New York.

Vail, L. and **White, L.** (1980) *Capitalism and Colonialism in Mozambique: A study of Quelimane District*. Heinemann, London.

Wallman, S. (ed.) (1996) *Kampala Women Getting By: Wellbeing in the time of AIDS*. James Currey, Oxford.

Webb, D. (1997) *HIV and AIDS in Southern Africa*. Pluto, London; David Philip, Cape Town.

Whiteside, M. (1998) Encouraging sustainable smallholder agriculture in southern Africa in the context of agricultural services reform, *Natural Resources Perspectives* [ODI] **36**.

World Bank (2000) *World Development Report 2000/2001*. Special edition: *Attacking Poverty*. Oxford University Press, New York.

World Bank (2002) *World Development Report 2003*. Oxford University Press, New York.

CHAPTER 2

Demographic change in eastern and southern Africa

Akim J. Mturi

Introduction

Over the past 20 years the countries in eastern and southern Africa have experienced some dramatic demographic changes. Demography can be defined as 'the scientific study of human populations, including their sizes, compositions, distributions, densities, growth, and other characteristics, as well as the causes and consequences of changes in these factors' (Haupt and Kane 1998). The emphasis in this chapter is on changes in fertility and mortality; significant population movements in the region as a result of rural–urban migration and the displacement of people by political violence have also occurred and are covered elsewhere in this volume (see Chapter 11 on Urbanisation and Chapter 9 on Geographies of Governance and Regional Politics).

Population size

The starting point for examining the population geography of this region is population size. Table 1.1 in the introductory chapter shows the UNDP's estimates for each country for 2000. It is useful to bear in mind that such estimates are based on projections from the population enumerated in the latest population census conducted in each country, and the more recent that census, the more reliable the estimates are likely to be. The latest published South African census was conducted in 1996, for example (although another was held in late 2001), but the last full census in Angola was conducted in 1970. In addition, in some countries the demographic data are less reliable than others or even unavailable. In this chapter the analysis of regional differences and similarities in demographic change will therefore tend to be based on those countries where the data are most reliable.

In terms of numbers of people, the largest nations in eastern and southern Africa include the Democratic Republic of Congo, South Africa, Kenya and Tanzania, all with over 30 million, and the other countries range down to Swaziland, Botswana, Namibia and Lesotho, all with very small populations of less than 2 million. Population numbers are, as mentioned, only the starting point for analysing a region's population geography (although they have direct relevance in terms of market size and political 'clout' in the international arena). Particularly in a region like this where, in the majority of the countries under consideration, most livelihoods are still based on the natural resource base, the land area and population : land ratio are other vital indices. As indicated in Table 1.1, country size is even more variable than population size, with the region including the second largest in sub-Saharan Africa (the Democratic Republic of Congo (DRC)), followed by Angola and South Africa. The smallest countries include Swaziland, Rwanda, Burundi and Lesotho. There is some congruence between ranking in country size and population (e.g. DRC, South Africa, Swaziland and Lesotho) but of far more significance are the cases where there is incongruence, translating into very high or low population densities. Particularly notable are tiny Rwanda and Burundi, where populations of around 6 million produce the highest national population densities in Africa of 229 and 292 people per square kilometre (three times higher than the next most densely populated countries, Malawi and Uganda), and the large countries of Angola and Zambia with populations less than twice those in Rwanda and Burundi, and very low population densities of only 10 and 14 people per square kilometre (and thus much scope for agricultural expansion and development). The even lower densities found in Botswana and Namibia, on the other hand, are largely due to the aridity of their environments which is a severe constraint on rural densification in the absence of irrigation. Direct comparisons of density are further complicated by very varied environments *within* some countries: in Kenya, for example, two-thirds of the land area is either semi-arid or arid, leaving only one-third suited to arable agriculture. Varied agro-ecological conditions within countries are often reflected in internal variations in crude densities; thus in Tanzania, for example, Kilimanjaro region had 84 persons per square kilometre in 1988, compared to only 10 in Lindi region (Bureau of Statistics 1994).

In terms of crude population density, a second group of countries in the region are relatively densely populated: Uganda, Malawi and Lesotho where population densities range from 67 to the high nineties.[1] At a very general level East Africa is more densely populated than southern Africa.

Fertility

Fertility refers to the number of live births in a society and is usually measured by the total fertility rate (TFR) – the average number of children

---

[1] Note that if only arable land is used, the population density of Lesotho increases from 70 to 588.

that would be born alive to a woman during her lifetime if she were to bear children at the prevailing age-specific rates. TFRs for each country in East and southern Africa are given in Table 2.1. They may not be strictly comparable because they refer to different time periods, but they do give an indication of the situation. Great care has been taken to use the most reliable sources available. These include the Demographic Health Surveys (DHS)[2] which have been run in many African countries and are an important source of demographic data where census data are unreliable or very out of date. Even for Angola, where war has rendered most development-related statistics guesstimates at best, two demographic surveys in the mid-1990s commissioned by the World Bank and UNICEF did produce some reasonable demographic data (Agadjanian and Prata 2001). As shown, South Africa has the lowest fertility rate and the evidence suggests that it is still declining quite steeply. For the period 1990–95 it was estimated at 4.0 (Kirk and Pillet 1998) and it had dropped to 2.9 by 1998. The second lowest current fertility rate is found in Lesotho. A national survey conducted in 1991/92 estimated a TFR of 4.8 (Mturi and Hlabana 1999) and the 1996 population census indicated further moderate decline to 4.1.

Table 2.1   Total fertility rates

Country	Total fertility rate	Census/survey year
Angola[a]	6.9	1996
Botswana[b]	4.3	1996
Burundi[c]	6.9	1987
DRC[d]	6.7	1984
Kenya[c]	5.4	1993
Lesotho[e]	4.1	1996
Malawi[c]	6.7	1992
Mozambique[g]	5.6	1997
Namibia[c]	5.4	1992
Rwanda[c]	6.2	1992
South Africa[f]	2.9	1998
Swaziland[d]	5.0	1988
Tanzania[c]	5.8	1996
Uganda[c]	6.9	1995
Zambia[c]	6.1	1996
Zimbabwe[c]	4.3	1994

Sources
[a] Prata (1999).
[b] Central Statistical Office (1999).
[c] Demographic and Health Surveys (1997).
[d] Gould and Brown (1996).
[e] Mturi and Hlabana (1999).
[f] Medical Research Council, Macro International Inc. and Department of Health (1999).
[g] Instituto Naçional de Estatistica and Macro International Inc. (1998).

[2] The DHS surveys are conducted by Macro International Inc. based in Maryland USA and are available on the web at www.macroint.com.

Perhaps the most documented fertility declines in sub-Saharan Africa during the 1980s and 1990s have been in Botswana and Zimbabwe where the pace has been faster than other countries in the region (with the exception of Kenya). As Table 2.1 indicates, fertility in eastern Africa is generally higher than in southern Africa. Nevertheless, Kenya's fertility rate of 5.4 in 1993 represents the most dramatic scenario of fertility decline in the region since, in the 1970s, Kenya had a record national fertility rate of 8.1 and the highest rate of natural increase in sub-Saharan Africa. Tanzanian fertility has also declined but at a more moderate rate, from 7.2 during the 1970s to 5.8 in 1996 (Hinde and Mturi 2000). These are the only countries in eastern Africa where significant fertility decline has occurred so far; all the other countries in the sub-region still have very high fertility levels and any measured decline has been minimal. Another category of countries observable in Table 2.1 includes those which were/are affected by civil wars and other conflicts. Fertility levels in Angola, Burundi, the DRC, Mozambique, Uganda and Rwanda are the highest among the countries covered.

Three conclusions can be drawn from this discussion. First, fertility is lowest in southern Africa – and South Africa is exceptional with a fertility regime for all races very unlike those typical of sub-Saharan Africa. Second, fertility decline is clearly under way in most of southern Africa (see Potts and Marks 2001) but is only firmly established in selected countries in eastern Africa. Third, countries that have experienced civil wars for many years are lagging behind in experiencing fertility decline (see, for example, Agadjanian and Prata 2001).

Fertility levels and trends can be influenced by a number of factors. Some affect fertility directly (usually referred to as proximate determinants of fertility), while others affect it indirectly (background determinants) by operating through proximate determinants. The major proximate determinants of fertility in sub-Saharan Africa include contraception uptake, marriage or union patterns, post-partum infecundability, and pathological or primary sterility (Jolly and Gribble 1993). These are discussed in turn below. The factors that affect fertility indirectly include socio-economic and demographic factors such as education, occupation, mortality, income, etc. The interplay of both sets of factors determines the levels and trends of fertility in a given population. For instance, high levels of maternal and paternal education in a society can be associated with lowered infant and child mortality and high contraceptive prevalence, both of which are associated with lowered birth rates. A change in fertility is thus associated with a decreasing or rising trend in one or more of the proximate determinants.

Proximate determinants of fertility in eastern and southern Africa

The evidence from sub-Saharan Africa is that the uptake of contraception is the *major* contributing factor to fertility declines (Caldwell *et al.* 1992; Gould and Brown 1996) and this has been very important in East and southern Africa. In this region, countries where substantial fertility decline has been observed, such as South Africa, Botswana, Kenya and

Zimbabwe, have contraceptive prevalence rates (the proportion of married women aged 15–49 currently using a contraceptive method) of well over 30 per cent. The other countries where fertility has also been declining, albeit at a slower pace (such as Tanzania and Lesotho), also have relatively high contraceptive prevalence rates (CPRs) compared to those where there is little or no fertility decline. Several countries in the region, including Burundi, the DRC and Mozambique, still have rates below 10 per cent, and such low levels of contraceptive prevalence in sub-Saharan Africa are usually associated with high levels of fertility. The strength of the negative association between contraceptive prevalence and fertility is illustrated for East and southern Africa in Figure 2.1. On average, contraceptive prevalence in southern Africa (47 per cent) is three times higher than that of eastern Africa (16 per cent).

The CPRs discussed in the preceding paragraph are based on what are often termed 'modern' and 'traditional' methods of family planning. The Demographic Health Surveys define modern methods as the pill, IUD, injectables, Norplant, vaginal methods (foam, jelly or diaphragm), condom, and female and male sterilization and traditional methods as the calendar (rhythm) method, mucus method and withdrawal. These are the methods used to compute CPRs in DHS surveys. In reality the distinction being made is between scientific and non-scientific 'modern' methods (or methods which are familiar to European societies) for, despite the nomenclature, 'traditional' often excludes various indigenous practices which have long been part of fertility control in African societies. These methods have received less attention in demographic literature (see, for example, Agadjanian 1998; Page and Lestaeghe 1981; Newman and Lura 1983; Mutambirwa 1979). Demographers regard them as methods whose mode of operation or use has no proven scientific or biological bearing on the reproductive process (Zulu 1998). They include the wearing of waist strings,

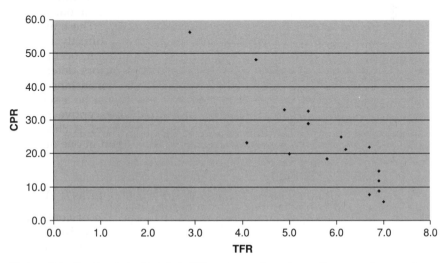

Figure 2.1   Scatterplot of total fertility rates and contraceptive prevalence rates
   Source: UN (1999a)

the position method, the use of traditional oral medicines to stop bearing children and various taboos and customs which vary between groups. With-drawal is also known to have been used 'traditionally' (e.g. in Zimbabwe). An overly 'scientific' analysis of the effectiveness of customary methods may, however, miss the point. Taboos against intercourse, for example, can be wholly effective as contraceptives *if* they are observed. Waist strings may signal to a man that a woman is not sexually available – perhaps because she is breastfeeding as there is a strong belief in many African societies that sperm can 'poison' breast milk and make the infant ill. Since traditionally many mothers breastfed infants for two years or longer, in effect, when observed, this taboo greatly increased child-spacing (with hugely beneficial results for infant nutrition, maternal health and child mortality) and lowered fertility. In the demographic literature the gap between a birth and the resumption of the mother's sexual activity is known as post-partum abstinence. There is some confusion about how to classify periodic and prolonged abstinence in terms of the DHS categories of 'modern' and 'traditional' since it is not 'modern' or 'scientific' contracep-tion but is evidently of proven scientific efficacy. In much of the literature on sub-Saharan African fertility, abstinence is discussed together with breastfeeding since these two are closely interlinked, as discussed above. Full breastfeeding in any case delays the return of ovulation, and has been one of the most important regulators of fertility in all human societies throughout history.

The timing of entry into first sexual union and the proportion of women in sexual union in any society indicate the degree to which women of re-productive age are exposed to the risk of becoming pregnant (see Box 2.2 for an illustration of the significance of this factor). Where age at first marriage is low and is coupled with low divorce (or separation) rates, the exposure to pregnancy is very high. Increases in age at first marriage can reduce that exposure. For example, changes in nuptiality patterns in Tanzania are implicated in the decline of fertility there from 7.2 in 1978 to 6.2 in 1989/90 and 5.8 by 1994–96 (Lejeune and Hinde 1999). Analysis of birth cohort data from a DHS survey in 1996 found a clear increase of more than two years in the median age at marriage for women born in the 1970s compared to those born in the 1940s, and postponement of marriage is argued to be the main factor in the decline of the TFR in the 1990s (Hinde and Mturi 2000).

However, there is a debate on whether the importance of *marriage* as a proximate determinant of fertility is still valid, given the rising trend in non-marital births (many of these being pre-marital), particularly in south-ern Africa. In Botswana, for example, marital births account for not much more than half of all births; in 1988 non-marital births increased the TFR by 89 per cent (Jolly and Gribble 1993) (see also Box 2.1). Research on South African and Namibian non-marital fertility has also indicated that the high incidence of non-marital childbearing in these countries means that the fertility-inhibiting effects of marriage are now small (Chimere-Dan 1997, 1999). Although this is an area that needs thorough assessment, it can be generalised that countries in southern Africa have higher rates of non-marital childbearing, later ages at first marriage and higher divorce

Box 2.1 Adolescent and extra-marital fertility in Botswana

Marriage is no longer common in Botswana. The 1991 Botswana population census data found that only 24 per cent of women aged over 12 were married compared to 56 per cent who had never been married. Eleven per cent were living with male partners but were not married. The proportion of never married women of childbearing age (15–49 years) was 58 per cent. This breakdown of the institution of marriage has been a major contributor to pre- and extra-marital childbearing in Botswana. In 1991, for 6 out of every 10 women who had given birth, childbearing had occurred outside of any formalised union.

One factor which has contributed to the breakdown of the institution of marriage in this country is the very long-standing tradition of male labour migration which has been such a common phenomenon in southern Africa. For generations, from the late nineteenth century, a significant proportion of the male labour force in Botswana migrated to South Africa to work in the mines. It has been argued that though this labour migration was an important element of Batswana household economies, it also weakened local family structures. It separated husbands and wives for long periods of time, leading to a breakdown of domestic control. As a result, children tended to do much as they liked and took little notice of their mother, especially if she had taken a lover. The disorganisation of the family and associated changes in sexual and reproductive behaviour contributed a great deal to the breakdown of the institution of marriage. Furthermore, men could delay the age at which they married until after their migrant careers, since local accumulation of wealth within Botswana was not contingent upon being married to a woman and a consequent allocation of arable land upon which she could cultivate. Instead, in semi-arid Botswana, wealth was accumulated in cattle holdings which could be tended by male relatives.

Women in Botswana are also argued to be educating themselves out of the marriage market. For example, it has been reported that 'numerous female teachers remain unmarried, stating explicitly that they are doing so rather than being beaten by foolish, illiterate husbands'. Many of these women go on to have children outside of marriage. Their education gives such women a degree of freedom from economic dependence on men.

Despite these changes in marriage patterns, sexual activity and childbearing start at an early age in Botswana. Women's desire to bear children remains very strong and girls and young women in Botswana will go ahead and have children whether they are married or not. Although such behaviour would have been socially unacceptable a few generations ago, the pattern of extra-marital fertility is now sufficiently long-standing, and so entrenched, that it is no longer frowned upon.

Based on: Gaisie (1998); Timaeus and Graham (1989); Gulbrandsen (1986).

rates than East African countries, although within southern Africa the patterns are quite variable (Mturi and Moerane 2001).

Despite the high rates of non-marital childbearing in much of southern Africa, total fertility there is still lower than it is in East Africa. Evidently, therefore, the factors responsible for reducing fertility in southern Africa are more significant than this factor which enhances it. Indeed some of the same factors which reduce fertility overall in southern Africa may simultaneously enhance non-marital fertility. These include urbanisation, higher educational attainment for women, and women's economic status. Urbanisation has been accused of eroding traditional practices that militated against pre-marital sexual activities and pregnancies. Improvements in girls' educational attainment prolong the period between menarche and marriage which may, in turn, increase the chances of an *unmarried* girl having sex. Older adolescent girls are also often at risk because of the attention of teachers and older boys at school, as well as other men outside the school environment. The latter can involve older men who 'buy' girls' affections with gifts and outings (the so-called 'sugar daddy' phenomenon). Young women who are working and have enough income to take care of themselves and their children, may also decide to have children whether married or not. This has been reported in Botswana, for example (see Box 2.1).

The duration, intensity and pattern of breastfeeding are further important proximate determinants of fertility as they have an impact on the length of lactational amenorrhea (the period after birth when a woman cannot conceive until her normal pattern of ovulation returns). Breastfeeding in East and southern Africa is still almost universal and prolonged, with the sole exception of South Africa (Medical Research Council, Macro International Inc. and Department of Health 1999). In Tanzania, for example, the median duration of breastfeeding is over 21 months (Hinde and Mturi 2000) and in Lesotho it is 20 months (Mturi and Hlabana 1999). The mean duration of breastfeeding in a range of other countries in the region in the early 1990s was reported as 19.2 months in Botswana, 23.9 months in Burundi, 20.1 months in Kenya, 19.1 months in Uganda and 18.2 months in Zimbabwe (Jolly and Gribble 1993). As noted already, breastfeeding has very significant fertility-inhibiting effects and it is probably the strongest of all the proximate determinants of fertility in eastern and southern Africa (with the exception of South Africa). However, because the trend in breastfeeding has not changed much over time, it has not contributed to the fertility *decline* observed in the region. Indeed, any trend in breastfeeding, particularly in parts of southern Africa, is probably towards a shortening of the period thereby enhancing fertility in the absence of modern contraception. However, evidently this impact has in the 1980s and 1990s been overwhelmed by the combined effect of all the other factors which are reducing fertility there.

In many African cultures, the resumption of intercourse is linked with weaning. Breastfeeding and sex, as discussed above, are frequently believed to be incompatible. Many cultures in this region therefore do not allow sexual intercourse while breastfeeding (e.g. Ndeti and Ndeti 1977), resulting in the prolonged duration of post-partum abstinence. In some cases such periods of abstinence last beyond the end of breastfeeding.

This happens when abstinence is used as a regulator of birth intervals. Traditionally, short birth intervals such as one year were not acceptable in many societies and a woman who conceived shortly after a birth would thus feel embarrassed. African fertility levels have tended historically to be significantly below the biological maximum and the primary factors involved were the long duration of breastfeeding and the long duration of post-partum abstinence, which had significant 'contraceptive' effects long before the introduction of 'modern' methods of contraception.[3] Other factors that traditionally reduced fertility levels below the biological maximum in many (but not all) African societies in this region included strong disapproval of sexual intercourse among unmarried people and of extra-marital fertility. Such 'morality' factors can sometimes have unforeseen consequences, however. For example, throughout the region unmarried women often fear to access modern contraception because they perceive (correctly) that they will meet with moral disapproval and, perhaps, that confidentiality will not be respected (e.g. Garenne *et al.* 2001; Mturi and Moerane 2001). This leads to many unwanted pregnancies today as historic sociocultural and economic norms which bolstered this 'morality' are rarely still in effect. Box 2.2 further illustrates the fertility-enhancing possibilities of 'moral' decisions relating to female fertility and sexuality in a specific Kenyan example (see also Gaitskell (1982) for a South African example).

Sterility and induced abortion are other proximate determinants of fertility. The unavailability of data on induced abortion makes it difficult to assess its impact on fertility in this region, however. In all the countries in East and southern Africa, with the exceptions of South Africa and Zambia, induced abortion is illegal unless performed to save the life of the mother. So, although it is known that induced abortion is not uncommon in the region, most of it goes unrecorded. Nevertheless, its impact on fertility levels (relative to the other proximate determinants discussed above) is generally considered to be minimal. Pathological sterility, caused mainly by pelvic inflammatory disease as a result of sexually transmitted infections such as gonorrhea (Gray 1982), is also of little significance for fertility in this region of Africa (Jolly and Gribble 1993) and in general sterility is not a major demographic factor.

Indirect influences on fertility

In the early 1990s, Caldwell *et al.* (1992) argued that the fertility transition in sub-Saharan African countries was different in some ways from that experienced elsewhere in the world. Their analysis rested on three countries in East and southern Africa – Kenya, Botswana and Zimbabwe – where significant fertility decline was observable. The common and relevant characteristics of these three countries included the following:

[3.] Bongaarts and Potter (1983) suggest that in the absence of contraception, if everyone marries (and if there is no marriage delay or breakdown) and there is no breastfeeding or abstinence, the number of children per woman in populations will vary from 13 to 17, with an average near 15.

Box 2.2   Fertility rises caused by modernisation

The pastoral Kipsigis of Kenya live mainly in Kericho district and belong to a linguistically related group of people called Kalenjin. In this ethnic group, children were traditionally highly valued members of a household for both emotional and economic reasons. In spite of a desire for many children, in the precolonial period women's fertility was not particularly high, and the TFR is estimated to have been under five from 1900 to 1925. Several Kipsigis' customs helped to restrain fertility then: late female initiation and age at marriage, breastfeeding of at least one year, long periods of post-partum sexual abstinence and a strong tradition of polygyny.

By the 1950s Kipsigis' cultural traditions surrounding the regulation of fertility had undergone considerable change and fertility had greatly increased. By 1948 the TFR was estimated at 9.0, and it increased further to a peak of 9.8 by 1962 – double the rate in the precolonial period. Several decades of colonialism had thus brought social, cultural and economic changes which tended to increase fertility levels.

The most important proximate determinant of increased fertility among Kipsigis women was reduction of the age at marriage to as low as 13 years. This decline was due to influence exerted by the missionaries and colonial government. The customary late age of marriage had been accompanied by some pre-marital sexual activity (and occasional infanticide, since pre-marital pregnancy was not sanctioned) which was greatly frowned on by the missions. By lowering the age of female initiation and marriage, youthful sex occurred within marriages and was, thus, deemed morally acceptable. The missionary influence had also decreased the incidence of polygyny, which among the Kipsigis was associated with lower fertility than monogamous marriages.

A number of indirect determinants of fertility were also responsible for the rise in the TFR. These included the move to a market economy and a change from pastoralism to mixed agriculture. Rising incomes brought about through agricultural innovation are thought to have allowed increased fertility among the Kipsigis. The initial stages of formal education for women were a further factor in the fertility rise.

The experience of the Kipsigis shows clearly that an assumption that 'traditional' African customs are always associated with very high fertility, and that 'modernisation' must be associated with a downward trend in fertility, is erroneous. Fertility trends among the Kipsigis in the post-colonial era have mirrored those of other Kenyan ethnic groups, and fertility began to fall by the 1970s.

Source: Lura (1985).

first, they alone in the region exhibited infant mortality rates below 70 per 1,000 live births; second, they had unusually high levels of education; and third, they were unique in their high levels of contraception. The relationship between contraceptive use (a proximate determinant) and fertility has already been explained. The influence of the other two indirect factors is considered below.

The mortality of children, especially infants, influences fertility in a number of ways. Three of these mechanisms are very important in the sub-Saharan African context. First, there is the *replacement effect*, where a couple who intend to achieve a certain number of children, attempt to have a replacement birth when a child dies before the end of reproduction. Second, there is the *insurance effect*, when a couple attempt to bear more children than their desired completed family size, because they expect that some may die. Third is the *physiological effect* when a child dies before the mother would normally have finished breastfeeding. In such a case, the period of infertility following a birth may be reduced and thus, in the absence of contraception, the interval between births may also be reduced, thus increasing the number of births over the woman's lifetime. All these fertility-enhancing effects are strongly positively associated with high infant mortality rates and therefore any reductions in these rates tend also to reduce fertility. Thus the improvements in infant mortality rates in Kenya, Botswana and Zimbabwe had facilitated their fertility transition.

The spread of education and literacy among women is believed to be fundamental to changes in reproductive behaviour (Caldwell 1980) but the impacts are quite complex and sometimes contradictory. In many developing countries the relation between the two variables is found to be curvilinear. That is, fertility tends first to rise with education and then to decrease sharply once a certain level of education is attained (United Nations 1995). The argument is that education is positively associated with improved health, lower levels of infertility, the abandonment of traditional constraints upon sexual behaviour and the practice of breastfeeding, all of which are known to raise fertility levels. As female education levels increase, however, marriage tends to be postponed which tends to reduce fertility and counteracts the initial effect of fertility increase. Moreover, educated women usually desire fewer children, have high contraceptive prevalence and have a higher chance of working outside their home. All of these factors are known to lower fertility levels. Research on the relationship between fertility and education in sub-Saharan Africa specifically, confirms that the two are generally inversely related there but has also found that fertility does not appear very responsive to small amounts of education (Cohen 1993). The evidence from Botswana, Kenya and Zimbabwe suggests therefore that, in those countries, female education has attained levels generally sufficient for the balance of forces influencing fertility to be profoundly negative.

Fertility regimes can also be influenced by male education. When a woman has a more educated husband (or partner) this tends to be associated with higher family income. On the one hand this can make children more affordable and possibly increase the demand for children, but on the other hand improved living conditions and status also tend to be associated with a desire for higher 'quality' children (who are more expensive) and thus a reduction in their number (Bulatao and Lee 1983).

Besides infant mortality and education, there are several other indirect demographic and socio-economic influences on fertility of significance in East and southern Africa. One is female employment resulting in improved income, although yet again this can have conflicting results. An initial effect of increased income may be to increase fertility levels by increasing the availability of health care and improving women's nutrition, both of which can result in increased fecundity (Lura 1985). There are, however, two major negative theoretical effects of women's employment on fertility which can be defined as the 'role incompatibility' and 'opportunity cost' hypotheses (United Nations 1985). The first argues that a woman may be expected to have fewer children if child rearing puts some restrictions on performing her duties as a worker. The second hypothesises that as labour market opportunities for women increase, the opportunity cost of rearing children (in terms of time and financial input) also increases, and that fertility therefore decreases. In East and southern Africa employment opportunities for women vary greatly between countries, hence the impact of such effects is also likely to be very variable. Furthermore, real employment opportunities for most women (i.e. those generating regular, reasonable incomes) have not been improving for decades in much of the region, so the 'opportunity cost' effect will usually not be operating in the way hypothesised. Nevertheless there has been a significant increase in women's employment in the informal sector in urban areas throughout East and southern Africa. Although much of this has been necessary to maintain, rather than improve, household incomes (Potts 1997; Rogerson 1997), both role incompatibility and opportunity cost effects on fertility are likely to be associated with this increased labour force participation.

Urban residence is generally associated with lower fertility than in rural areas and is another indirect determinant. The strength of the association in African countries has been illustrated in research by Cohen (1993). Higher education, higher work status, a more modern environment and aspirations for higher living standards are among the factors which can lead to fertility decline among urban women. Also, it is generally safe to assume that urban women have better knowledge of, and access to, modern methods of contraception than rural women. As can be seen, there are overlaps between the influence on fertility of female employment per se, and that of urbanisation.

The final factor considered here is polygny. This is generally argued to enhance child spacing in most African societies and thereby reduce fertility. Female abstinence, for instance, can be maintained more easily in a society that practises polygyny and where women's independence is restricted. This appears to be illustrated by evidence from Tanzania where pregnancies are more frequent in monogamous unions than polygynous unions (Henin 1979). Research by Lura (1985) on the history of fertility change among the Kipsigis of Kenya also supports this association (see Box 2.2). On the other hand, however, it has been argued that polygyny is negatively associated with contraception (Caldwell and Caldwell 1981). In part this may be because women married to the same man feel they must compete to bear children, particularly in societies where a woman's status depends on the number of her surviving children. The importance of bearing children for women's status still remains central in most East African and southern African societies (Potts and Marks 2001; Harrison and Montgomery 2001; Upton 2001). As with so many of the indirect determinants of fertility considered in this section, the impact of polygyny in this region is thus not straightforward.

Mortality

Comparisons of mortality between countries or regions usually use two indices. The first is the infant mortality rate (IMR), defined as the number of children dying before their first birthday, per 1,000 live births (the child mortality rate – the proportion dying before age five – is also often used). The second is life expectancy at birth which is the average number of years a newborn child is expected to live if current mortality patterns were to continue for the rest of the child's life. Childhood mortality is of particular interest to demographers because, when high, it tends to be largely due to poor nutrition and sanitation and inadequate or untimely preventive and curative health intervention. Childhood mortality can therefore be taken as a very useful indicator of the socio-economic status and the quality of life in a country.

Table 2.2 indicates that there is a gap in terms of IMRs between South Africa, Botswana, Namibia, Zimbabwe (all countries at the southern tip of the region) and the other countries although, as with fertility rates, Kenya stands out as somewhat exceptional in East Africa with lower rates than all its neighbours. At a rate of 37, Botswana had the lowest IMR in the 1990s, followed by South Africa (45), Zimbabwe (53), Namibia (57) and Kenya (62). These are known to be the most 'developed' countries in the region using other standard economic measures of development, and these IMRs compare with those experienced in west European countries in the 1950s.[4] It should be remembered, however, that in South Africa, and to a lesser extent in Namibia, these average rates mask sharp differences between the experiences of different races, with whites enjoying contemporary European rates, while black Africans have higher rates than the averages suggest. The average child (under five) mortality rate in South Africa in 1998 was 59, for example, below the African population group average of 64 and way above the white average of 15. Residing in the rural areas of a former homeland in South Africa, where socio-economic conditions are very poor due to the legacies of apartheid, can sharply increase African child mortality to levels of well over 100 (Mencarini 1999) which

[4.] The IMR in 1999 was estimated at 5.9 infant deaths per 1,000 live births for the UK and at a mere 3.6 in Sweden (Population Research Bureau 1999). However, the IMR in the UK at the beginning of the 1950s was 28 per 1,000, compared to 41 in Ireland, 60 in Greece and Italy, 62 in Spain and 45 in France and Belgium (United Nations 1986).

Table 2.2 Infant and child mortality rates, life expectancy at birth, and AIDS prevalence rates (countries ranked by IMR in 1990s)

Country	Infant mortality rate per 1,000 live births[a] (1960–65)	Infant mortality rate per 1,000 live births[b] (recent estimates)	Child mortality rate per 1,000 live births			Life expectancy at birth (2000–5)[c]		HIV/AIDS prevalence (ages 15–49), end 2001[d] (%)
			1960	1980	1998/99	With AIDS	Without AIDS	
Botswana	115	37 (1996)	160	100	60	36	70	38.8
South Africa	130	45 (1998)	130	90	70	47	66	20.1
Zimbabwe	106	53 (1994)	160	120	90	43	69	35.0
Namibia	150	57 (1992)	210	110	70	44	64	22.5
Kenya	118	62 (1993)	200	130	120	49	66	15.0
Swaziland	150	72 (1999)	230	140	90	38	63	33.4
Lesotho	145	74 (1996)	200	160	130	40	64	31.0
Burundi	145	75 (1987)	260	190	180	41	52	8.3
Uganda	125	81 (1995)	220	180	130	46	54	5.0
Rwanda	142	85 (1992)	220	190	180	41	51	8.9
Tanzania	143	88 (1996)	240	180	140	51	59	7.8
Congo (DRC)	146	106 (1999)	300	200	210	52	58	4.9
Zambia	130	109 (1996)	210	160	200	42	60	21.5
Mozambique	173	134 (1999)	320	270	200	38	49	13.0
Malawi	204	134 (1992)	360	270	210	39	53	15.0
Angola	200	166 (1996)	350	260	290	46	48	2.3[e]

Note: In column 2, figures in parentheses indicate the year the data were collected.
Sources: Child mortality rates: UNDP, *Human Development Report*; UNICEF, *State of the World's Children*; World Bank, *African Development Indicators*; World Bank, *World Development Report* each for various years.
[a] United Nations (1986).
[b] Demographic and Health Surveys (1997) except Angola (Prata 1999), Lesotho (Bureau of Statistics 1998), South Africa (Medical Research Council, Macro International Inc. and Department of Health 1999), DRC, Mozambique and Swaziland (Population Research Bureau 1999), Botswana (Central Statistics Office 1999).
[c] UNDP (2002).
[d] UNDP (2002).
[e] UNAIDS (2000).

are more typical of the rest of the region. Similar racial differences can be found in Zimbabwe, but there the white population is now such a small proportion of the total population (less than 1 per cent) that it has little impact on national demographic indices.

At the other end of the spectrum, five countries still recorded tragically high IMRs of over 100 per 1,000 live births in the 1990s. Of these, however, three (Angola, Mozambique and the DRC) have been, or are still, affected by civil wars. The other two (Malawi and Zambia), neighbouring land-locked countries, have a history of notably high infant and child mortality rates in the region due to economic difficulties and undernutrition.

Generally speaking, infant and child mortality levels are higher in sub-Saharan Africa than elsewhere in the world. But a declining trend was observed from the 1960s up to around 1990, as indicated by comparing the IMRs for those two time periods in Table 2.2. In the early 1960s all countries in East and southern Africa had IMRs in triple figures. In Angola and Malawi the rates were over 200, and the child mortality rates there and in Mozambique were over 300. The subsequent decline in many countries was encouraging. In many instances the 1960s IMR had been approximately halved by the 1990s. In Tanzania, for example, an IMR of 160 per 1,000 live births in 1967 had declined to 115 by 1988, to 92 by 1991/92, and to 88 by 1996 (Mturi and Curtis 1995). Zimbabwe achieved a halving of its rate in the decade of the 1980s alone. In the absence of war, a downward trend in mortality rates *up until the early 1990s* was fairly typical for the region. The exception was Zambia where infant and child mortality rates were recorded to have increased in the 10–15 years before the 1992 DHS survey. In 1980 the IMR was 92 (UNICEF 2002) but had increased to 109 by 1992. The factors contributing to this situation were widespread undernutrition, deteriorating economic conditions, very poor medical services and the increasing prevalence of AIDS (Demographic and Health Surveys 1997).

Tragically, during the 1990s, these generally encouraging trends in mortality have now decelerated, stagnated or even reversed in many other countries in the region besides Zambia. By far the most significant contemporary factor influencing these changes has been the HIV/AIDS epidemic, which has been cataclysmic for East and southern Africa. Because of the time lag between infection with HIV and death from AIDS, the impact of the epidemic on death rates and demographic patterns has usually only begun to have a really significant national statistical impact towards the end of the 1990s – this is why Zambia's increasing child mortality before this, where, as indicated above, other factors were also at play, was remarkable. Yet preliminary data from Zambia's census in 2000 indicate that its population had still grown at about 3 per cent per year from 1990 to 2000 (CSO 2000). National population growth rates in southern African countries are now being markedly reduced by AIDS deaths: in some cases rates of zero are now anticipated. In South Africa this scenario is projected for 2011, reducing from 1.9 per cent per year in 2000 (a rate then little affected by AIDS deaths) (*Financial Times* 2001).

By the turn of the century AIDS deaths were really having a marked effect on both child mortality (because of children infected through their

HIV+ mothers) and adult mortality, with the impact worst in the countries of southern Africa. The effect is dramatically shown by changes in life expectancy. In essence the gradual gains in life expectancy which had been such a positive marker of the post-colonial era in most of the region have been swiftly dissipated in many countries by HIV/AIDS, as shown by the differential life expectancy columns in Table 2.2 (see also Box 2.4). As indicated, in the absence of AIDS, Botswana (70), Zimbabwe (69), Kenya (66) and South Africa (66) would now have life expectancies at birth which are high by sub-Saharan African standards. But the HIV/AIDS epidemic has undermined their regional comparative advantage. When taking AIDS into account, Botswana now has the region's lowest life expectancy at birth, and only the DRC (which before AIDS had one of the lowest life expectancies) and Tanzania have life expectancies of over 50 years. The epidemic, therefore, is by far the most serious developmental and welfare issue facing the region today, particularly in the countries furthest to the south.

The adult AIDS prevalence rates estimated at the end of 2001 for the region are also shown in Table 2.2. They range from 2.3 per cent in Angola to a horrendous 38.8 per cent in Botswana. The next most badly affected countries, all with rates over 30 per cent, were Zimbabwe, Swaziland and Lesotho, followed by South Africa, Zambia and Namibia with rates over 20 per cent. South Africa, with 4.2 million infected people, has the sad distinction of having the largest number of people living with HIV/AIDS in the world (UNAIDS 2000).[5] As the table shows, southern Africa's rates are clearly worse than those in East Africa. This has not always been the case however: the worst-hit region in the late 1980s and early 1990s was in East Africa, comprising a core area including parts of Uganda, and districts of Tanzania and the DRC around Lake Victoria (see Box 2.3). It is of great significance therefore that Uganda, which in the early 1990s was the country most affected by the HIV/AIDS epidemic, has brought down the prevalence rate with strong prevention campaigns (see Box 2.3). UNAIDS (2000, 2002) reported that Zambia showed signs of following Uganda's experience, with rates falling among young women in particular, although the general adult prevalence rate for 2001 (Table 2.2) was still higher than that for 2000, which was 20 per cent. It does appear, however, that rates in East African countries by the end of the 1990s were stabilising, with the 2001 UNAIDS estimates generally slightly lower than those for 2000 (except in Kenya).

The current debates on what should be done to combat the HIV/AIDS epidemic focus on dealing with the stigma attached to HIV/AIDS. In sub-Saharan Africa HIV is primarily spread through heterosexual behaviour, but discussions about sex and sex-related issues are still a taboo in nearly all the societies of East and southern Africa. Indeed the global silence on HIV/AIDS based on ignorance, fear and denial has been accused as the major factor leading to the spread of the epidemic in such a devastating

---

[5.] Another alarming point about South Africa is how swiftly the prevalence of HIV-infected people accelerated. In a period of two years (1997–99), the rate increased from 12.9 to 19.9 per cent (UNAIDS 2000).

Box 2.3   AIDS in the East African 'core' region

The countries most affected by the early stages of Africa's HIV/AIDS epidemic in the 1980s were Uganda, Rwanda, Tanzania, Congo, Zaire, Zambia and Malawi. However, in the 1990s, HIV infection rates in southern Africa overtook those in this 'core' region. West African rates remain somewhat below those in East Africa. Within East Africa urban centres have been more affected than rural areas. In Uganda, for example, the HIV seroprevalence rate at the end of the 1980s was 24.1 per cent in the large towns and 12.3 per cent in rural areas. In Rwanda the respective rates were 20.1 per cent against 2.2 per cent.

The main mode of transmission in eastern Africa is heterosexual relationships. Transmission from mother to child and via blood transfusion have also played a part, but to a lesser extent. A significant factor in heterosexual transmission is labour migration by men to work in towns, mines and on farms (a factor which is probably even more significant in southern Africa's epidemic). This encourages extra-marital affairs due to prolonged periods of separation between husbands and wives. The circulation of commercial sex workers between cities and between urban areas and their rural homes is also known to enhance the transmission rate of HIV. Yet another vector are truck drivers who, by the nature of their work, are highly mobile. These drivers are the main clients of commercial sex workers at truck stops along their customary highway routes. Truck drivers have been cited as highly significant vectors of HIV along the highway linking Mombasa–Nairobi–Kampala–Kigali, for example.

Despite being one of Africa's worst affected countries by the epidemic in the 1980s, with a seroprevalence rate among the sexually active of 28 per cent in 1990, Uganda is now frequently acclaimed as an HIV/AIDS success story because it has succeeded in significantly reducing its infection rates by behavioural changes. One of the explanations for this achievement was active government commitment. As early as 1986, the government of Uganda acknowledged the HIV/AIDS crisis and began mobilising both domestic and international support to combat it. The President, Yoweri Museveni, was in the forefront of this campaign. He publicly (if reluctantly) announced that Ugandans should use condoms to protect themselves – a highly significant step in a region where condom use is highly contentious. It is apparent that the President's influence was very important. There has also been a strong campaign in Uganda stressing the need for women to have enhanced social status and a more equal relationship with men, since sexual inequality is felt to be a central issue enhancing the spread of HIV/AIDS. By 1997 the HIV prevalence rate in urban areas in Uganda had fallen to 15 per cent and was probably less than 10 per cent in the rural areas and, as indicated in Table 2.2, had fallen further by 2001 when it was estimated at 11 per cent nationally.

Sources: Barnett and Blaikie (1992); Opolot (2000); Potts and Marks (2001); UNAIDS (2000); UNDP (2002).

way. The vital necessity of changing this situation was highlighted at the 13th International AIDS Conference held in Durban, South Africa in July 2000, where the conference's main theme was *Break the Silence*.

Political commitment is another area that needs to be addressed very seriously. The few countries like Uganda which have managed to reverse the trend of the spread of HIV/AIDS have succeeded because their governments were very committed to tackling the issue, including promoting the need for people to alter their sexual behaviour – an issue which is evidently hugely sensitive and notoriously hard to change. Many governments are wary of entering into this policy arena. There is also a debate about where to get the resources needed to combat the epidemic. The expansion of the health sector now needs more money than ever before, yet structural adjustment programmes throughout the region have imposed austere fiscal constraints on government expenditure. Some successes have been achieved in forcing drug companies to provide cheaper drugs to prolong the lives of people with AIDS, or allow local production of much cheaper, generic versions of those drugs. However, for nearly all the countries in the region, the expense of these drugs is still completely out of reach, and even in the slightly wealthier countries the health infrastructure is too weak to deliver and monitor the complex drug regimes needed.[6] The developmental problems associated with the epidemic are legion. For example, the number of AIDS orphans who need care is increasing daily throughout the region, and poverty in AIDS-affected households is escalating because people who were supposed to generate income have died or are ill.

Age–sex structure

The age–sex composition or structure of a country's population is the proportion of males and females in each age group, and is usually illustrated by population pyramids. The shape of these pyramids is determined by the cumulative result of past trends in fertility, mortality and migration in that country. Hence it is possible to infer a good deal about a country's demographic history by examining its population pyramid. United Nations population data have been used to construct population pyramids for each of the countries in eastern and southern Africa for the years 1975, 1985 and 1995 (United Nations 1999c). Those for Malawi, Kenya and South Africa, all representing rather different demographic trends over the past 30 years, are presented in Figures 2.2–2.4. Before turning to these specific cases, however, three general observations can be made for the region from analysis of the entire set. First, all 16 countries have a young population

---

[6.] This is one of the arguments used by the South African government, under the leadership of Thabo Mbeki, against the provision of these drug regimes for its HIV-affected population. However, there are other complex factors at play in this country (whose CBOs (Community Based Organisations) and NGOs led the lobby to force changes in the world's drug companies on HIV drug pricing and policies), including Mbeki's concerns about drug toxicity and the role of poverty in facilitating the spread of HIV. Sadly these concerns have proved to be a real hindrance to policies which, in relatively wealthy South Africa, could be making some improvements to the situation.

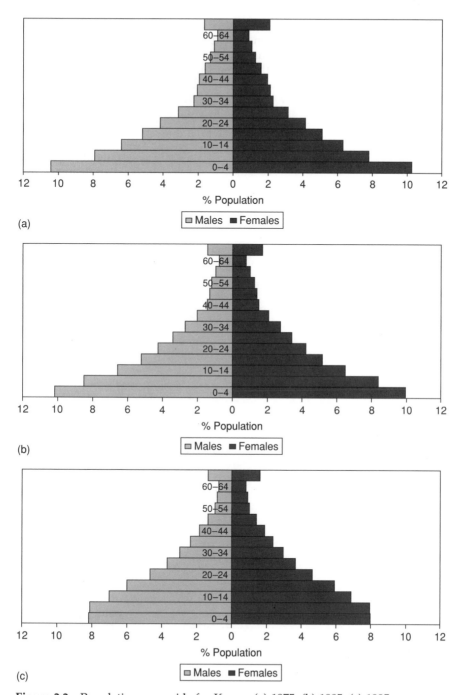

Figure 2.2 Population pyramids for Kenya: (a) 1975; (b) 1985; (c) 1995

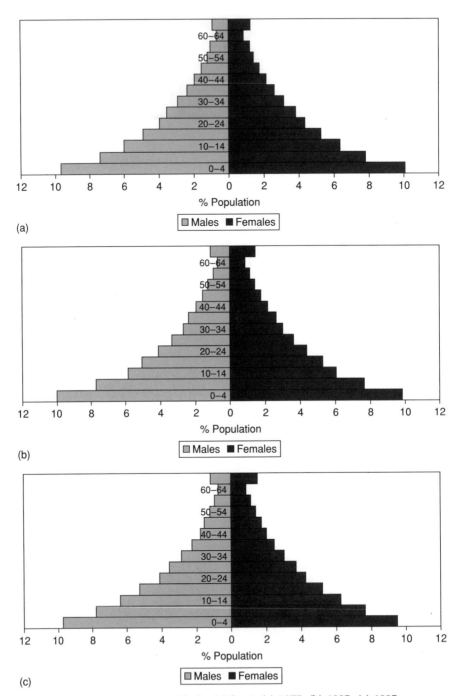

Figure 2.3 Population pyramids for Malawi: (a) 1975; (b) 1985; (c) 1995

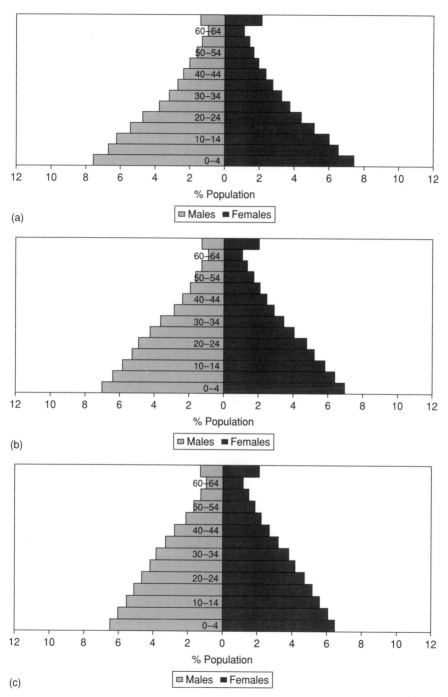

Figure 2.4 Population pyramids for South Africa: (a) 1975; (b) 1985; (c) 1995

with small proportions of old people, and thus have a pyramid with a wide base and a very narrow top. In all cases, the population aged over 65 years old is well under 5 per cent of the total and, with the exception of South Africa, over 40 per cent of the total population is aged under 15. Second, the proportion of the population decreases gradually as age increases in each country. Third, there has been no *major* change in age–sex structure during the period 1975–95.

Another observation from the population pyramids is that some countries in southern Africa have a deficit of males compared to females from age group 20–24 or 25–29 onwards.[7] These countries include Botswana, Lesotho, Malawi, Swaziland and Zambia. The deficit of males from age 20 to around age 50 could partly be attributed to emigration (labour migration) to other countries (e.g. South Africa and Zimbabwe), a system which has long, historical roots in the southern African region. However, there have been very significant changes in labour migration patterns since the 1970s when South Africa began a process of internalising its mine contract labour (i.e. using domestic South African labour) (Crush and Wilmot 1995). As a result the flows of labour from its neighbours have been very significantly reduced, although the impact on Lesotho was less marked until the 1990s. In Zimbabwe the need to use foreign labour was greatly reduced by the 1980s and this will also have reduced in-migration. Botswana's strong economic growth has also meant that migrant flows there have tended to shift to internal rural–urban movements. The impact on these countries' population pyramids of legal, contracted labour emigration has therefore been reduced and, except in Lesotho and perhaps Swaziland, would not be particularly marked by 1995. Comparison of the 1975 and 1995 pyramids for Malawi shows that the male : female imbalance in the 20–24 age group in 1975 had largely disappeared by 1995, in line with the ending of labour recruitment by other countries. On the other hand there has been a significant rise in casual (often 'illegal') movement to South Africa (Crush 1997) from surrounding countries (including, now, Zimbabwe), and it may be that this type of emigration is being picked up in the 1995 population pyramids. Yet there is a significant involvement of women in this casual migration, which should not translate into a marked deficit of males in the population pyramids.

At advanced ages the imbalance in sex ratios found in the population pyramids is caused by excess male over female mortality. Adult mortality in sub-Saharan Africa is generally higher than other parts of the world and male adult mortality is higher than that of females (Timaeus 1993). In all 16 countries the proportion of males aged 65 years or older is lower than that of their female counterparts.

The three sets of population pyramids in Figures 2.2–2.4 illustrate well three different demographic scenarios and trends in the region for 1975 to 1995. Kenya and Malawi started off with almost identical pyramids, with very wide bases and concave sides, typical of countries with very high

---

[7.] It is common for more boys than girls to be born. However, male mortality among infants tends to be higher, and in the oldest popuation cohorts. A greater degree of gender parity in the cohorts between these ages is usual.

fertility and mortality. Malawi has experienced little change in the shape of its pyramid over the 20 years and represents countries which have maintained this general demographic regime. Kenya, by contrast, represents those countries which have experienced massive changes in their fertility over the period – indicated in its case by the sudden shrinking of the 0–4 age group to the proportions of the 5–9 group by 1995. In South Africa, fertility was already lower in 1975 than the regional norm, as is evident from its narrower base then than Kenya or Malawi. The very different shape of South Africa's pyramid by 1995 compared to 1975, or to either of the other countries, illustrates graphically the points made earlier about its pre-eminence in the region in terms of its position in the fertility transition. The differing trends are also indicated by the following data: by 1995 the proportion of South Africa's population which was under five was 13 per cent, compared to 15 per cent in 1975. By comparison the proportions in Kenya and Malawi were, respectively, 16 per cent compared to 20.5 per cent, and 19 per cent compared to 20 per cent.

Family structure

The Population Research Bureau defines a family as a group of two or more persons residing together and related by birth, marriage or adoption (Haupt and Kane 1998). Families are of central importance to demographic study as traditional institutions for reproduction and child rearing. In sub-Saharan Africa, a traditional family is created by the marriage of two opposite sex adults (monogamy) and in many societies men are allowed to marry more than one wife (polygyny). These may later be joined by their own children (born as a result of marriage or adoption) to form a nuclear family or joined by other relatives to form an extended family. Other family structures are created when marriages end because of death, divorce, separation or desertion. If a marriage breakdown is immediately followed by remarriage, as used to be almost universally the case in the region, the family structure will remain either nuclear or extended. Nowadays, however, the female-headed households are a very important type of family structure. Single parents, especially women, have increased because marital breakdowns are less often followed by swift remarriage and there has also been an increase in pre-marital childbearing. Another important cause of female-headed households in southern Africa is absenteeism of the male head because of migration. Usually these men remain the *de jure* head of their households but the wives at home take care of day-to-day household activities (i.e. they are the *de facto* head).

A gender-based study conducted in 10 countries in East and southern Africa in the 1990s allows genuine comparisons to be made across the region on family structure issues such as household size and female headship. It covered Malawi, Botswana, Zimbabwe, Uganda, Mozambique, Kenya, Tanzania, Namibia, Zambia and Swaziland. Average household size ranged from 4.5 to 6.0,[8] with the smallest in Malawi and Botswana

---

[8.] Note that a *household* sometimes includes someone who is not a member of the family.

and the largest in Swaziland and Zambia (Regional Gender Statistics Project 1995). Female-headed households were more common in southern than East Africa and were most prevalent in Botswana, accounting for over 50 per cent of households, followed by Swaziland and Namibia. These patterns in southern Africa are in part a legacy of the long-standing and deep-rooted migrancy systems there. A breakdown or undermining of the institution of marriage in this region is also a factor with, as already indicated, many women having children outside of marriage. The two factors are often interrelated as, for example, absentee men abandon their rural wives or choose to delay marriage (Government of Lesotho and UNICEF 1994). Gaisie (1998), for example, identifies labour migration and the decline of polygyny as important influences in extra-marital fertility and subsequent female-headed households. Other, broader socio-economic factors are also at work such as education and employment patterns.

Regional differences can also be identified in relation to gender-based practices within the family. Broadly speaking, in East Africa girls tend to be somewhat disadvantaged. This is clear in terms of access to education, as statistics show that boys are more likely to be enrolled in schools than girls. The explanation is that, when there are limited resources in the family, parents prefer to educate boys in order to assist them to become economically independent by the time they become adults. Girls, on the other hand, are expected eventually to marry and join another family. So investing in boys is seen as more beneficial to the family than investing in girls. This imbalance has often been worsened by structural adjustment programmes as schooling has become much more expensive across the region as governments have imposed user fees. Although these fee re-gimes have subsequently been reversed in some cases when the extremely adverse effects on enrolments were recognised, most schools are still having to impose levies on parents because government funding is now completely inadequate; in these circumstances girls are frequently more likely to be taken out of school than boys.

On the other hand, in some parts of southern Africa girls may receive preferential access to education. In Lesotho, for example, gender discrim-ination can start at an early age with girls being sent to school and boys staying at home to herd animals. This can happen particularly when there is only one boy in the family. An important factor in this differential treatment is that, traditionally, during early adulthood most Basotho males have gone to work in the South African mines doing jobs that do not require many educational qualifications. Another effect of this pattern is that, even if they are enrolled in schools, it has been noted that male adolescents in Lesotho tend not to work as hard as their female counter-parts because they have traditionally had this alternative avenue for earning money. Similar issues have been noted in Botswana and Swaziland. The very major changes which have taken place in South African mine recruitment practices over the past 30 years mean that such traditions and assumptions often do not accord with reality, as foreign novices now find it very difficult to get contracts and there has been massive retrenchment of mine workers. The 'modernisation' of attitudes may also be changing this behaviour but for many people, especially in rural areas, it will take time to reverse these practices which have been operating for many generations.

Family structures have also been profoundly affected by the HIV/AIDS epidemic. Previously rare structures are becoming more prevalent. The number of AIDS orphans (see Box 2.4) is increasing daily. When both parents die or become too ill to work and earn a living, the oldest children may start work to earn money and support their siblings, long before the age at which they might have entered the labour force in the past. This is a major cause of under-age children working in Lesotho (Kimane and Mturi 2000). Another type of family created when both parents die often involves one or both grandparents caring for their grandchildren, especially if they are too young to take care of themselves. The HIV/AIDS epidemic is thus creating families which comprise only children (sometimes all of whom are too young to bear family responsibilities) or very old people with children. The trauma borne by children who are AIDS orphans is thus multiplied:

> A child whose mother or father has HIV begins to experience loss, sorrow and suffering long before the parent's death. And since HIV can spread sexually between father and mother, once AIDS has claimed the mother or father, the children are far more apt to lose the remaining parent. Children thus find themselves thrust in the role of mother or father or both – doing the household chores, looking after siblings, farming, and caring for the ill or dying parent or parents, bringing on stress that would exhaust even adults.
>
> (UNAIDS and UNICEF 1999: 5)

Concluding remarks

Fertility and mortality rates have been declining in East and southern Africa, although they remain high in comparison to most parts of the world. Mortality started to decline during the 1960s and 1970s. However, this decline stagnated in the 1980s and in some cases reversed in the 1990s due to a variety of factors including deteriorating economic conditions, civil warfare, undernutrition, inadequate medical services and, in particular, the spread of the HIV/AIDS epidemic. It is projected that most of the improvements in mortality in the region made in the 1960s and 1970s will vanish in the near future. AIDS is now reported as the primary cause of death in sub-Saharan Africa, with the most affected countries in East and southern Africa (UNAIDS 2000); nevertheless other causes of death have also contributed to this terrible situation.

Declining fertility in southern Africa in particular is now clearly established and documented. It seems probable that family planning programmes, deteriorating economic conditions and the spread of HIV/AIDS will assist further decline. The AIDS epidemic appears to reduce fertility through a variety of behavioural adaptations according to research in Zambia (Baylies 2000) and Zimbabwe (Gregson *et al.* 1998; Grieser *et al.* 2001). It is possible that the future rate of fertility decline in those countries in the region which, as yet, have hardly begun the fertility transition, may be quite rapid.

Box 2.4   Socio-demographic impact of HIV/AIDS

The most dramatic social and economic effects of HIV/AIDS occur among children who have been orphaned as a result of the epidemic. AIDS orphans are classified as children under 15 years old who have lost one or both parents. Uganda has the largest number of AIDS orphans with an estimated 1.7 million by the end of 1999. Other countries in the region with large number of AIDS orphans are Tanzania (1.1 million), Zimbabwe (0.9 million), Kenya (0.73 million), the DRC (0.68 million), Zambia (0.65 million), South Africa (0.42 million) and Malawi (0.39 million). A typical pattern for people with HIV/AIDS in the region is that they become infected with HIV before they are 25, and acquire AIDS and die by the time they turn 35, leaving behind a generation of children to be raised by their grandparents or left on their own in child-headed households. Tragically, AIDS orphans are known to be at greater risk of malnutrition, illness, abuse and sexual exploitation than children orphaned by other causes.

The long-term demographic impact of HIV/AIDS is expected to be significant in all the countries affected by the epidemic. The population pyramids – currently typically with a broad base which tapers off steadily with increasing age – will change to what is called a 'population chimney'. The base of the pyramid becomes less broad, and the population of women above their early 20s and men above their early 30s shrinks radically. As only those who have not been infected survive to older ages, the pyramid has straighter sides and starts to resemble a chimney. The economic implications are that there is only a small number of young adults to support a large number of young and old people. Many of these young adults will themselves be debilitated by AIDS and thus may even require care from their children or elderly parents rather than providing it.

The fall in life expectancy at birth due to AIDS is a huge demographic issue. The United Nations has estimated two sets of life expectancy (LE) at birth (e_0) statistics for each country: one set projecting what was expected to happen in the absence of the AIDS epidemic, and the other accounting for the expected impact of AIDS. The gap between the two sets of statistics is extraordinary. Botswana, one of the most affected countries, had an estimated gap of 21 years in 2000: in the absence of the HIV/AIDS epidemic, its e_0 would have been 68 years, but once the prevailing situation of HIV/AIDS was taken into account the estimate was reduced to 47 years (the estimate of e_0 in Botswana is therefore referred to as 68–47). Other countries have suffered similar drastic reductions although with somewhat narrower gaps. They include Zambia (58–40), Malawi (51–39), Kenya (64–52), Uganda (52–40) and Zimbabwe (64–44). Estimates for 2000–5 indicate the gaps continue to grow (see Table 2.3). It is worth remembering that both sets of LE data are based on various assumptions, in fact, since registration of births and deaths is very poor in most of the region, and some believe that the computer software designed to incorporate AIDS into LE tables somewhat overestimates mortality levels. Nevertheless, it does appear that that the HIV/AIDS epidemic is eliminating much, and perhaps all, of the improvement in mortality achieved in the past several decades in southern and eastern Africa.

Based on: UNAIDS (2000); UNAIDS and UNICEF (1999); United Nations (1999b).

There is no doubt that the spread of HIV is the most threatening catastrophe in the region. Unfortunately there are a number of cultural traditions that inhibit behavioural changes which could slow the rate of spread. Traditionally, women have lower status in families than men. Once married, men automatically become the heads of their households and women are expected to be submissive to their husbands. This makes it very difficult for women to negotiate use of family planning, especially condoms, which can help to stop the spread of HIV. Furthermore many men in the region take the attitude that if a woman asks for a condom to be used that this means that she is promiscuous (Kimane *et al.* 1999; Maharaj 2001; Potts and Marks 2001). This presents women with the impossible dilemma of choosing between their reputation and a relationship which may be central to their livelihood, and safe sex. The success, or otherwise, of policies to change such attitudes and practices will be a major determinant of demographic patterns of mortality and population growth in the region, particularly in southern Africa, in the near future (see Box 2.4).

References

Agadjanian, V. (1998) Women's choice between indigenous and western contraception in urban Mozambique, *Women and Health* **28**(3): 1–17.

Agadjanian, V. and **Prata, N.** (2001) War and reproduction: Angola's fertility in comparative perspective, *Journal of Southern African Studies*, special issue on Fertility in Southern Africa **27**(2): 291–309.

Barnett, T. and **Blaikie, P.** (1992) *AIDS in Africa: Its present and future impacts*. Belhaven Press, London.

Baylies, C. (2000) The impact of HIV on family size preference in Zambia, *Reproductive Health Matters* **8**(15): 77–86.

Bongaarts, J. and **Potter, R.G.** (1983) *Fertility, Biology and Behavior: An analysis of the proximate determinants*. Academic Press, New York.

Bulatao, R.A. and **Lee, R.D.** (1983) An overview of fertility determinants in developing countries, in Bulatao, R.A. and Lee, R.D. (eds) *Determinants of Fertility in Developing Countries: Fertility regulation and institutional influences*, vol. 2. New York, Academic Press: 757–87.

Bureau of Statistics [Lesotho] (1998) *1996 Population Census Analytical Report*, vol. IIIA: *Population Dynamics*. Maseru, Government of the Kingdom of Lesotho.

Bureau of Statistics [Tanzania] (1994) *1988 Population Census of Tanzania: The analytical report*. Dar es Salaam, President's Office, Planning Commission.

Caldwell, J.C. (1980) Mass education as a determinant of the timing of fertility decline. *Population and Development Review* **6**(2): 225–55.

Caldwell, J.C. and **Caldwell, P.** (1981) Cause and consequence in the reduction of postnatal abstinence in Ibadan City, Nigeria, in Page, H.J. and Lesthaeghe, R. (eds) *Child-Spacing in Tropical Africa: Traditions and change*. London, Academic Press: 181–99.

Caldwell, J.C., Orubuloye, I.O. and **Caldwell, P.** (1992) Fertility decline in Africa: a new type of transition?, *Population and Development Review* **18**(2): 211–42.

Central Statistics Office (1999) *The 1996 Botswana Family Health Survey III.* Gaborone, Republic of Botswana.

Chimere-Dan, O. (1997) Non-marital teenage childbearing in Namibia, *African Population Studies* **12**(2): 87–95.

Chimere-Dan, O. (1999) Marriage and fertility transition in South Africa. Paper presented at the 3rd African Population Conference, Durban, South Africa, 6–10 December 1999.

Cohen, B. (1993) Fertility levels, differentials, and trends, in Foote, K., Hill, K. and Martin, L. (eds) *Demographic Change in Sub-Saharan Africa.* National Academy Press, Washington, DC: 6–67.

Crush, J. (1997) *Covert Operations, Clandestine Migration, Temporary Work and Immigration Policy in South Africa.* Southern African Migration Project, Migration Policy Series; 1, Cape Town.

Crush, J. and **Wilmot, J.** (eds) (1995) *Crossing Boundaries: Mine migrancy in a democratic South Africa.* IDASA, Rondesbush and IDRC, Ottawa.

CSO (2000) *2000 Census of Population and Housing, Preliminary Report.* Central Statistical Office, Lusaka, Zambia.

Demographic and Health Surveys (1997) *DHS Newsletter* **9**(1), Calverton, Macro International Inc.

Financial Times (2001) AIDS ravages South African truck drivers, 3 May.

Gaisie, S.K. (1998) Fertility transition in Botswana, *Journal of Contemporary African Studies* **16**(2): 277–96.

Gaitskell, D. (1982) Wailing for purity: mothers, daughters and Christian Prayer Unions in South Africa, in Marks, S. and Rathbone, R. (eds) *Industrialisation and Social Change.* London, Longman, Harlow.

Garenne, M., Tollman, S., Kahn, K., Collins, T. and **Ngwenya, S.** (2001) Understanding marital and pre-marital fertility in rural South Africa, *Journal of Southern African Studies* **27**(2): 277–90.

Gould, W.T.S. and **Brown, M.S.** (1996) A fertility transition in sub-Saharan Africa? *International Journal of Population Geography* **2**: 1–22.

Government of Lesotho and **UNICEF** (1994) *The Situation of Women and Children in Lesotho.* Ministry of Planning, Economic and Manpower Planning, Maseru.

Gray, R. (1982) Factors affecting natural fertility components: health and nutrition, in Bulatao, R.A. and Lee, R.D. (eds) *Determinants of Fertility in Developing Countries.* National Academy Press, Washington, DC.

Gregson, S., Zhuwau, T., Anderson, R.M. and **Chandiwana, S.K.** (1998) Is there evidence for behaviour change in response to AIDS in rural Zimbabwe? *Social Science and Medicine* **46**(3): 321–3.

Grieser, M., Gittelsohn, J., Shankar, A., Koppenhaver, T., Legrand, T., Marindo, R., Mavhu, W. and **Kenneth Hill, K.** (2001) Reproductive decision-making and the HIV/AIDS epidemic in Zimbabwe, *Journal of Southern African Studies*, special issue on Fertility in Southern Africa, **27**(2): 225–43.

Gulbrandsen, O. (1986) To marry or not to marry: marital strategies and sexual relations in Tswana Society, *Ethnos* **15**(1–2): 7–28.

Harrison, A. and **Montgomery, E.** (2001) Life histories, reproductive histories: rural South African Women's narratives of fertility, reproduction, health and illness, *Journal of Southern African Studies* **27**(2): 311–28.

Haupt, A. and **Kane, T.T.** (1998) *Population Handbook*, 4th international edition. Population Reference Bureau, Washington, DC.

Henin, R.A. (1979) *Effects of Development on Fertility and Mortality in East Africa: Evidence from Kenya and Tanzania*. Population Studies and Research Institute, University of Nairobi, Kenya.

Hinde, A. and **Mturi, A.J.** (2000) Recent trends in Tanzanian fertility, *Population Studies* **54**(2): 177–91.

Instituto Naçional de Estatistica and **Macro International Inc.** (1998) *Mocambique Inquerito Demografico e de Saude 1997*. Maputo, Mozambique and Calverton, Md.

Jolly, C.L. and **Gribble, J.N.** (1993) The proximate determinants of fertility, in Foote, K., Hill, K. and Martin, L. (eds) *Demographic Change in Sub-Saharan Africa*. National Academy Press, Washington, DC: 68–116.

Kimane, I., Molise, N.M. and **Ntimo-Makara, M.** (1999) *Socio-Cultural Phenomena Related to Population and Development in Lesotho*. Report submitted to UNFPA, Lesotho.

Kimane, I. and **Mturi, A.J.** (2000) *Rapidly Assessing Children at Work in Lesotho*. Government of Lesotho and UNICEF, Maseru.

Kirk, D. and **Pillet, B.** (1998) Fertility in sub-Saharan Africa in the 1980s and 1990s, *Studies in Family Planning* **29**(1): 1–22.

Lejeune, A. and **Hinde, A.** (1999) The impact of pre-marital fertility, marital fertiliy and nuptiality on recent trends in Tanzanian fertility, in Union for African Population Studies, *The African Population in the 21st Century*, vol. 2. Proceedings of the Third African Population Conference, Durban, 6–10 December, Dakar: 227–43.

Lura, R. (1985) Population change in Kericho District, Kenya: an example of fertility increase in Africa, *African Studies Review* **28**(1): 45–6.

Maharaj, P. (2001) Male attitudes to family planning in the era of HIV/AIDS: evidence from KwaZulu-Natal, South Africa, *Journal of Southern African Studies*, special issue on Fertility in Southern Africa, **27**(2): 245–57.

Medical Research Council, Macro International Inc. and **Department of Health [South Africa]** (1999) *South Africa Demographic and Health Survey 1998: Preliminary Report*.

Mencarini, L. (1999) An analysis of fertility and infant mortality in South Africa based on 1993 LSDS data, in Union for African Population Studies, *The African Population in the 21st Century*, vol. 1. Proceedings of the Third African Population Conference, Durban, 6–10 December, Dakar: 109–27.

Mturi, A.J. and **Curtis, S.L.** (1995) The determinants of infant and child mortality in Tanzania, *Health Policy and Planning* **10**(4): 384–94.

Mturi, A.J. and **Hlabana, T.A.** (1999) Fertility in Lesotho: recent trends and proximate determinants, in Union for African Population Studies, *The African Population in the 21st Century*, vol. 1. Proceedings of the Third African Population Conference, Durban, 6–10 December, Dakar: 145–62.

Mturi, A.J. and **Moerane, W.** (2001) Non-marital childbearing among adolescents in Lesotho, *Journal of Southern African Studies* **27**(2): 259–75.

Mutambirwa, J. (1979) Traditional Shona concepts on family life and how systems planned on the basis of these concepts effectively contained the population growth of Shona communities, *Zimbabwean Journal of Economics* **1**(2): 96–103.

Ndeti, K. and **Ndeti, C.** (1977) *Cultural Values and Population Policy in Kenya*. Kenya Literature Bureau, Nairobi.

Newman, J.L. and **Lura, R.** (1983) Fertility control in Africa, *Geographical Review* **73**: 396–406.

Opolot, E. (2000) Uganda's AIDS fightback, *News Africa* **1**(6) (Aug. 14): 53.

Page, H. and **Lestaeghe, R.** (eds) (1981) *Child Spacing in Tropical Africa: Traditions and change*. Academic Press, London and New York.

Population Research Bureau (1999) *World Population Data Sheet*. Washington, DC.

Potts, D. (1997) Urban lives: adopting new strategies and adapting rural links, in Rakodi, C. (ed.) *The Urban Challenge in Africa: Growth and management of its large cities*. United Nations University Press, Tokyo: 447–94.

Potts, D. and **Marks, S.** (2001) Fertility in Southern Africa: the silent revolution, *Journal of Southern African Studies*, special issue on Fertility in Southern Africa, **27**(2): 189–206.

Prata, N. (1999) Fertility in Angola: what do we know today? paper presented at Workshop on Fertility in Southern Africa, SOAS, London, September 11–13.

Regional Gender Statistics Project (1995) *Women and Men in East, Central and Southern Africa: Facts and figures*. Central Bureau of Statistics, Nairobi.

Rogerson, C. (1997) Globalization or informalization? African urban economies in the 1990s, in Rakodi, C. (ed.) *The Urban Challenge in Africa: Growth and management of its large cities*. United Nations University Press, Tokyo: 337–70.

Timaeus, I. (1993) Adult mortality, in Foote, K., Hill, K. and Martin, L. (eds) *Demographic Change in Sub-Saharan Africa*. National Academy Press, Washington: 218–55.

Timaeus, I. and **Graham, W.** (1989) Labour circulation, marriage and fertility in southern Africa in Lestaeghe, R. (ed.) *Reproduction and Social Organisation in sub-Saharan Africa*, University of California Press, Berkeley.

UNAIDS (2000) *Report on the Global HIV/AIDS Epidemic*. June.

UNAIDS (2002) *Factsheet 2002: sub-Saharan Africa* [www.unaids.org/barcelona/presskit/factsheets/FSssafrica_en].

UNAIDS and **UNICEF** (1999) *Children Orphaned by AIDS: Front-line responses from Eastern and Southern Africa*. Division of Communication, UNICEF, New York.

UNICEF (2002) *Statistics, Child Mortality* [www.childinfo.org/cmr/revis/db1.htm].

United Nations (1985) *Women's Employment and Fertility: A comparative analysis of world fertility survey results for 38 developing countries*. Department of International Economic and Social Affairs, New York.

United Nations (1986) *World Population Prospects: Estimates and projections as assessed in 1984*. Department of International Economic and Social Affairs, New York.

United Nations (1995) *Women's Education and Fertility Behaviour: Recent evidence from the demographic and health surveys*. Population Division, Department of Economic and Social Information and Policy Analysis, New York.

United Nations (1999a) *Levels and Trends of Contraceptive Use as Assessed in 1998*. Population Division, Department of Economic and Social Affairs, New York.

United Nations (1999b) *The Demographic Impact of HIV/AIDS*. Report on the Technical Meeting of 10 November 1998. Population Division, Department of Economic and Social Affairs, New York.

United Nations (1999c) *World Population Prospects: The 1998 revision*. Vol. II: *The Sex and Age Distribution of the World Population*. Population Division, Department of Economic and Social Affairs, New York.

United Nations Development Programme (2002) *Human Development Report 2002*, Oxford University Press, New York.

Upton, R. (2001) Infertility makes you invisible: gender, health and the negotiation of fertility in Northern Botswana, *Journal of Southern African Studies* **27**(2): 349–62.

Zulu, E. (1998) What can family planning programs learn from traditional reproductive beliefs and practices? A case study from Malawi, paper presented at the IUSSP Seminar on Reproductive Change in Sub-Saharan Africa held at Nairobi, Kenya, 2–4 November.

CHAPTER 3

Structural adjustment in eastern and southern Africa: the tragedy of development

Colin Stoneman

Structural Adjustment: Why it Wasn't Necessary and Why it Did Work.

(Bracking 1999)

The Threat of a Good Example?

(Melrose 1985)

A great tragedy has been enacted in eastern and southern Africa, as in much of the 'developing world'. It came out of a conflict between the post-colonial aspiration for national economic development (on the model of other late developers) and the hardening aspiration of the international financial institutions (IFIs) to establish a world market. Most people in the region are now poorer, and live shorter and less fulfilled lives than they did 10 or 20 years ago, and in some cases 40 years ago, before independence.

This chapter is headed with quotations from the titles of two works spanning almost the whole period during which structural adjustment has been the orthodoxy for developing countries. They encapsulate the main hypothesis of the chapter: that structural adjustment was not economically necessary, and that its motivations were primarily political. This may seem overstated, and incapable of adequate testing in so short a chapter, but I would argue that, given more space, an even stronger case could be justified: that structural adjustment has proved almost completely harmful, and that for some countries that were succeeding with non-market policies, *it was intended to be*.

Over the last two decades some intemperate rants against the World Bank and the IMF have been published, but there have been many more damning indictments of their record in the 'developing world' by analysts using intellectually respectable procedures, while hiding their anger against the destruction of livelihoods and lives that has often been one of the consequences. This chapter falls in the latter category. I do not seek to underplay the mistakes, inappropriate policies and corruption of many countries, and agree that these needed addressing. However, such factors can also and, I believe, more significantly, should be seen as providing the pretext for the imposition by the IFIs of policies that were even more inappropriate in a developing country context. Some observers have seen these institutions as experimenting rashly with developing countries, but learning from their mistakes, so that before long they may be exerting a beneficial influence after all.[1] Again I see this as too weak, in that it assumes that policy development has proceeded through purely technical considerations, thus neglecting the fact that the policies imposed benefited Western interests unequivocally – in lowering commodity prices, opening up opportunities for Western corporations, and bailing out Western banks which had made rash loans during the 1970s. These were predictable, even inevitable, consequences of structural adjustment, while belief in the benefits to developing countries amounted to a 'leap in the dark in three senses' (Mosley *et al.* 1995: I, 305) despite which the IFIs insisted that 'there is no alternative' to their policies.

False dawn: a brief history of the first two decades

Most countries in eastern and southern Africa gained independence in the 1960s. Many of the new leaders were inspired by the analysis of Kwame Nkrumah, the first President of Ghana, which became independent in 1957. He saw political independence as only the first step in the construction of viable nations, and was suspicious of the relatively bloodless relinquishing of control by colonial powers.[2] Britain and France, in particular, no longer gained from direct colonialism but might gain more from what he called a 'neocolonial' relationship with their former colonies, divided into many small countries with little individual bargaining power. France in fact split its former French West Africa into half a dozen new countries, but then cajoled nearly all of them (only Guinea refused to join), along with other French ex-colonies, into a 'French Union' and a monetary area tied to the French franc. 'Divide and rule' seemed still to be the order of the new dispensation; even today, Africa, with about 50 countries, has barely half the population of India.

In response, Nkrumah proposed pan-Africanism, and the Organisation of African Unity (OAU) was established in 1963 with the ultimate aim of unifying the continent, maybe in a United States of Africa (Mazrui 1967: 219–42). But in the interim he attempted to establish a base for a more

---

[1] The major work (Mosley *et al.* 1995), involving detailed case studies of nine countries undergoing structural adjustment, may fall into this category. Although I take a more sceptical line towards the IFIs than these authors, most of their findings, which I draw on particularly in the cases of Kenya and Malawi, are also consistent with a more radical interpretation.

[2] In French and British colonies it was only where there was a substantial settler community, with interests divergent from those of the metropolitan power (as in Algeria and Zimbabwe in particular), that independence required sustained armed action. Portugal, however, a much poorer European power, continued to benefit from its colonies and there was relatively little division of interest between Lisbon and the settlers.

independent economy in Ghana alone. This involved nationalist policies that were far from revolutionary, having been applied in many countries attempting to catch up, from Germany in the nineteenth century, through Latin America beginning in the 1930s, to the contemporary policies in what later came to be called the newly industrialising countries (NICs). In the absence of either a significant domestic entrepreneurial class or markets in many parts of the economy, these policies necessarily involved a central role for the state. A shift away from overemphasis on production of export commodities (in Ghana's case cocoa) that had distorted the colonial economy, and protection for domestic firms, especially those that would form the basis for industrialisation, were key policies.

With minor modifications, these elements provided the basis for economic strategy in nearly all countries in eastern and southern Africa as they became independent. Some countries weakened the neo-colonial grip by direct means, especially through nationalisation of productive assets; others welcomed foreign investment despite an economic nationalist framework (on the model of import-substituting industrialisation in Latin America). But the state nearly always played a central role, as it did in most successful late developers, including Germany, Japan and the NICs.

Furthermore, except in states which allied themselves with the Soviet Union or China, the strategy was not seriously challenged by the West. Development aid, after all, is more cheaply given in large quanta and its effects are more visible in large projects; but in any case there was rarely any significant private sector to support instead. Thus the World Bank itself encouraged the very policies that, after 1980, it attacked as anti-market manifestations of misguided or corrupt, and certainly overpowerful, states.

And, by and large, the economic record of the 1960s was encouraging: many African states experienced growth rates of over 5 per cent per annum; indeed before 1970 an analyst would not automatically have predicted that east Asia rather than Africa was going to produce the new economic miracle (see Table 3.1).

Both groups of countries had started in the 1950s with per capita incomes of a few hundred US$, and both had been growing fairly steadily since, with fairly similar policies. Most of the reasons that the World Bank and IMF applied with hindsight to explain Africa's subsequent failure could have been applied to the NICs had they failed: a dominant state

Table 3.1 GDP annual average growth rates, 1950–69: Asia and Africa

Region	GDP AAGR (%)		
	1950–60	*1961–65*	*1966–69*
Africa	4.0	4.3	4.3
East Asia	4.7	5.5	7.4
South Asia	3.6	3.5	5.2

Sources: Pearson (1969: 358); World Bank/IDA (1970: 66–7).

role, including direct involvement in productive industry, high protection of domestic enterprise, growing corruption, distortion of market signals and so forth. Of the differences, these tell in both directions: the NICs forced export growth rather than the domestic market (as did some African countries such as Côte d'Ivoire and Malawi), but they were *more* hostile to foreign capital and had far more pervasive control over financial markets, with banks remaining in state ownership until the 1990s.

The turning point for Africa came in 1974 with the first oil-price hike, followed by the collapse in prices of many other commodities. This affected Zambia worst of all, as it was still dependent for over 90 per cent of its export earnings on copper. Some have seen the relative resource poverty of the NICs as their good fortune in these circumstances, and indeed resource-poor African countries such as Tanzania were less affected, continuing to achieve economic growth of about 5 per cent per annum until the threefold rise in oil prices in the second hike in 1979. Whereas the NICs by this time had substantial industrial capacity with growing export capability, Africa had not yet freed itself from commodity dependence, having invested too heavily in infrastructure (which might, given time, have been the correct strategy) with manufacturing industry lagging. Thus the NICs could pay for higher fuel costs by higher export earnings, much as the industrialised countries did, while Tanzania, for example, found that fuel costs, previously taking 40 per cent of export earnings, were now three times higher, while its mainly commodity earnings (principally from sisal) had fallen. As commodities are relatively homogeneous, exporters can rarely influence the price received significantly, as can industrial exporters through design and quality changes. So overnight the Tanzanian economy was shifted by a change in world market prices from a path that most had thought was leading to long-term growth, into a situation of complete insolvency.

Five years before this, Zambia had had the ground cut from beneath its feet in similar fashion, but because its commodity, copper, was still of long-term importance to the world economy, and because the World Bank had already supported major infrastructural projects designed to begin the process of reducing dependence on copper through industrialisation,[3] it was reasonable to expect Zambia to borrow heavily so as to complete and maintain these projects, in the expectation of a recovery in copper prices (Fundanga 1989: 142–3). This recovery never came, and Zambia's crisis demonstrates in extreme form how the debt crisis arose in most African countries: they borrowed cheap 'petrodollars' at the instigation of the West,[4] which then caused a further slump in commodity prices by the

---

[3.] This was lagging as compared with the NICs, partly because independence only came in 1964, partly because the country was much smaller, and partly because the World Bank did not support productive industries directly, seeing industrial investment as the province of private investors in response to market signals.
[4.] Western banks were swamped with the extra (dollar-denominated) earnings from OPEC countries after the first oil-price hike; they were under extreme pressure to on-lend these funds even at an interest rate as low as 5 per cent, which was negative in real terms as inflation was then running at over 10 per cent in industrial countries.

deflationary policies of the early 1980s. There were thus no earnings to repay the debts, which themselves became larger in real terms as interest rates became positive when world inflation fell. Zambia's predicament was greatly exacerbated by its landlocked position and the massive opportunity costs incurred in transport after Rhodesia declared UDI (Unilateral Declaration of Independence) in 1965, and Zambia struggled to use or develop alternative transit routes to the sea, in line with international sanctions against Rhodesia's white minority regime (see, for example, Osei-Hwedi and Osei-Hwedi 1990).

Triumph of market-Leninism over Marxist-Leninism

Although few countries in eastern and southern Africa professed openly Marxist-Leninist[5] policies (the main exceptions were Angola and Mozambique from 1975 to the late 1980s), policies of state-led self-reliance, coupled in many cases with active membership of the Non-Aligned Movement, had long prompted political suspicion in the West. The lack of overt opposition from the IFIs may be related to the disingenuous official position that they were not concerned with internal politics. But, in any case, they probably saw such policies as merely transitional, as they had been in other cases of capitalist development.

With the change in political climate following the elections of Margaret Thatcher in the UK and Ronald Reagan in the USA, this more relaxed attitude was overruled, and the World Bank, in particular, moved from defending to attacking the development policies of the previous two decades. It appeared to forget the integral role that it itself had played in supporting and developing these policies.

The key document was the so-called 'Berg Report' (World Bank 1981), which identified internal factors as almost exclusively responsible for the deterioration in the economic situations and prospects of African countries. Neither the oil-price hikes (the main cause of the debt crisis) nor any other external economic factor was regarded as being an important cause. The report focused almost exclusively on what it saw as the overextended role of the state, overvalued currencies and the neglect of market signals generally. The clear implication was that the main thing Africa needed to do was 'to get the prices right'. This could be achieved by reducing the role of the state to supervising the efficient operation of markets and improving the infrastructure and education, where conditions were not yet favourable for private capital to do this.

The criticism of the Berg Report was extensive, and the two subsequent World Bank reports on Africa in the 1980s (World Bank 1984, 1989) made progressive strategic retreats on the *interpretation* of the causes of the

---

[5.] 'Leninist' implies the assumption by the party of a dictatorial role, acting 'on behalf' of the proletariat rather than allowing the free action envisaged by Marx. In practice it meant the elimination of all opposition or consideration of alternative policies.

economic deterioration, now allowing a contribution from negative exogenous factors, but without altering the *prescription* in any way. The harmful social and distributional effects caused by its first programmes were not taken as evidence against their efficacy, for these factors were not of interest to the programme framers who focused on *policy* indicators such as the freeing of currencies, trade, labour and goods from state inter-ference; some programmes were judged as successful under these criteria, despite the fact that *real economy* indicators, such as export growth, invest-ment and GDP growth either showed little response or reacted negatively along with the social and distributional indicators (for example, Nigeria; see Olukoshi 1989). The only concession was to admit that there might be costs as well as benefits, losers as well as gainers, from adjustment, and that some of the losers might be the very poor, rather than the unde-serving 'rent-seekers' (the unproductive category of people who benefited from the price distortions and physical controls caused by government intervention, such as licences, trade quotas and exchange control). Pro-grammes to ameliorate these effects for the poor were thus introduced. These were generally assessed as having very little impact for a variety of reasons; one example was the much criticised Social Dimensions of Adjustment programme in Zimbabwe (Moorsom *et al.*, 1997).

The ideological, rather than technical, nature of the Bank's strategy in the 1980s is demonstrated by the fact that much the same policy was applied to widely different countries in very different economic circum-stances, not just in Africa but across the world. The reports cited above treat Africa as a whole with few references to individual countries and no case studies (see Hoogvelt *et al.* (1992) for a detailed critique). The Bank's individual country reports did, of course, pay attention to the detail of countries' economic situations, but little effect on the prescriptions was discernible. Basically all countries were assumed to have the structure of the worst: pervasive and inefficient government involvement in the econ-omy, seriously overvalued exchange rates, an anti-export bias, 'feather-bedding' of inefficient industry and a bias against agriculture, and non-functioning markets generally. Examples from the region where such distortions attracted early pressure from the Bank and Fund included Zambia (from the late 1970s) and Tanzania (from the early 1980s), although the worst manifestations only became clear after the oil-price hikes of the 1970s (and so had, at least in part, exogenous causes). The less than objec-tive nature of this pressure is demonstrated by the cases of Cold War client countries like Zaire and Malawi which, despite having comparable or worse distortions, were then under little pressure.[6] Even where all of these assumed negative characteristics were operating, there were still arguments that could have been made against the Bank's policies, but their prescrip-tions could be really destructive in countries where many domestic policies were in fact succeeding. For a time in the 1980s it seemed that, for the

[6.] Another example concerns the countries in the franc zone, where the overvalued currency was supported by France, and this distorting factor somehow in these cases failed to surface as a cause for concern.

Bank, any government role was by definition bad, so that it had to be reduced, even if, as in the NICs, it was demonstrably having a dynamic effect on economic structure. The most dramatic case of this in Africa was in Zimbabwe, where arguably the Bank's attitude can be made explicable only in terms of a hidden agenda of wishing to avert a 'threat of a good example' (Melrose 1985). Other aspects of its attack on Zimbabwe's successful 'heterodox' policy in the 1980s are discussed in Box 3.3.

Stabilisation or adjustment?

At this stage it is important to distinguish not only between the IMF's stabilisation programmes and the World Bank's structural adjustment programmes (SAPs), but also between countries in serious macro-economic imbalance and those more nearly in balance. As described above, Tanzania moved from viability to serious imbalance in its trading account as a result of the second oil-price hike. A country previously balancing its payments (even if helped by fairly predictable aid or other foreign capital inflows) cannot continue without major changes if an extra 80 per cent of GDP is added to its import costs, least of all in the context of falling earnings. In exceptional cases, states have been bailed out by proportionate aid flows, for instance South Korea after the end of the war with the North, or Mozambique after the end of South African destabilisation and the rapprochement between Frelimo and Renamo. But for most countries in such circumstances, imports have to be cut drastically and efforts have to be made to raise exports. Cutting essential imports in Tanzania meant that a vicious downward cycle occurred as factories producing millions of dollars of output closed for lack of a few hundred dollars' worth of spare parts or fuel (Biermann and Campbell 1989). Attempts to minimise the pain by protecting the exchange rate (so keeping imports cheap) reduced the incentive to export, and when devaluation was finally forced, the currency traded at a fraction of its former value. It is such macro-economic imbalances – in the trade account, in the exchange rate, and in high levels of inflation with attendant growth in the money supply – that require action, whether caused by bad internal policies, exogenous events, or a combination of the two.[7] The original justification for the IMF was to provide short-term (about a year's) support for countries facing such problems, and usually it imposed conditions relating to a country's ability to achieve balance thereafter. The lending to Zambia in the 1970s came at first from commercial banks, but increasingly from the IMF to allow repayments of commercial loans during what was generally seen as a temporary dip in the price of copper. The IMF is ill-equipped to deal with

---

[7] Protection of infant industry is usually associated with some currency overvaluation (typically 10–30 per cent). Although this is opposed by the World Bank and the IMF, it may be defended as part of a deliberate strategy. In South Korea's case for example, it has been argued that the essence of its economic success was that it found the right way 'to get prices wrong'! (Amsden 1989: 13–14). On the other hand, supporting an exchange rate that has become unrealistic – overvalued by 50–100 per cent or more – is not defensible.

problems of this type: of what to do in the longer term if the copper price does *not* rise or the import price of oil does *not* fall. All it can do is insist on austerity measures which, although possibly appropriate for responding to short-term shocks, are likely to be both socially and developmentally negative in the medium to long run. The IMF is concerned with short-term stability, not long-term development.

Nevertheless, *stabilisation* of economies in macro-economic imbalance is not really controversial: in the absence of massive aid inflows[8] it will either happen painfully under something like the IMF model (see the Zimbabwe case study in Box 3.3), or very painfully (as in Tanzania, see Box 3.2) after trying to carry on as if nothing had happened. On the other hand, *structural adjustment*, as advocated by the World Bank, should be seen as one, highly controversial, way of trying to change economic structures, either to accelerate a sluggish growth rate (their argument in Zimbabwe) or to create a new structure where (as in Tanzania) the old one had become unviable. As we have seen above, the type of structural adjustment favoured by the World Bank emphasises markets, and sees states as intrinsically hostile to the efficiency-generating characteristics of free markets. In its most extreme form, *any* government action is argued to result in suboptimal outcomes, although government roles in conserving free markets (for example through anti-monopoly legislation) and in such functions as education, are usually accepted. For a time in the 1980s even this began to be challenged to some extent with the introduction of conditions requiring some 'cost recovery' through school fees. This was soon largely abandoned in the face of evidence of the fall in school attendance by the rural poor and girls in particular. It is of interest that by the early 1990s, the Bank was urging its latest recruit, Zimbabwe, which was keen to curry favour with it, *not* to impose fees in rural areas.

A World Bank SAP thus included the following elements: trade liberalisation (successively moving from quantitative controls and outright bans to equivalent tariffs, then to non-differential tariffs and finally to low or zero tariffs); removal of export subsidies; exchange liberalisation (allowing market determination of the exchange rate for the currency, with a minimal role for the Central Bank); removal of discrimination against foreign firms; removal of price controls; removal of wage and employment controls; reduction and removal of subsidies; withdrawal of the state from the economy through privatisation and closure of marketing boards, etc.; and reduction of government expenditure so that the budget deficit is progressively reduced to at most 3 per cent of GDP (see, *inter alia*, Cornia and Helleiner 1994; Engberg-Pedersen *et al.* 1996; van der Hoeven and van der Kraaij 1994; Gibbon 1993; Himmelstrand *et al.* 1993; Hope 1998; Mohan *et al.* 2000; Simon *et al.* 1995). Some of these elements overlap with the IMF's stabilisation measures, but the IMF is more concerned, for instance, that government expenditure should be low so that the economy is in balance, whereas the World Bank wants it low so that market forces can

---

[8] At the levels received by, for example, Israel, or more recently Mozambique (but only *after* structural adjustment was agreed in the latter case (Hanlon 1991: 248, 270)).

have free rein. Some contradictions have been pointed out: the pursuit of balance by the IMF may lead to austerity measures that destroy World Bank programmes of economic development by, for example, reducing the market for the output of small farmers and new entrepreneurs.

In summary, what could be seen even in the 1970s as a project of the World Bank and the IMF to create a single world market, irrespective of the harm that this might do to small developing corners of it, was given its perfect pretext in the 1980s in Africa and Latin America by exogenous factors.[9] The few countries which had avoided serious debt and were less vulnerable to IFI leverage found themselves pressured in a number of controversial or contradictory ways. For example, Zimbabwe was denied further financial support for an already successful export promotion scheme.

Arguably such pressure was ideological in nature, as countries with unorthodox policies could not be seen for too long to be succeeding while the World Bank's success stories were so unconvincing. In the late 1980s Zimbabwe achieved an average growth rate of just over 4 per cent per annum. It achieved this while servicing its debts fully (at a time when almost all other countries in Africa needed to have their debts rescheduled) despite an aid freeze from major donors, with the result that it had to shoulder a net capital outflow of 5 per cent of GDP. Yet the World Bank regarded Zimbabwe negatively, and regarded Ghana's comparative growth rate of 5 per cent, after a period of deep depression, as a success story even though there this was accompanied by a net capital *inflow* of 5 per cent, largely from the IFIs (Stoneman 1992: 104; Haynes 1989).

Wider *choice* for consumers to select which products they preferred in commodity markets were buzzwords of the 1980s. Choice of economic systems or even policies was another matter: there was 'no alternative'[10] – or if there was it had to be destroyed. Market-Leninism was triumphant.

A taxonomy of countries in the region

In this section the countries of the region are grouped in three categories: failures and victims; market 'successes'; and threatening successes. Lesotho, Swaziland and Namibia are excluded, not because they are small, but because they have not experienced SAPs of any significance. This is also true of South Africa, but its special economic significance within the region and the relatively 'developed' economy inherited by the post-apartheid government in 1994 clearly placed it in the 'threatening success' category.

Failures and victims

In this first category are countries needing stabilisation for one reason or another. These are the DRC, Malawi, Tanzania and Zambia, which

---

9. In Latin America SAPs did seem partially to address the economic crisis there in the 1980s, but in sub-Saharan Africa this decade witnessed further, radical deterioration in the socio-economic situation (see Simon 1995).

10. TINA – 'there is no alternative' – was the common slogan of the orthodoxy of the time.

suffered from adverse world conditions and poor policy (although each of them at one time or another has been praised by the IFIs); and Angola, Mozambique, Rwanda and Burundi, in all of which civil wars have played a decisive role in destroying what (in some cases at least) may have been promising developmental strategies. These countries are considered briefly in turn below, except for Tanzania and Mozambique which are discussed in detail in Boxes 3.1 and 3.2.

The **DRC** (Democratic Republic of the Congo, formerly Zaire) was perhaps the most extreme case where cold-war politics allowed a rapacious ruling class or dictator to exploit a country's resources and international aid for private rather than national ends. In 1993 (when the unreliability of their statistics caused the World Bank to stop listing the country in the *World Development Report*) its per capita GDP at US$117 was one-third the level of 1958, just before independence. Although the IFIs attempted to impose conditionality on financial support over a period of two decades this was almost completely ineffective. President Mobutu Sese Seko continued to salt away a high proportion of the funds received to add to his private fortune in Swiss banks and similar untraceable accounts. This was

---

Box 3.1   Mozambique: adjusting during and after war

The case of Mozambique illustrates in extreme form how the IFIs' global agenda dominates even the poorest country's interests. Their involvement began while the country was in the throes of a civil war between the Frelimo government and Renamo rebels (originally encouraged by white-ruled Rhodesia and then sustained by apartheid South Africa). The basic attack on government spending amounted to an almost unprecedented constraint on a government fighting a war. Had similar policies been imposed in the Second World War they would have vitiated the successful prosecution of the British or American war efforts. In the words of Mozambican government ministers, 'During World War II in Europe, every country intervened in the market. But the IMF tells us we cannot' (Hanlon 1991: 128). One of the IFIs' fundamental aims – to shift countries towards free-market policies – was similar, in some ways, to the aims of the very enemy that the Frelimo government was fighting, namely Renamo and South Africa.

The first five years of independence until 1980 were times of great hope and achievement, as the country recovered from the initial devastation caused by the precipitate exodus of almost the entire white population, taking all they could carry and sabotaging what they could not. Education and health services made significant advances and economic growth resumed. But a combination of the worsened external conditions that affected all countries in the region and the start of South African destabilisation, spearheaded in Mozambique's case by Renamo, made the next five years a period of collapse, effectively exacerbated by the IFIs. In 1983, for example, when Frelimo appealed for food aid because of a drought, aid was reduced. Three demands were then made by Mozambique's creditors: that Mozambique sign the 'Nkomati Accord' with South Africa (thus agreeing to stop supporting the ANC); that Mozambique join the World Bank and the IMF; and finally that Mozambique allow foreign NGOs to engage in independent aid and development work. Only then was food aid given. When Mozambique in 1986 again needed food aid, this was only released after a fourth agreement was signed: to accept a structural adjustment programme.

With their political aims achieved, the IFIs became quite generous, with total aid jumping from US$360m. in 1985 to an average of US$1,100m. from 1990 to 1994 (Hanlon 1996: 164), making Mozambique the largest recipient in sub-Saharan Africa. Unfortunately, much of this aid was wasted through lack of coordination between the burgeoning NGOs, or kept idle in forced additions to reserves. This paradox of idle aid money occurred because a condition of the stabilisation programme was to limit government expenditure (which included much of the aid) and cut inflation. The IMF was thus imposing what has been called a 'tax on aid' and even though the government had the money available, it could not even repair bridges and roads damaged in the war, or restore health and education services. Even some World Bank reconstruction loans were affected, causing friction between the two institutions. The consequence of this was that the government effectively lost control over many parts of the country to the NGOs as well as to Renamo (until the ceasefire in 1992). Trade liberalisation was also pursued dogmatically – in 1995 the World Bank forced a highly controversial liberalisation policy on the cashew nut exporting industry which effectively destroyed nut processing (the fourth largest employer) in Mozambique. Hanlon (1996: 37) argues that 'the issue had little to do with nuts, and was more related to the need for the Bank to show its power and prevent a government alternative being accepted'.

By the late 1990s the World Bank enjoyed 'a virtual monopoly on policy advice' (Hanlon 1996: 48) to Mozambique. Money borrowed had to be spent following this advice, and it had to be repaid in full even if the advice turned out to be wrong. A notorious example is the case of the Roads and Coastal Shipping programme costing almost US$1bn, described by some as a fiasco due to its totally inappropriate technological standards for a desperately poor, mainly rural, country. The Bank, however, maintained that it could not 'be held accountable for "incorrect" advice' (Hanlon 1996, citing a World Bank staff member in Mozambique).

Despite the aid inflows, there was no 'peace dividend' for adjusting Mozambique: under IMF demands, credit fell from US$438m. in 1990 to US$150m. in 1995; industrial production fell from 50 per cent of its 1980 level in 1989 to under 30 per cent in 1993; state agricultural marketing boards were abolished without the promised safeguards to peasant producers, who then suffered at the hands of monopoly traders. By 1995, per capita GDP had fallen to US$91 from US$108 in 1990, making Mozambique the *poorest* country in the world, despite half its GDP being provided as aid.

As a postscript it should be added that the World Bank now regards Mozambique as a great success story on the strength of its subsequent improved economic growth rates of 8 per cent (including 11.2 per cent in 1998). This is welcome evidence of a reversal of the earlier trend, possibly thanks to a more relaxed attitude to government spending confirmed in an IMF statement in June 1999 allowing more aid funds to be spent. However, it will take a few more years to judge whether Mozambique is on a sustainable growth path or merely experiencing a partial recovery from a devastating slump. Growth rates dropped sharply again at the beginning of the twenty-first century, although these were not helped by two years of devastating floods and a regional drought in the 2001/2 crop year. Analysis is not helped by the unreliability of the statistics: in 1997 there were four different official estimates of GDP per capita, ranging from the World Bank's US$120 to the IMF's US$207 (the latter implying a seemingly implausible doubling in two years from US$91 in 1995). A further example of such statistical difficulties can be seen from an estimate that the output from a major new aluminium smelting plant (Mozal) near Maputo will alone raise Mozambique's GDP by 10 per cent in the near future (SAPA, 2000). Hence the macro-economic indices are highly susceptible to fluctuations due to events which in larger economies would hardly cause a blip. It is equally clear that such fluctuations tell us little or nothing about trends in the welfare of the population as a whole.

Indeed the impact of structural adjustment policies need not only to be analysed at the national, but also at the subnational level. In Mozambique the improvements in national economic performance in the late 1990s have had little positive impact on the livelihoods of the rural and urban poor. Instead, many of the policies, such as liberalisation of trade and government cutbacks which brought about many formal sector job retrenchments, and the squeeze on public investment which has hampered the health and education sectors which were in disastrous shape after the war, have caused further impoverishment. There is even evidence from a trade union study that many formal sector jobs in Mozambique now yield such poor incomes and conditions that retrenched workers feel there is little to choose between such jobs, and farming or informal sector work. Even in Maputo, which appears to be attracting the lion's share of 'liberalised' economic activity, the urban poor are often having to develop new rural bases, or redevelop old ones, merely to subsist. Yet rural conditions in Mozambique remain exceptionally difficult. Of course, according to the central argument of this chapter – that SAPs were not only not necessary for, but also antithetical to, 'development' (in its normative sense) – this evidence is unsurprising.

Perhaps the greatest cause for optimism concerning Mozambique's prospects should derive from its having qualified for debt relief under the recent HIPC (Highly Indebted Poor Countries) programme. Although it had in any case previously been unable to service its total debt (which the World Bank estimated at 4.28 times GDP in 1996), actual debt service is expected to fall from an average of US$114m. per year to US$73m. per year.

Sources: Cramer (2001), Disch (1993), Jenkins (forthcoming), Hanlon (1991, 1996, 1998, 1999, 2000), SAPA (2000).

permitted by the West as a result of his skilful exploitation of the strategic position of Zaire in Africa in the face of supposed Soviet ambitions.[11]

Economic growth averaged about 6 per cent over the 1962–74 period of economic nationalism. But declining copper and other commodity prices at a time of rising prices of imported fuel following the first oil-price hike in 1974 undermined progress. The first in a series of stabilisation programmes began in 1976, and these continued until 1987, with both the IMF and the World Bank putting their own people in key posts in the Bank of Zaire, the Customs Office and the Ministries of Finance and Planning in the later stages (Parfitt and Riley 1989: 82). It was during this period, and in particular 1984–89, that Zaire entered a phase of 'generalised predation' and extortion perpetrated by those close to the presidency. Despite the IMF and World Bank presence, it is estimated that about US$1bn was drained from the state treasury in this period (United States Institute of Peace 1997: 9).

In 1991 the IMF declared Zaire ineligible for further loans as it was over US$6bn in arrears on repayments; the World Bank and African Development Bank followed suit in 1995 when foreign debt stood at US$12.3bn (over US$300 per capita). Although misgovernment and corruption were almost certainly the main causes of this situation, it was compounded by the same world market factors which overtook Zambia. Since 1995, of course, the civil war has further worsened the situation; when it is resolved it will not be a question of recovery from the war, but rather a complete reconstruction of the country (Collins 1996).

Malawi is another example of a country with a fairly good growth record under policies of economic nationalism in the 1960s and 1970s – annual GDP growth averaged 5.5 per cent between 1967 and 1979 – followed by a poor record in the 1980s under World Bank SAPs, when growth fell to 1.2 per cent on average in the 1980s. According to an independent assessment, in the eyes of the World Bank Malawi provided

> a favourable testing ground for its new mode of programme lending. The country possessed a strong, highly-centralised political regime and an economic ideology which appeared receptive to the types of structural reforms advocated by the Bank, while the macro-economic disequilibrium experienced by Malawi in the late 1970s and early 1980s, and the structural weakness consequently exposed, appeared to be moderate.
> (Harrigan 1991: 201)

After three adjustment programmes in the 1980s, however, 'there was an air of disillusion amongst Bank staff and the Malawian authorities'. The assessment at that time, which was substantially unchanged at the end of the century, was that the export base remained narrow, vulnerability to external shocks remained high, the public sector deficit was still large, and

---

[11.] Mobutu's support from the West dated back to the early 1960s when he received covert support from Belgium and the CIA to oust the socialist Patrice Lumumba, the country's first elected Prime Minister, who was then assassinated in 1961, again with covert Western involvement.

investment and GDP per capita had fallen. According to Harrigan (1991: 202), although there may have been some bad luck (as the Bank claimed), more important was the inappropriateness of some of the policies imposed by the Bank. What can be drawn out of this experience is that *either* the poor economic record in the 1980s and 1990s was caused by external factors (in contradiction to the Berg Report's and the World Bank's general rationale for imposing structural adjustment) *or* the World Bank's reforms had had a negative effect. In late 1999 the IMF admitted that Malawi's debt service payments were 'onerous' – despite the fact that the country was not eligible for debt relief under the current Heavily Indebted Poor Countries (HIPC) Initiative because its debt service payments were considered too small (IMF 1999).

Zambia is the classic case of a country where the economic structure has been distorted by production of a single dominant commodity for export. After independence in 1962 an attempt was made to begin a process of more balanced development, and in the 1960s and early 1970s GDP growth averaged almost 6 per cent per annum, much of it deriving from growth of manufacturing industry. The collapse in copper prices in 1973 and 1975, well before the country could have expected to have developed significant alternative exports, threatened this strategy, and Zambia borrowed heavily during what was generally believed to be a temporary fall in commodity prices, so that infrastructural and productive investments could be completed and yield a return. Thus was laid the basis for an external debt reaching US$7bn by the 1990s, almost US$700 per head or about twice the annual GDP. Undoubtedly the economy needed stabilisation after 1975, and the resistance to such measures made matters worse. A particular problem in the Zambian case was the indiscriminate protection of inefficient parastatals for many years (Karmiloff 1990: 300–3). Structural adjustment programmes have been in place since the mid-1980s but, after a short and weak recovery in 1987–88 (during a 'self-help' programme after the IMF programme was suspended following maize-price riots), growth rates turned negative until 1996. The stabilisation measures introduced have probably been beneficial, but structural adjustment measures have deepened poverty, raised mortality rates and harmed education. Efforts to sell off parastatals included the copper industry, the nationalisation of which had been a mark of national pride in the 1970s (Kashambuzi 1995; *Financial Times Survey* 1992, 1997). The sale of the four state-owned copper mines proved difficult and controversial however, as they were run down and were in any case reaching the end of their 'lives' (i.e. the best copper had been mined out). Eventually Anglo-American Corporation of South Africa, having negotiated 'lavish' and 'swingeing' concessions, took most of them over at a very low price and on terms disadvantageous to the workforce (*Africa Confidential* 2000a, b).

Angola is perhaps the prime example, along with Mozambique, of a country where war has dominated all other considerations. The immediate post-independence economic policy which delivered greatly improved access to healthcare and education has been defended as well as criticised, but the costs of civil war, supported by both South Africa and the USA from 1975 until the end of the Cold War, destroyed the economy. Attempts to restore macro-economic stability began in 1987, with most subsidies being abolished in 1994 when a new stabilisation programme began following discussions with the IMF and the World Bank. In 1994 the external debt, of which over a half was in arrears, was US$11.2bn, about US$880 per capita, more than double the annual GDP (Hodges 1995). The civil war reached new heights in the later 1990s, again rendering development policies, of whatever ideological flavour, virtually irrelevant (except perhaps in the few cities whence fled millions of internally displaced people – but even these, with the exception of the capital, Luanda, could be periodically cut off from the government or otherwise affected by military activity). At the same time the MPLA government found itself greatly financially advantaged by high oil prices and explorations indicating significant new offshore oil potential which led to an 'oil rush' as European and American companies bid for new licences. Unfortunately there has been little attempt to use government revenues to improve the desperate circumstances of Angola's people (albeit that the war imposed serious constraints), leading to the unusual situation where the IMF (which is usually associated with cutting government budgets) was insisting on improved health-related and other welfare expenditure in 2000 (IMF 2000) as a condition for loans under a so-called 'Staff Monitored Programme' (SMP). There is clear evidence that MPLA politicians, many of whom have long since eschewed their Marxist ideological roots (see, for example, Hodges 2001), are involved in corrupt financial dealings on a scale which could seriously undermine any development policies (*New York Times* 2000). In 2001, for example, one estimate made was that around US$1.4bn (one-third of state revenues) had gone missing (*Financial Times* 2002).[12] The lack of financial transparency in the government budget and the oil accounts were primary reasons for the IMF suspending the SMP in 2001. Since the war finally ended in 2002, there is some hope that development prospects will improve, but quite how IFI conditional policies will combine with this internal financial corruption and the country's extreme need for significant investment to rehabilitate the economy (see Box 3.2 on Mozambique) remains, at the time of writing, to be seen.

Rwanda and **Burundi** are both dominated by desperate internal instability, which has spilled over into genocide at different times in both countries, most terribly in 1994 in Rwanda. Discussion of the impact of IFI policies has not been restricted to whether these policies were inappropriate or beneficial as in other countries: it has also been argued that they have had a causative role in exacerbating communal dissension. Dependence on coffee exports meant that after market pressures destroyed the International Coffee Organisation in 1991, the collapse in prices affected these countries as seriously as the collapse in copper prices had affected Zambia. However, in this case the crisis in the peasant economy sharpened inter-ethnic divisions. The immediate imposition of stabilisation measures

---

[12] A billion dollar bank account in the Virgin Islands with two signatories close to the Angolan President was also identified.

Box 3.2 Structural adjustment in Tanzania

Tanzania's case is illustrative of the situation of many very poor, rurally based countries which attempted in their first decades of independence to adjust the structures of their economies away from the colonially distorted form in which only one or two primary commodities were exported, and almost no manufactured goods were produced let alone exported. This either involved import substitution starting with production of light manufactures or, as in Tanzania's case, to attempt to force the pace through developing basic (usually heavy) industries and infrastructure. Like many other countries, it was blown off course by the world market disruptions of the late 1970s and early 1980s.

Its economic growth rate from independence in 1960 to the end of the decade was probably about 6 per cent; from 1967 to 1975 it was 4.3 per cent. Unlike Zambia it just about weathered the first oil-price rise, growing by 3 per cent per annum until 1980. GDP then fell and only recovered its 1980 value in 1985, so per capita income must have fallen by over 10 per cent. In this period Tanzania tried to do the impossible and resisted the needed stabilisation measures following the second oil-price rise in 1979, rejecting IMF loans because of the attached conditionality. It refused to devalue, so that the real exchange rate of the currency doubled over the five years as inflation averaged about 30 per cent, and savings and investment dropped sharply. Declining aid flows and continuing pressure from the IMF and the World Bank eventually forced the country into both stabilisation and structural adjustment in 1985. From 1988 to 1998 the growth rate was about 2.9 per cent.

It is interesting to quote from a World Bank report of 1990 which discusses the first two decades of Tanzania's independent growth path. According to the Bank,

> Tanzania's unprecedented access to concessionary flows of external capital ... allowed it ... to maintain a high rate of largely ill-conceived and uneconomical industrial investment.... The Bank's continuous exhortation of the donor community to provide assistance ... sustained a constant inflow of official aid that helped maintain irrational domestic policies.
>
> (*Financial Times* 1999: unpaginated)

The Bank's former support for Tanzania's early development path is evident, as is its attempt to distance itself from those policies, given its subsequent neo-liberal conversion. In addition, close examination of the actual aid inflows over the past 40 years and of the macro-economic outcomes of the so-called 'internal policies' of the 1960s and 1970s compared to the era of SAPs reveals some interesting challenges to the assumptions behind this statement. First, manufacturing output as a percentage of GDP rose from 10.7 per cent in 1967 to a peak of 13 per cent in 1976, since when it has fallen almost continuously to 6.8 per cent in 1998. But maybe we can accept the Bank view that this was 'artificial' or uneconomic investment. What, however, did this 'unprecedented access' to aid amount to? The percentage of GDP contributed by aid did indeed rise steadily from 2.3 per cent in 1967 to 7.2 per cent in 1974 and

11.7 per cent in 1975 before falling to about 9 per cent in the next two years. After 1980 it was progressively squeezed under IFI pressure until it was only 6.9 per cent in 1985. It then climbed steadily as Tanzania was rewarded for adopting structural adjustment policies, reaching 49.5 per cent in 1992. It is hard to escape the conclusion that the World Bank, or some World Bank experts at least, regard aid-supported industrial development at any level as 'ill-conceived and uneconomical' and 'irrational' in an economy where existing comparative advantage is in agriculture, but that aid support at five times such 'unprecedented' levels in pursuit of de-industrialisation and the raising of agricultural exports is rational. Under this policy exports have indeed risen from a trough of 4.8 per cent of GDP in 1985 to 14.8 per cent in 1992, but only in 1993 did they again reach 22.2 per cent, about the level of the late 1960s and early 1970s.

Tanzania's external debt has been rescheduled on five occasions since 1986 when it was 146 per cent of GDP; in 1998 it was still 94 per cent of GDP at US$7.4bn or US$230 per capita. Without debt relief, servicing of this debt would have been US$800m. annually, about 24 per cent more than total export earnings in 1998. It is estimated that without HIPC debt reduction the debt would grow to US$10bn by 2005.

Sources: Nyoni (1997), World Bank (1999b), *Financial Times* (1999).

then put extra burdens on the poorest, further intensifying the political tensions. Perhaps the 1994 genocide would have occurred in any case, but it is widely accepted that the adjustment programme exacerbated tensions and helped to create the environment in which the genocide occurred (Hanlon 1998).

Market 'successes'

There are very few economic success stories to point to in eastern and southern Africa (or anywhere in Africa) beyond a few periods of less than a decade, mostly in the 1960s and early 1970s, and rarely since 1975 (when market policies began to take over).

Botswana is the main exception to this generalisation. It not only equalled, but at times exceeded, the growth rates of the NICs, averaging GDP growth of 10.7 per cent between 1977 and 1987, although the record since is less impressive at 4.8 per cent. Its GDP per capita of US$4,860 in 1998, nearly three times that of South Africa, now places it in the upper-middle-income countries. Yet only 4.8 per cent of income derives from manufacturing industry (down from over 7 per cent in the 1970s despite some real growth, as diamond production has grown much faster), and two-thirds of exports are diamonds. Given its enormous diamond wealth, Botswana has, of course, needed no stabilisation programme and has no external debt. Its policies have been nominally free-market, although complicated investment regulations control the ineffective industrialisation policy (heavily constrained by membership of the Southern African Customs Union, SACU). But the same wealth has allowed lavish and inefficient programmes of state support for enterprises of almost any type under the Financial Assistance Policy (FAP). For instance, just one provision of the FAP, the Unskilled Labour Grant, is estimated to be equivalent to a subsidy of about 25 per cent over a five-year period (Granberg and Parkinson 1988: 17–18). Such subsidies would have fallen foul of the IFIs elsewhere. but in Botswana they have no leverage through which to express opposition effectively. No conclusions one way or the other can therefore be drawn about the merits or otherwise of World Bank policies in this case. Whether the economy's dynamism would survive a long-term fall in the market price of diamonds is doubtful.

Kenya is one of the countries that was seen in the 1960s and 1970s as an advertisement for the success of free-market capitalism (which for a time it called 'African socialism') in contrast to the state socialist policies being pursued in Tanzania in particular. However, its average annual growth rate at 6.6 per cent from independence in 1963 to 1973 was little more than Tanzania's at 6 per cent. One particularly encouraging factor was the more rapid development of industry which grew at 10 per cent annually between 1964 and 1978. In fact its policies involved many market interferences, albeit not in pursuit of socialist objectives. Everything began to go wrong in the late 1970s with the dissolution of the East African Community in 1977, the collapse in the commodity boom (in particular the sharp fall in the price of coffee in 1979) and the second oil-price hike. The World Bank, however, chose to attribute the major blame to rapid growth in the public and parastatal sector, with the public share in GDP rising from 24.7 per cent in 1976–7 to 35.4 per cent in 1980–81. There were 147 parastatal organisations and the government owned shares in a further 176 companies, many in manufacturing industry (Tostensen and Scott 1987: 250). But perhaps as significant as the scale of the public sector (whose influence was also pervasive in the NICs)[13] was Kenya's failure to exert any effective financial discipline over parastatal operations.

The first IMF loans were made in 1975 and 1978, but since 1980 Kenya has undergone a series of stabilisation and structural adjustment programmes, with increasing policy conditionality, which should not have been necessary (on World Bank assessments) had it been the free-market model claimed. In fact the experience of the 1980s and the 1990s was of a series of IMF standby loans and World Bank programmes introduced and then cancelled as Kenya failed to meet the conditions imposed. The growth rate declined to 3.9 per cent in 1977–87 and 2.3 per cent in 1988–98, while manufacturing industry growth fell to 4.9 per cent and 2.9 per cent in the same periods. Total external debt rose from US$1.7bn in 1977 (37 per cent of GDP) to US$7bn in 1998 (61 per cent of GDP and US$245 per capita) (World Bank 1999a; Mosley 1991; Swamy 1994). The IFIs may well argue

---

[13] One of the World Bank's prime 'free-market' model countries, Taiwan, is estimated to have had public sector final demand at about 33 per cent of GDP at this time (Wade 1990: 173), equivalent to Kenya's.

that Kenya's failure in the last two decades was caused by its failure to meet their conditions; this may have some weight in that it was reluctant to *stabilise* fully following external shocks. There is less evidence that its slowness to adjust away from an economic *structure* which earlier had delivered much higher growth was a major cause.

Uganda is the one clear success story in recent years for a country in the region undergoing a SAP. After considerable growth in the 1960s the country went into decline under Idi Amin and the civil war following his downfall. From 1987, however, recovery began, with the next 10 years registering average growth of 6.7 per cent (although this barely restored the income level of 1971). This is the more impressive because the price of the main export, coffee, had collapsed in the interim; earnings from coffee fell from US$394m. in 1986 to US$98m. in 1992 (recovering to US$404m. in 1996). Stabilisation was important in the early stages, but structural adjustment has since begun, with cuts in subsidies, the budget deficit, deregulation of marketing boards and a privatisation programme. Desperate for a success story, the World Bank has embraced Uganda eagerly, but proving causality is another question, given the similarity of these policies to those which have failed nearly everywhere else. There are other ways of explaining the record: at US$1.4bn (about US$75 per capita), debt was not so serious a problem in 1986 at the start of recovery although it had risen to US$3.5bn (US$185 per capita) in 1996, about 58 per cent of GDP. Perhaps most significant is the level of external support that Uganda has received. The government reported in 1998 that 49 per cent of the budget in the previous financial year was covered by external support, down only marginally from 53 per cent in 1995/96. This support financed, according to the World Bank, about 80 per cent of public investment. Gross aid flows in the mid-1990s totalled about US$550m., more than twice the value of merchandise exports. (Only Tanzania and Mozambique – also after adopting structural adjustment – received aids flows at comparable levels (Nyoni 1997: 2–3; Hanlon 1996: 164).) So all that this 'success story' demonstrates for certain is that stabilisation is easier when the books are balanced by external donations. It also suggests that structural adjustment *may* lead to economic growth, in the context of massive aid inflows (Laishley 1993; *Financial Times Survey* 1996, 1998).

Threatening successes

Only two countries in the region, South Africa and Zimbabwe, have enjoyed long periods of economic growth which have led to them developing significant manufacturing industries, accounting for about a quarter of GDP. Only they, therefore, can yet be seen as having laid the basis for sustained development which may enable them eventually to attain developed status. Neither of these experiences has owed anything to orthodox market policies, and both countries have been markedly less successful economically since embracing them. Indeed the industrialisation process seems now to have been completely aborted in Zimbabwe.

South Africa developed under nationalist economic policies, beginning in the 1920s and accelerating during the Second World War, and then again in the 1950s and 1960s. Interpretation of this experience is made more difficult because the economic nationalism co-existed with the super-exploitation of black labour under apartheid, and this may well have been a factor contributing to rapid capital accumulation in this period (see Lipton 1985 and Sampson 1987 for different views). All that need concern us here, however, is that rapid growth and industrialisation occurred under conditions in which market forces were not dominant and the state played a key role in forcing industrialisation, as it did in Japan and the NICs. In the 1970s and 1980s the regime began a process of liberalisation, in part in response to growing inefficiencies in the apartheid system, in part as an attempt at a rapprochement with the West. As majority rule approached, measures akin to structural adjustment, including privatisation of successful parastatals, began. This formed part of a successful attempt by the outgoing minority government to constrain the incoming ANC government by removing many key resources from state control. These had been used by the apartheid state to benefit *its* main constituency (Afrikaans-speaking whites), but it did not want these to be available to benefit the ANC's main black constituency. Although the need to undo the extraordinary distortions of apartheid arguably provided grounds for high levels of state expenditure for a decade at least, the ANC immediately took the decision to move to more market-oriented policies. The market was thus allowed largely to confirm the income and asset distribution engineered by apartheid. A key lesson of the first three decades of Afrikaner rule was ignored: that even in a context of economic growth, state action had been needed to reverse the relative disadvantage of Afrikaans-speaking, as compared with English-speaking, whites. Few in that period would have expected market forces to bring about this reversal, yet such a faith on behalf of the black majority is now relied on. Why the incoming ANC government reversed state-oriented policies that it had proclaimed at least as far back as the Freedom Charter of the 1950s is a matter of debate which cannot be pursued here (but see Bond 2000; Marais 2000 for critiques). Nor can the issue of why it failed to learn from the Afrikaner experience. But it is clear that an opportunity was thereby lost for South Africa to provide an example of the merits of heterodox (from the point of view of the IFIs) economic policies to the rest of Africa.

Zimbabwe can also be characterised as once having been a 'threatening success'. Its fastest periods of growth were in periods of relative detachment from the world economy, during the Second World War, under sanctions between 1965 and 1975 (falling off during the height of the liberation war) and from 1980 to 1990. A detailed analysis of its economic development before and after structural adjustment is provided in Box 3.3.

Conclusion

Almost all countries in eastern and southern Africa experienced much faster economic growth before 1980 than after. This is true of countries following state-led industrialisation programmes as well as ones which were always closer to World Bank prescriptions, some of which were

Box 3.3 Zimbabwe: opting for structural adjustment

The case of Zimbabwe is particularly interesting, in that here the World Bank and the IMF confronted a country that had been quite successful with its earlier policy, enjoyed fair macro-economic stability, and retained an *option* as to whether to adopt their liberalisation policies or not. All the adjusting countries we have discussed so far (apart from South Africa) had, in the end, no choice but to take the medicine, whether because their policies had been wrong, poorly or corruptly executed, or aborted by war or world market storms. Recognising Zimbabwe's relative economic strength and stability, the World Bank's resident representative in Zimbabwe stated in early 1991 that the prospects for ESAP were bright and that 'Zimbabwe could be the first African country to succeed with such reforms' (Economist Intelligence Unit 1991: 13, citing Christian Poortman), a revealing recognition that the Bank's numerous earlier claims for success were spurious.

Zimbabwe had been most successful in economic terms before 1975 when, under sanctions, the illegal white minority regime was forced to run the economy through comprehensive controls and planning. The high level of protection, coupled with a battery of measures to promote exports and prevent serious exploitation of monopoly positions, produced a growth rate of over 7 per cent from 1967 to 1975; in the early 1970s the rate exceeded 10 per cent, resulting in some overheating and the famous phrase that 'we must slow down to a gallop'. Manufacturing industry developed fast, with its contribution to GDP rising from 17 to 27 per cent over this period. When the World Bank surveyed Zimbabwe's industry in 1982 and again in 1985, it assumed throughout its commentaries that most of this industry must be extremely inefficient. However its own data showed otherwise: the average DRC (domestic resource cost) was only 1.27, much lower than in any other World Bank survey of a developing country, implying that a small devaluation – as little as 21 per cent – would have been required to make it internationally competitive on average.[1] The prescription of liberalisation, however, had clearly been decided on before the surveys were carried out, and these results were not allowed to influence the conclusions in any way.[2] In a 'heads we win, tails you lose' equation, the message that emerged was that inefficient industries deserved to be 'rationalised' by liberalisation, whereas efficient industries (which theoretically should not have existed at all) would benefit from it. As later experience showed, both categories in fact suffered. Of course some industries were indeed inefficient (whether through overprotection or

---

[1.] The DRC is related to the protection afforded to industry by the amount of overvaluation of the domestic currency (i.e. in this case 21 per cent – 27/127).
[2.] The ideological character of the Belli Report in particular is seen in the disjuncture between the frequent reference to 'surprisingly efficient' industries or companies (with a total absence of any attempt to explain how their efficiency came about) and the immediate attribution of any inefficiency to the existence of protection.

underinvestment), but many were highly competitive. Effectively rejecting World Bank recommendations in the 1980s, Zimbabwe decided not to close the former but to seek to improve them through investment.

After independence in 1980, bilateral aid flows and World Bank loans prompted a degree of liberalisation and, during the drought of 1982–84, support was sought from the IMF. In March 1984, however, Zimbabwe terminated the programme after drawing down only one-third of the funds, mainly because the IMF's conditions were aborting the ambitious social and economic development programmes, particularly in health and education, almost before they had started. Furthermore, the balance of payments position had been worsened by the partial liberalisation of trade (the IMF had promised it would improve). From 1984 to 1990 Zimbabwe operated its own home-grown stabilisation programme (but with very little structural adjustment), and achieved a growth rate of over 4 per cent annually. Modest though this may seem, it was three times the African average in this very difficult decade. Furthermore, although industry grew slightly more slowly than GDP at about 3 per cent, it was reorienting itself to export markets and, by the last third of the decade, overall exports were growing by about 9 per cent per annum, with 'non-traditional' *manufactured* exports exceeding 15 per cent growth. All this was achieved, almost uniquely in Africa, without a single restructuring of a substantial debt burden, and by 1988 Zimbabwe had reduced its debt-service ratio from over 40 per cent to below 25 per cent of exports. The circumstances under which this record was achieved included not only significant destabilisation by apartheid-ruled South Africa[3] but also, as there was no IMF programme and as the resistance to structural adjustment meant very little balance of payments support, restriction of aid to project aid (i.e. no programme aid). As a result, net foreign flows in this period amounted to a 5 per cent outflow. A growth rate of 4 per cent under these circumstances was truly remarkable. Indeed in 1985 the Bank forecast near zero growth under existing policies, but promised 3 per cent following structural adjustment; thus Zimbabwe's own policies achieved a higher economic growth rate than the Bank's policies promised.

Nevertheless, dissatisfaction with the modest growth of the 1980s, the severe constraint on imports, and low growth in formal employment, made Zimbabwean politicians and industrialists vulnerable to World Bank pressure about the benefits to be had from structural adjustment. Overoptimistically, having succeeded with its own stabilisation, it attempted to introduce its own structural adjustment programme containing the standard World Bank and IMF ingredients. The result in the early and mid-1990s was a succession of economic failures despite substantial support from the IMF and the World Bank.

Let us now look in a little more detail at the heterodox policies of the 1980s, partly because they correspond fairly closely to the successful policies operated in the NICs and partly because they were also not too different from those

---

[3.] Defence expenditure, most of it devoted to keeping the 'Beira Corridor' open in face of Renamo attacks, amounted to at least 10 per cent of GDP for most of the 1980s.

advocated by Mosley *et al.* (1995) after their close scrutiny of World Bank policies. First, the policies were orthodox in their search for macro-economic stability, i.e. they involved home-grown stabilisation, manifested in a resolution to keep the exchange rate of the currency realistic, unlike the case in many other countries attempting to resist the IMF. Thus the currency was adjusted monthly on the basis of a trade-weighted basket of currencies and, although it was widely perceived to be overvalued, even the World Bank generally thought that this was by no more than 10–20 per cent.[4] Inflation was held down (and was well below the rate in South Africa, Zimbabwe's main supplier) by price and wage controls, and the money supply was controlled. The main failure occurred in government expenditure where the budget deficit remained obstinately around 10 per cent of GDP. There were, however, some very good reasons for this: South African destabilisation (primarily through support of Renamo in Mozambique) required 10,000 Zimbabwean troops to be in Mozambique to keep the cheapest trade route through the Beira Corridor open; even so, most trade had to travel the three to four times greater distances through South Africa. Thus destabilisation alone could have accounted for nearly all the budget deficit. Beyond that, as has been mentioned, Zimbabwe serviced its debts without rescheduling in face of restrictions on aid inflows, while maintaining its social programme.

Protectionist trade policies continued, relying on quantitative controls, but any bias against exporting was cancelled out by an export subsidy of 9 per cent and two other export-incentive measures, in addition to an export-revolving fund which meant that exporters, unlike other manufacturers, did not suffer from a foreign-exchange shortage for their inputs. Subsidies continued for the poor, most notably on basic foodstuffs, while education and healthcare remained free for nearly all.

None of this impressed the IFIs; or perhaps more accurately it caused them to redouble their efforts to be rid of this 'threat of a good example'. They were aided by the USA, which suspended all aid in 1986 on a flimsy pretext, and the UK which, although it continued with project aid, refused programme aid on the grounds that Zimbabwe did not 'have an IMF programme'. As such programmes are intended for countries in difficulties, it was disingenuous to penalise Zimbabwe for not needing one! The World Bank also contributed to the pressure. In 1981 it had loaned Zimbabwe the funds to set up the 'export-revolving fund'[5] which, as explained, had greatly benefited manufacturing exporters. The success prompted Zimbabwe *and* local World Bank officials to negotiate an extension to the fund in 1985–86 so that it could benefit all sectors of the economy. Yet although these negotiations were successfully completed, they then languished in pending trays in Washington for nearly two years,

---

[4] The correctness of this view is borne out by the fact that the real exchange rate following liberalisation was little lower, corresponding to a devaluation ranging from 0 to 20 per cent at different times.

[5] Manufacturers who needed to import inputs necessary for the fulfilment of an export order could borrow foreign exchange from the fund. They thereby avoided the forex constraint suffered by other producers.

before officials there decided to make a further loan *conditional on trade liberal-isation*, i.e. on a policy under which an export-revolving fund would become irrelevant. The problem was that while local Bank officials simply saw that the fund had worked and that a larger fund would work better, those in Washington whose ideological convictions were not challenged by observing pragmatic successes plainly regretted that it had worked at all and feared an extension would be a further 'threat of a good example'.

The advent of Zimbabwe's own ESAP in 1990 was economically paradox-ical and must be explained mainly by political considerations. The economic position was much stronger than in 1987, the debt crisis had largely been resolved and there was the immediate prospect of an end to destabilisation by South Africa. Since Zimbabwe had managed 4 per cent growth in the late 1980s and more than double that rate of growth in exports, what might it have achieved in the potentially much more favourable political environment of the 1990s? Instead ESAP was introduced and the World Bank promised un-conditional support ('no conditionality' was promised to President Mugabe in 1990) in financial terms. However it then proceeded to drive a hard bargain, demanding irreversible advances in trade liberalisation before it actually delivered the financial support needed to take the immediate risk out of such measures, so that control of the balance of payments was lost almost at once. This forced the country into approaching the IMF (which it had vowed not to do), and the needed funds came at the expense of a massive devaluation and commitments to accelerate trade liberalisation, and the dismantling of wage, price and currency controls. Inflation took off, investment and exports fell, and all the good work of the 1980s in reducing the debt burden was undone. GDP growth over the 1990s was probably under 2 per cent; the share of manufacturing industry fell to 17 per cent of GDP in 1998 (down from 25 to 27 per cent in the 1980s) and has fallen further since;[6] inflation rose from an average of under 15 per cent in the 1980s to over 30 per cent in the 1990s and was about 50 per cent in 1999; and the budget deficit remained obstinately around 10 per cent of GDP. Even exports fell at first, although there was a recovery in the middle of the decade, but with the collapse of the textile industry (precisely the one promised to do better under trade liberalisation), exports are now more 'colonial' in structure, with the country ever more vulnerable to swings in the world tobacco market.

The impact on the people of Zimbabwe has been deeply negative as real incomes have fallen for those in work, while unemployment and informal sector employment have soared, and the social welfare systems have been undermined. A cynic might say that the main aim of the IFIs' policy was not to produce an economic success story but to ensure that the country was so deeply beholden that it had no way back to its former policies.

The end of the 1990s and beginning of the new century in Zimbabwe saw deepening opposition to the government as the economy and social

---

[6.] The index of manufacturing output fell 6.7 per cent in the first half of 2000, when it was running at 5.4 per cent below the level of 1980, having been 40 per cent higher at the start of ESAP (CSO 2001).

conditions worsened. The ruling party's increasingly desperate, and violent, attempts to hold on to power have further harmed the economy as economic support by the IMF and most donors has been suspended. The gains of the 1980s have now been entirely lost. This complex political crisis cannot be analysed here, but its best known aspect outside Zimbabwe – the mass occupation of white-owned commercial farms – has been linked explicitly to Zimbabwean people's increasing poverty under ESAP by Moyo, Zimbabwe's main analyst of land reform.

Sources: Belli (1986), Bijlmakers *et al.* (1996), CSO (2001), Economist Intelligence Unit (1991), Jansen (1983), Mosley *et al.* (1995), Kanji and Kajdowska (1993), Moyo (2001), Potts and Mutambirwa (1998), Stoneman (1989).

already regarded as models.[14] Thus the original explanation, relying exclusively on internal factors, that the World Bank offered in the 'Berg Report' for the decline in the late 1970s seems implausible. The weakest conclusion must be that a common worsening of world market conditions and a growth in indebtedness following the oil-price hikes of the 1970s and fall in commodity prices in the early 1980s affected *all* countries in the region adversely, and this is the most likely cause of the general decline in performance. Harder to prove, but increasingly likely, is that World Bank structural adjustment policies adopted by, or forced on, nearly all countries, made matters worse rather than better. The Bank's analysis was *wrong*: external factors rather than inefficient states were the major problem. The Bank's prescription was *wrong*: making states weaker (and usually therefore less efficient and often more corrupt) did not help markets to do the job more efficiently. The Bank, and the Fund, may nevertheless not have been much disappointed with the outcome, as they were able to point to numerous successes in terms of *indicators* of liberalisation, such as freeing of exchange rates and trade, privatisations and removal of subsidies and controls. A more general failure in *outcomes*, such as GDP growth, investment, access to education and healthcare and poverty reduction were no doubt regrettable, but possibly only secondary considerations.

Much research has shown in detail why there was this general failure (see, for example, Mosley *et al.* 1995), but a little knowledge of the history of the development of the now developed states would have sufficed: market forces can provide an automatic efficiency mechanism, but only after basic economic structures are in place and sociopolitical structures command widespread assent. Before that, market forces are more likely to

---

[14.] As well as Kenya (see above), Malawi was often regarded as a model (Acharya 1981). Elsewhere in Africa the Côte d'Ivoire was a particular favourite (den Tuinder 1978). These World Bank assessments were to prove something of an embarrassment in the 1980s.

direct resources outside the country. This is why all successful countries have in early stages of development used the state to shape industrialisation and other policies: quite generally the state share in economic decision-making has followed an inverted U-curve as wealth has increased. The World Bank, observing the advantage in efficiency (if not equity) terms of freeing markets in wealthy countries (i.e. those on the downward part of the inverted U) decided, without evidence, to argue that countries on the *upward* part of the curve should also be made to go down! Some in the World Bank, no doubt, were honest (but bad) technicians, with no political programme. Others were pragmatists, and continued to support state policies, as in Mozambique and Zimbabwe, where it was clear that they were working, but the market ideologues were the ones who prevailed. It now seems clear that African countries *did not need* adjustment (they needed more efficient state-led programmes), but the World Bank needed them to adjust as part of a wider project in which they were pawns, and it *did not need*, and it therefore made sure that it *did not get*, any heterodox success stories.

References

Acharya, S.N. (1981) Perspectives and problems of development in Sub-Saharan Africa, *World Development* **9**(2): 109–47.

Africa Confidential (2000a) Copper quarrels, **41**(16).

Africa Confidential (2000b) Copper politics, **41**(17).

Amsden, A. (1989) *Asia's Next Giant: South Korea and late industrialization.* Oxford University Press, Oxford.

Belli, P. (1986) *Zimbabwe: An industrial sector memorandum.* World Bank Report 6349-ZIM, of a Mission to Zimbabwe in October–November 1985, led by Pedro Belli. World Bank, 10 December.

Biermann, W. and **Campbell, J.** (1989) The chronology of crisis in Tanzania 1974–1986, in Onimode, B. (ed.) *The IMF, the World Bank and African Debt.* Vol. 1: *The Economic Impact.* Zed Books, London.

Bijlmakers, L., Bassett, M. and **Sanders, D.** (1996) *Health and Structural Adjustment in Rural and Urban Zimbabwe.* Nordiska Afrikainstitutet, Uppsala.

Bond, P. (2000) *Elite Transition: From apartheid to neoliberalism in South Africa.* Pluto Press, London.

Bracking, S. (1999) Structural adjustment: why it wasn't necessary and why it did work, *Review of African Political Economy* **80**(26): 207–26.

Central Statistical Office (CSO) (2001) *Stats-Flash (December 2000)*, Harare, February (and earlier issues).

Collins, C.J.L. (1996) Zaire's economic crisis deepens, *Africa Recovery* **10**(1): 14–17.

Cornia, G.A. and **Helleiner, G.K.** (eds) (1994) *From Adjustment to Development in Africa: Conflicts, controversy, convergence, consensus?* St. Martin's Press, New York.

Cramer, C. (2001) Privatisation and adjustment in Mozambique: a 'hospital pass', *Journal of Southern African Studies* **27**(1): 79–103.

den Tuinder, D.A. (1978) *Ivory Coast: The challenge of success.* Johns Hopkins University Press for the World Bank, Baltimore.

Disch, A. (1993) Country strategy for Mozambique. Internal paper, World Bank, Maputo, 25 January.

Economist Intelligence Unit (EIU) (1991) *Zimbabwe, Malawi: Country report No. 2.* EIU, London.

Engberg-Pedersen, P., Gibbon, P., Raikes, P. and **Udsholt, L.** (eds) (1996) *Limits of Adjustment in Africa.* James Currey, London.

Financial Times (1999) Debt: Running a 6-year race [ft.com/ftsurveys/ country/sc5835.htm, 31 March].

Financial Times (2002) Report on Angolan oil industry urges freezing of stolen assets, 25 March.

Financial Times Survey (1992) *Zambia,* 17 December.

Financial Times Survey (1996) *Uganda,* 25 April.

Financial Times Survey (1997) *Zambia,* 4 March.

Financial Times Survey (1998) *Uganda,* 24 February.

Fundanga, C. (1989) The role of the IMF and World Bank in Zambia, in Onimode, B. (ed.) *The IMF, the World Bank and the African Debt.* Vol. 1: *The Economic Impact.* Zed Books, London.

Gibbon, P. (ed.) (1993) *Social Change and Economic Reform in Africa.* Nordiska Afrikainstutet, Uppsala.

Granberg, P. and **Parkinson, J.R.** (eds) (1988) *Botswana: Country study and Norwegian aid review.* Chr. Michelsen Institute, Bergen.

Hanlon, J. (1991) *Mozambique: Who calls the shots?* James Currey, London.

Hanlon, J. (1996) *Peace without Profit: How the IMF blocks rebuilding in Mozambique.* James Currey, London.

Hanlon, J. (1998) Global roots of civil war. Paper presented at the Conference: *Africa and Globalisation: Towards the Millennium.* Preston, Lancashire, 24–26 April.

Hanlon, J. (1999) Mozambique gains an extra US$28m. per year from HIPC debt relief, Jubilee 2000, 5 July [www.jubilee2000uk.org/ main.html].

Hanlon, J. (2000) Power without responsibility: the World Bank and Mozambican cashew nuts, *Review of African Political Economy* **83**, March: 29–45.

Harrigan, J. (1991) Malawi, in Mosley, P., Harrigan, J. and Toye, J. (eds) *Aid and Power: the World Bank and policy-based lending.* Vol. 2. Routledge, London: 201–69.

Haynes, J. (1989) Ghana: indebtedness, recovery, and the IMF 1977–87, in Parfitt, T. and Riley, S.P. (eds) *The African Debt Crisis.* Routledge, London.

Himmelstrand, U., Kinyanjui, K. and **Mburugu, E.** (eds) (1993) *African Perspectives on Development: Controversies, dilemmas and openings.* James Currey, London; EAEP, Nairobi; Baobab, Harare; Fountain Publishers, Kampala; Mkuki na Nyota, Dar es Salaam; St. Martin's Press, New York.

Hodges, T. (1995) Angola on the road to reconstruction, *Africa Recovery* **9**(4), December: 22–8.

Hodges, T. (2001) *Angola: From Afrostalinism to petro-diamond capitalism.* James Currey, Oxford.

Hoogvelt, A., Phillips, D. and **Taylor, P.** (1992) The World Bank and Africa: a case of mistaken identity, *Review of African Political Economy* **54**(July): 92–6.

Hope, R.K. (ed.) (1998) *Structural Adjustment, Reconstruction and Development in Africa.* Ashgate, Aldershot.

IMF (International Monetary Fund) (1999) *Enhanced Structural Adjustment Facility Policy Framework Paper 1998/99–2000/01* (ESAF PFP). 5 January [www.imf.org/external/np/pfp/malawi/index.htm].

IMF (2000) The IMF's Staff Monitoring Program for Angola: the Human Rights Implications (a backgrounder by Human Rights Watch, New York, 22 June 2000, updated 25 September 2000).

Jansen, D. (1983) Zimbabwe: government policy and the manufacturing sector, mimeo.

Jeffries, R. (1992) Urban popular attitudes towards the economic recovery programme and the PNDC government in Ghana, *African Affairs* **91**(363): 207–26.

Jenkins, P. (forthcoming) Image of the city in Mozambique: civilisation, parasite, engine of growth or place of opportunity, in Bryceson, D. and Potts, D. (eds) *African Urban Economies: Viability, vitality or vitiation of major cities in East and southern Africa?*

Kanji, N. and **Kajdowska, N.** (1993) Structural adjustment and the implications for low-income urban women in Zimbabwe, *Review of African Political Economy* **56**: 11–26.

Karmiloff, I. (1990) Zambia, in Riddell, R. (ed.) *Manufacturing Africa: Performance and prospects of seven countries in Sub-Saharan Africa.* James Currey, London: 297–336.

Kashambuzi, E. (1995) Poverty still deepening in Zambia, *Africa Recovery* 9(1): 13.

Laishley, R. (1993) Uganda: turning growth into prosperity, *Africa Recovery* 7(2), October: 16–21.

Lipton, M. (1985) *Capitalism and Apartheid.* Gower Press, Aldershot.

Marais, H. (2000) *South Africa: Limits to change: the political economy of transition.* Zed Books, London.

Mazrui, A.A. (1967) *Towards a Pax Africana: A study of ideology and ambition.* Weidenfeld and Nicolson, London.

Melrose, D. (1985) *Nicaragua: The threat of a good example?* Oxfam, Oxford.

Mohan, G., Brown, E., Millward, B. and **Zack-Williams, A.** (eds) (2000) *Structural Adjustment: Theory, practice and impacts.* Routledge, London and New York.

Moorsom, R.J.B. with **Sachikonye, L.** and **Matanga, J.** (1997) *Evaluation of the Social Development Fund in Zimbabwe: A pilot study.* Research Report R1997: 9, Chr. Michelsen Institute, Bergen.

Mosley, P. (1991) Kenya, in Mosley, P., Harrigan, J. and Toye, J. (eds) *Aid and Power: The World Bank and policy-based lending.* Vol. 2. Routledge, London: 270–310.

Mosley, P., Harrigan, J. and **Toye, J.** (1991, 1995) *Aid and Power: The World Bank and policy-based lending* (2 vols). Routledge, London.

Mosley, P., Subasat, T. and **Weeks, J.** (1995) Assessing 'Adjustment in Africa', *World Development* 23(9): 1459–73.

Moyo, S. (2001) The interaction of market and compulsory land acquisition processes with social action in Zimbabwe's land reform, paper presented at London School of Economics Seminar, April 2001.

New York Times (2000) Angolan paradox: oil wealth only adds to misery, 9 April.

Nyoni, T.S. (1997) *Foreign Aid and Economic Performance in Tanzania.* African Economic Research Consortium, Research Paper 61, March, Nairobi.

Olukoshi, A. (1989) Impact of IMF-World Bank programmes on Nigeria, in Onimode, B. (ed.) *The IMF, the World Bank and the African Debt.* Vol. 1: *The Economic Impact.* Zed Books, London.

Osei-Hwedi, B. and **Osei-Hwedi, K.** (1990) *The Tanzania–Zambia Railroad: An analysis of Zambia's decision-making in transportation.* Lawrence, Virginia, Brunswick.

Parfitt, T. and **Riley, S.P.** (1989) *The African Debt Crisis.* Routledge, London.

Pearson, L.B. (1969) *Partners in Development: Report [to the World Bank] of the Commission on International Development.* Pall Mall Press, London.

Potts, D. and **Mutambirwa, C.** (1998) Basics are now a luxury: perceptions of ESAP's impact on rural and urban areas in Zimbabwe, *Environment and Urbanization*, special issue on Beyond the Rural–Urban Divide, **10**(1): 55–76.

Sampson, A. (1987) *Black and Gold: Tycoons, revolutionaries and apartheid.* Hodder and Stoughton, London.

SAPA (2000) Mbeki to Attend Mozal Launch in Maputo, Pretoria, 20 September [in *ANC Daily News Briefing*, 21 September 2000].

Simon, D. (1995) Debt, democracy and development: sub-Saharan Africa in the 1990s, in Simon, D., van Spengen, W., Dixon, C. and Närman, A. (eds) *Structurally Adjusted Africa: Poverty, debt and basic needs.* Pluto Press, London and Boulder.

Simon, D., van Spengen, W., Dixon, C. and **Närman, A.** (eds) (1995) *Structurally Adjusted Africa: Poverty, debt and basic needs.* Pluto Press, London and Boulder.

Stoneman, C. (ed.) (1988) *Zimbabwe's Prospects.* Macmillan, London.

Stoneman, C. (1989) The World Bank and the IMF in Zimbabwe, in Campbell, B. and Loxley, J. (eds) *Structural Adjustment in Africa.* Macmillan, London.

Stoneman, C. (1992) Policy reform or industrialisation? The choice in Zimbabwe, in Adhikari, R., Kirkpatrick, C. and Weiss, J. (eds) *Industrial and Trade Policy Reform in Developing Countries.* Manchester University Press, Manchester.

Swamy, G. (1994) *Kenya: Structural adjustment in the 1980s.* World Bank Policy Research Working Paper 1239, January.

Tostensen, A. and **Scott, J.G.** (eds) (1987) *Kenya: Country study and Norwegian aid review.* Chr. Michelsen Institute,

United States Institute of Peace (1997) Zaire: Predicament and prospects, *Peaceworks* **11**, January.

van der Hoeven, R. and **van der Kraaij, F.** (eds) (1994) *Structural Adjustment and Beyond: Long-term development in sub-Saharan Africa.* James Currey, London.

Wade, R. (1990) *Governing the Market: Economic theory and the role of government in East Asian industrialisation.* Princeton University Press, Princeton.

World Bank (1981) *Accelerated Development in Sub-Saharan Africa.* World Bank, Washington.

World Bank (1984) *Towards Sustained Development in sub-Saharan Africa: A joint program of action.* World Bank, Washington.

World Bank (1989) *Sub-Saharan Africa: From crisis to sustainable growth.* World Bank, Washington.

World Bank (1999a) *Kenya at a Glance,* 9 September [www.ifc.org/html/ABN/cic/kenya/english/glance.htm].

World Bank (1999b) *Tanzania at a Glance,* 22 September [www.ifc.org/html/ABN/cic/tanzania/english/glance.htm].

World Bank/IDA (1970) *Annual Report.* World Bank, Washington.

CHAPTER 4

The persistence of poverty

Anthony O'Connor

Several chapters in this book are concerned not just with African characteristics but with African problems, or problems which are especially
severe in Africa: but none more so than this chapter. The people of eastern
and southern Africa may, of course, experience as much happiness and
joy as people anywhere else: such things are unmeasurable. However, this
chapter is concerned with the least happy aspects of their lives, many of
which are encapsulated by the word 'poverty'. This region is often considered to be part of 'the developing world', but by what criteria most of it
can really be regarded as 'developing' over the past 20 years is far from
clear. We are dealing mainly with materially poor people in materially
poor countries, whose circumstances have not improved over the past two
decades, and indeed whose lives have in many cases become more difficult.

A widely respected assessment of where countries stand in terms of the
human condition is the Human Development Index (HDI) produced
annually by UNDP. Table 1.1 (in the introductory chapter) provides the
HDI value for 2000 for each country covered in this volume and their
world rank. Two-thirds of the countries in eastern and southern Africa
were ranked in the lowest quartile of global rankings with HDI values less
than 0.535. Only Zimbabwe, Botswana, Swaziland, Namibia and South
Africa had better indices. Burundi and Mozambique fall almost at the
bottom of the world list – only Niger and Sierra Leone had lower indices.
The precise HDI figures hardly matter given the margin of error involved,
especially for countries as conflict-ridden as Angola and the Democratic
Republic of the Congo (DRC). If South Africa were excluded, the poverty
situation considered here would be broadly similar to that in both Western Africa and the Horn of Africa, i.e. the world's worst; and many issues
raised here apply across the continent from Mozambique as far north as
Eritrea and as far west as Mauritania. Most existing writing on this theme

that is not confined to individual countries (e.g. Chege 1995; Iliffe 1987; O'Connor 1991; Oxfam 1993; White and Killick 2001) adopts this broader near-continental framework.

This chapter tries neither to present a mass of detailed factual information nor to advance or even test theory. It does aim to increase awareness, promote thought, and perhaps deepen understanding about material poverty in eastern and southern Africa. Its concern is with poverty in absolute terms rather than with inequality, and so attention will be focused on eastern and south-central Africa rather than on South Africa. There is of course much extreme hardship and some total destitution in South Africa; but in contrast to Iliffe's (1987) book, *The African Poor*, this chapter does not concentrate on such destitution. The main concern here is with the less intense poverty that afflicts the great majority of the population in countries such as the DRC, Zambia and Malawi, but not further south.

No comprehensive explanation of this poverty is attempted, for it results from a highly complex array of past and present factors, some internal to Africa and some external, all interacting with each other. Attention is confined to the present extent, nature and spatial pattern of poverty, and some of the internal forces contributing to its current persistence. The concern is both with poor people, households and families, and with poor countries administered by poor governments (or over large tracts by rebel movements). Outsiders (e.g. UK students) often suggest that African governments ought to provide better social services, and even provide 'decent' houses for people; but most are in no position to do this, and even within countries such as Uganda or Mozambique, both with better macro-economic performance than most in the 1990s, few people seriously expect that they will.

The chapter is not based on any single theoretical approach, but on the belief that diverse approaches have contributions to make to understanding; and also that the factors contributing even just to the persistence of poverty (whatever its origins) are too numerous, their combinations too complex, and their spatial variations too great, for any single fundamental cause to be identified. Furthermore, no recipe for poverty alleviation is offered here, but only reminders of issues that must be taken into account by any policy-makers.

Poverty in context

Clearly there are links between poverty and the topics discussed in almost every other chapter. Some of these topics provide vital parts of the context for African poverty, so they should at least be mentioned here also. Africa's colonial history is highly relevant to a full understanding of its poverty, but less so to its persistence and even intensification in the 1980s and 1990s. Global economic and political structures and forces operating today are more clearly relevant to this (see Chapters 9 and 10); but here the focus is on the internal characteristics of eastern and southern Africa. Three of these are a generally harsh physical environment, rapid population growth and widespread political instability.

In the 1960s and 1970s many renowned writers on development and poverty, including some who profoundly disagreed on these matters, such as Rostow and Frank (Potter *et al.* 1999), seemed to share the view that the environment was of little relevance. The influence of the physical environment on development is now once again appreciated, although without resurrecting crude environmental determinism. Anyone familiar with the impact of drought across vast tracts of Africa (Benson and Clay 1998), or with the ravages of malaria (Spielman and D'Antonio, 2001), must recognise the importance of the environment (see also the influential American historian Landes (1998)). A succinct summary of key African poverty-related environmental issues has been provided by Grove (1991); many aspects are examined in Ahmed and Mlay (1998); and Adams (1992) has demonstrated the importance of one aspect, water management, now complemented by Christie and Hanlon (2001) on the Mozambique floods.

The fact that Africa has the world's highest population growth rates is also highly relevant, sometimes linked to environmental issues, as where population pressure forces farmers onto more marginal land. In South Africa the annual increase has fallen below 2 per cent, with a fertility rate below 3 (see Table 2.1): but in Angola, the DRC, Malawi, Zambia, Uganda, Rwanda and Burundi fertility rates remained between 6 and 7 in the 1990s, and annual growth in that decade was still averaging 2.5–3 per cent (despite increasing AIDS deaths – although as noted in Chapter 2 this factor is now feeding through into significantly reduced population growth rates in southern African countries). One consequence is that in many areas half the people are under the age of 18, a fact which became more economically significant from the 1960s to the 1980s as school enrolments increased. Another consequence is that those who do most of the farmwork in these countries are generally either pregnant or working with a baby on their back. It is widely noted that fertility has fallen universally as economic development has proceeded, but across much of this region 'development' is not occurring so the relevance of this interrelationship for many eastern and southern African countries is limited. The UNDP's *Human Development Report 2001* actually gives higher fertility figures than in earlier years both for sub-Saharan Africa in general and for countries such as Zimbabwe. Sai (1988) has provided a succinct outline of these population growth issues.

Political upheaval may also directly contribute to poverty, as the example of Uganda in the 1970s and 1980s clearly demonstrated (Jamal 1998). Without civil war the intensity of poverty in Angola would surely be less; its ending in Mozambique in 1992 brought some poverty alleviation; and peace since independence has contributed to relative prosperity in Botswana. There is no direct correlation, however, for the intensity of the violence in Rwanda has not proportionately changed the poverty situation there, while relative stability in Tanzania and Zambia has not brought prosperity to either country. It is perhaps arguable for Tanzania that the huge effort put into ensuring political stability and promoting national integration in the 1970s and 1980s (Yeager 1989) was partly at the expense of economic advance.

The political context of poverty in Africa is not only a matter of instability and conflict. Another feature of many countries is highly personalised power relations, and what many political scientists term 'neopatrimonialism'

(Chabal and Duloz 1999). This operates at all levels from kleptocratic heads of state to poor people who see no chance of improved well-being other than through personal links with powerful local leaders. A related issue is the undoubted intensification of corruption in eastern and south-central Africa in the 1980s and 1990s (Hope and Chikulo 2000). Practices once more characteristic of West Africa, though intense also in the DRC, are now helping to perpetuate poverty in countries like Kenya, Zambia and Zimbabwe.

Another element in the context for poverty is the urbanisation process discussed in Chapter 11. There is no clear evidence in this region that rapid urbanisation has either increased or reduced poverty, but it has certainly changed both its location and its character (Jones and Nelson 1999; Potts 1997). For example, poor urban dwellers tend to be more dependent on cash transactions, and less dependent on subsistence activity, than rural dwellers. This applies to water supply, fuel and housing, as well as basic foods. Another difference is that poor people in cities are often living within sight of affluence, and are nearly always well aware of it, whereas in most rural areas people are not made aware of their relative poverty to the same extent.

The huge scale of post-independence refugee flows is also highly relevant to poverty in Africa (Blavo 1999). In the case of the movement of almost 1 million people from Mozambique to Malawi in the 1980s repatriation is now essentially complete, but deep scars remain in both countries. A long-standing refugee situation is that of first- and now second-generation Burundians in north-west Tanzania: more dramatic movements in the late 1990s, devastating for many of the people involved, were those between Rwanda and the DRC (Nafziger *et al.* 2000). Equally significant for poverty intensification are the massive internal displacements which occurred in Mozambique in the 1980s, and Angola until 2002, and which continue in the DRC, Burundi and Rwanda.

The final contextual issue that must be mentioned here (relevant but little-studied in refugee situations) is the human devastation caused by HIV/AIDS (see Chapter 2). While this is a global issue, the areas most severely affected are in eastern and southern Africa (Baylies and Bujra 2000; Hope 1999; Poku 2001; Webb 1997). The region accounts for about half the world total of 40 million people probably living with HIV, and well over half of AIDS deaths to date. By the end of 2001 estimates of adult HIV prevalence had reached 20 per cent in Namibia and Zambia, over 30 per cent in Lesotho, Swaziland, Zimbabwe and Kwa-Zulu/Natal Province of South Africa and almost 40 per cent in Botswana (see Table 2.2). Further north in Uganda, where the crisis struck earlier, the spread of infection has mercifully been checked, but in the 1990s it was there that the poverty implications in terms of prolonged illness of the breadwinners and proliferation of orphans were most apparent (Barnett and Blaikie 1992; Wallman 1996). The links with poverty are both ways: AIDS is devastating families and putting immense pressure on healthcare services, while poverty at both household and national level rules out any high-cost forms of medication. Prospects for less expensive drugs, affordable by some at least in southern Africa, was one of the best pieces of news for this region in 2001.

Poverty measurement

Poverty is extremely hard to measure, although attempts at international comparisons are now made each year by the World Bank and by various United Nations bodies. The 1998 *Human Development Report* included for the first time a Human Poverty Index, but even after revisions this seems more open to question than the HDI (Box 4.1). No use is made here of figures for the number 'below the national poverty line', since such a line is always arbitrary and rarely comparable from one country to another.

The problems of measurement are both conceptual and practical. The conceptual problems include the relative importance to be given to wealth,

---

Box 4.1   Poverty and inequality indicators

Sample surveys have been undertaken in most countries to discover how evenly or unevenly income is distributed among households, but they vary greatly in reliability, and some published data are totally implausible. Thus the 1999 UNDP *Human Development Report* (p. 148) gave real annual GDP per capita for Zambia of $US216 for the poorest 20 per cent and $US2,797 for the richest 20 per cent, with equivalent figures of $US217 and $US1,430 for Tanzania. Yet the same table informed us that 85 per cent of Zambians lived on less than one US dollar a day, but only 16 per cent of Tanzanians. A selection of the more plausible figures available is given below:

Country (date)	Income share of:		% of population with PPP income per day	
	Lowest 20%	Highest 20%	Below $1	Below $2
Mozambique (1996)	7	45	40	80
Uganda (1992)	7	45	40	80
Kenya (1994)	5	50	30	60
Zimbabwe (1991)	4	60	40	60
South Africa (1993)	3	65	10	40

Source: World Bank (2000).

The UNDP also now provides a Human Poverty Index that builds in distribution aspects more fully than in the Human Development Index. Again, however, some rankings are open to question. For example, in the 2001 HPI, Tanzania implausibly ranks more favourably in terms of poverty incidence than Namibia (or Morocco), and on a par with Egypt. Zimbabwe ranks more poorly largely due to the anticipated *future* impact of AIDS via current life expectancy at birth.

---

to income and to social welfare. For instance, some pastoralists with large herds of family-owned cattle might rank much higher on the first criterion than on the others (Anderson and Broch-Due 1999), whereas many culti-vators have few personal assets when the land they work is communally owned. Some specific poverty dimensions such as hunger are more straight-forward conceptually than poverty overall, but very difficult to measure. In general, the poorer the country the weaker is its information base (O'Connor 2002). Many governments cannot afford data-gathering exer-cises, while much activity may be of types that cannot be quantified. Who is to document how much water is drawn from streams, or even wells? And what value should be placed upon it? Access to healthcare depends on the availability and effectiveness of traditional herbalists as well as more documented dispensaries. Furthermore, in situations where the adage 'knowledge is power' applies in an extreme way, even when information has been gathered its disclosure may not be in the interests of those who hold it.

Nearly all the figures provided in this chapter therefore indicate only broad orders of magnitude. Where spurious precision has been offered in the original sources (because that is what someone is expected or paid to provide, regardless of their validity), the figures have generally been rounded. In fact, of course, if our main purpose is understanding of issues, precision is rarely needed; it is only when, say, absolute data are confused with per capita data that we really need to worry!

Income measures of poverty

The most widely used indicator of both prosperity and poverty at national level is gross national product or gross national income per capita, con-verted into US dollars. In every country some attempt is made to calculate this, though with variations in the extent to which estimates of the value of subsistence production and services are included. Figures for the world's richest countries are in the range US$20,000–30,000, while those for many middle-income countries, including Botswana and South Africa, are around US$3,000–4,000. As shown in Table 1.1, estimates for other eastern and southern African countries range from almost US$2,000 in Namibia to around US$100 in the DRC and Burundi. This clearly indicates far more pervasive poverty in some parts of our region than in others, whatever the limitations of this measure.

Some UN agencies now prefer to use World Bank data with GNP figures adjusted to purchasing power parity (PPP), especially since currency ex-change rates can become very distorted. This measure suggests average income in the richest countries only 30 times greater than in the poorest, rather than 100 times greater. This does not greatly change poverty rankings in the region, but it does help to identify four country groups: Botswana, Namibia, South Africa and Swaziland; Lesotho and Zimbabwe; Kenya and Uganda; and a remaining group of more impoverished countries (see Table 1.1). Although Angola's per capita PPP value was over $2,000 by 2001, there is no doubt that is highly misleading with regard to its relative ranking in terms of the incidence of poverty, due to the dual impact of war and corrupt use of government revenues. It is therefore peculiarly hard to categorise; its HDI ranking is thus the best guide to its relative position with regard to human poverty.

If we are concerned specifically with the *geography* of poverty, it would be helpful to have reliable figures disaggregated by region: countries of greatly varying size within ex-colonial boundaries are often not the most appropriate units for analysis. Generally, however, such figures simply do not exist. Even attempts to distinguish between rural and urban areas have led to much controversy. Thirty years ago average incomes in most cities and towns were far higher than in most adjacent rural areas, and the incidence of poverty was far lower. Undoubtedly disparities have since narrowed, both through migration of poor people into cities and through the differential impact of structural adjustment programmes (SAPs) (see Chapters 11 and 3). However, claims that over large parts of Africa the disparities have almost disappeared, and in some areas reversed (Jamal and Weeks 1993), seem dubious, especially if they are based on a nominal minimum urban wage and on earnings in relatively prosperous, rather than typical, rural areas.

In Uganda, for instance, material poverty is far more evident in the capital city, Kampala, today than in early post-independence years; but it is still even more pervasive in the almost wholly rural western and northern fringes of the country. Similarly in Zimbabwe, extremely low incomes, even allowing for subsistence production, are still much more characteristic of many rural areas than of Harare or Bulawayo. In terms of cash income alone, some of the widest urban–rural disparities may be in the DRC, where subsistence and barter livelihoods still predominate over vast areas (especially when old currency notes are suddenly declared null and void). Even, say, in Angola, where urban squalor is extreme, city dwellers tend to be more nimble, flexible and resourceful in respect of cash incomes than most rural dwellers.

Meanwhile, among the rural areas of countries as diverse as Kenya, Malawi and South Africa, there are huge variations in both average incomes and the incidence of extreme poverty. However, while it is clear that incomes are much higher in the Kikuyu areas of central Kenya than in areas with a similarly dense small-farmer pattern of settlement in the far west, it is much more difficult to compare either of these with the areas characterised by wage-labourers on large farms, or those occupied by nomadic pastoralists. One more complicating factor, especially in South Africa and Lesotho, is the extent of temporary outmigration from many rural areas, and hence the extent to which the local cash economy may depend on remittances.

Consumption measures of poverty

Absolute material poverty is sometimes defined in terms of failure to satisfy the most basic human material needs, such as a supply of clean water. Reliable water consumption data exist for countries where it is supplied through large organisations, but not for those where most people rely on streams or wells. However, enough local surveys have been undertaken to permit us to say that average per capita domestic water consumption in countries like Malawi, Tanzania and Uganda is between 3 and 5 per cent of that in western Europe.

Only the vaguest estimates exist for *total* per capita energy source consumption, including the fuelwood on which most African rural dwellers must depend. However, in respect of *commercial* energy use a clear picture emerges from World Bank and World Resources Institute sources. Even in South Africa, per capita commercial consumption is only half that in western Europe and a quarter of that in North America: yet it is there six times higher than in Kenya and probably 20 times higher than in Rwanda.

Much more is known about consumption of goods which have to be brought into African countries from outside; and these are of interest also because they indicate how far these countries are or are not tied into the set of processes now widely labelled 'globalisation'. Countries like Malawi and Uganda are often said to be highly 'dependent' on the industrialised world, but perhaps this really applies to only very small sectors of their populations. Even allowing for some smuggled and other unrecorded imports, Box 4.2 suggests that consumption of imported goods in most countries of eastern and south-central Africa is tiny. The US$15 worth per

---

Box 4.2   Globalisation or marginalisation?

South Africa, for some years very isolated for political reasons, is now at least in part closely integrated into the 'global economy', but this certainly does not apply to all of southern and eastern Africa (see table below). Microscopically small imports clearly reflect both national poverty and household poverty, rather than self-sufficiency. Telephones may be a personal consumer item only for the elite, but lack of them is relevant for poorer people since they can be used institutionally to serve the public at large. The same applies to motor vehicles, of which there are about 120 per 1,000 people in South Africa, but only 4 in Uganda, 3 in Rwanda and 2 in Mozambique, according to the World Bank's *African Development Indicators*.

Country	Recorded imports US$ per capita 1999	Telephones per 1,000 people 1999
South Africa	580	270
Zimbabwe	150	36
Kenya	90	11
Tanzania	50	6
Burundi	15	3

Sources: UNDP, *Human Development Report 2001*; World Bank, *African Development Indicators 2001*.

head annually in Burundi compares with US$12,000 worth in Belgium. This points much more clearly to marginalisation from the so-called 'global' economy than to extreme dependency.

It may also be appropriate to consider new communications technologies which are often said to be extending to every part of the world, with potential for contributing to poverty eradication. A mass of television aerials is indeed a feature of even the poorest housing areas in the cities of middle-income countries, from Mexico to Manila, via Casablanca and Tunis – as it is also in Soweto. However, this is not true of Kigali, Kampala or Dar es Salaam, while television sets are extremely rare in most rural areas of Rwanda, Burundi, Uganda, Tanzania and Malawi. The 1998 national figure for Burundi was two sets per 1,000 people, compared with 30 in both Botswana and Zimbabwe, and 130 in South Africa (UNDP 2000).

The rather similar figures for telephones in Box 4.2 largely speak for themselves. We are of course not considering telephones as an individual household item in most of our region, but the point to be stressed is that there is not even a telephone in most villages, nor in institutions such as schools or health centres. Furthermore, cost reductions almost bringing about 'the death of distance' (Cairncross 2001) over much of the world are far less evident in Africa – where geography matters as much as ever. Thus, while a telephone call from New York to Geneva or Tokyo now costs little more than one to Toronto, one to Bujumbura still costs many times more. The contrast in respect of rural Switzerland and rural Burundi is even starker, and even harder to eliminate through twenty-first-century technology. The fact that it is now possible to 'surf the net' not only in Johannesburg, but also in some parts of Lusaka and Nairobi, does not mean that the people of eastern Africa are all now part of the 'global village' and hence being drawn into mass-consumption society.

Welfare measures of poverty

The UNDP HDI, mentioned earlier, is intended to reflect social welfare as well as income or prosperity, and life expectancy data (as well as income and education factors) are incorporated into it. This is particularly significant for eastern and southern Africa where life expectancy has been drastically cut in recent years by HIV/AIDS (see Chapter 2). There are many ways in which poverty contributes to the rapid spread of AIDS, while the affliction certainly intensifies poverty, both for communities and for individual families. However, although poverty can exacerbate behaviour and mobility patterns which facilitate the spread of HIV, especially when women's status is low (as in almost all of the region), it is not in itself a dimension of poverty to the same extent as most childhood diseases or other adult diseases that arise from situations such as inadequate sanitation (see Box 4.3). Thus it is questionable whether rapid deterioration in Botswana's life expectancy ranking in the 1990s, because of the spread of HIV, really meant increasing poverty.

Table 4.1 provides data on infant and child mortality and life expectancy, with the countries ranked by child mortality in 1999. These indices are also discussed in Chapter 2 in relation to demographic trends (see

---

Box 4.3   Misleading data: access to sanitation

Access to satisfactory sanitation is an important social welfare issue for everyone, and is perhaps the dimension of poverty that is most evidently intensified by rapid urbanisation. But what counts as 'satisfactory' and what constitutes 'access'? Country differences in definition have produced ridiculous figures, which are nonetheless reproduced in highly reputable publications, and even used to guide policy. Consider the figures below (which are not misprints, as similar figures appeared in the 1998 UNDP Report):

	Population 'without access to sanitation' (%)
Tanzania	14
Mozambique	46
Zimbabwe	48
Malawi	97

Source: UNDP, *Human Development Report 1999*: 148.

It is not credible that Tanzania and Malawi are at opposite ends of the scale, or that the situation in Zimbabwe is no better than that in Mozambique. Furthermore, the 1996 UNCHS *Global Report on Human Settlements* indicated (p. 512) that in Malawi 84 per cent had 'access to sanitation', compared to 32 per cent in Tanzania. Hence we have two diametrically opposed versions of contrasting situations in what are really two rather similar countries.

The 2001 *Human Development Report* provides totally different figures for 'population using adequate sanitation'. These now include 77 per cent for Malawi – but only 8 per cent for Rwanda. So does this revision leave us any the wiser?

Source: UNDP, *Human Development Reports*.

Table 2.2), but here they are being judged as indicators of poverty. The disjuncture between child mortality rates and life expectancy rates (due to AIDS, as discussed above) indicates that child mortality is now more appropriate as a poverty indicator, especially *current* poverty, than life expectancy. The geographical pattern of child mortality is broadly similar to that of incomes, except that the relative positions of Zimbabwe, Uganda and Tanzania are rather better, while the situation in Zambia is worse, and just as bad as in Malawi and Mozambique. The figures for Angola are worse still – almost 250–300 deaths per 1,000 children under 5 according to both UNDP and UNICEF, a higher rate than for any country in the world apart from Sierra Leone (and contrasting with 60–70 per 1,000 for Bangladesh and India, often regarded as similar to most of Africa in respect

Table 4.1 Social welfare indicators for 1999 (countries ranked by child mortality)

Country	Child (under 5) mortality per 1,000	Infant (under 1) mortality per 1,000	Life expectancy at birth
Angola	250	150	47
Malawi	220	130	40
Mozambique	200	130	42
Congo (DRC)	190	120	48
Burundi	190	110	42
Rwanda	190	110	41
Zambia	190	110	41
Uganda	150	90	43
Tanzania	140	90	47
Lesotho	140	90	48
Kenya	120	80	50
Zimbabwe	110	70	43
Swaziland	100	60	50
Namibia	90	60	48
Botswana	80	60	42
South Africa	70	50	52
World	80	60	65

Note: The figures above represent the average of figures that are in some cases very disparate, partly due to different estimates of the impact of AIDS (see Table 2.2 and related discussion in text for detailed analysis of the impact of AIDS on life expectancy). The infant mortality rates here are generally higher than those in Table 2.2 which relies more on Demographic Health Survey sources which usually find lower rates than those cited in UN and World Bank sources.
Sources: Global Coalition for Africa, *African Social and Economic Trends*; UNDP, *Human Development Report 2002*; UNICEF, *State of the World's Children 2001*; World Bank, *African Development Indicators*; World Bank, *World Development Report 2000/2001*.

of poverty). The HDI and child mortality rankings are, not surprisingly, very similar (compare Table 4.1 with Table 1.1).

One social welfare issue that is particularly sensitive to variations in material poverty at both national and family levels is maternal mortality. This has now fallen to below 10 per 100,000 births in some of the world's richest countries, while it stands at around 100–200 in middle-income countries like Brazil and Egypt, and 400 in India. It was also 400 in Zimbabwe in the mid-1990s, but the situation is very much worse in countries like Mozambique, for which both UNDP and UNICEF provide a figure of 1,100 maternal deaths per 100,000 births.

Other widely quoted social indicators reflecting the extent of poverty across national populations concern access to healthcare, clean water and sanitation. In general terms these again indicate much more extensive deprivation in eastern and south-central Africa than in southern Africa, and also a worse situation than in South Asia; but as Box 4.3 demonstrates, problems of varying definitions severely limit the value of such data for specific country, or urban–rural, comparisons.

Poverty trends

In static terms there is at present little difference between most of sub-Saharan Africa and South Asia with respect to the incidence and intensity of material poverty: but there are huge differences with respect to recent trends. The phrase 'developing countries' can appropriately be applied to most in South and South-East Asia but, as suggested earlier, it is now highly inappropriate for most in Africa and it will remain so until fundamental changes occur.

Per capita income rose steadily in most parts of Africa in the 1960s and early 1970s, and the geographical pattern of this advance was examined in detail in O'Connor (1978). The annual GNP growth rate for many countries was then around 4–5 per cent, compared with population growth of 2.5–3 per cent (see Chapter 3). Not all of this income growth was concentrated in the hands of a few, as some writers have argued: there was modest improvement in the standard of living for the majority in both urban and rural areas, and some real reduction of absolute poverty almost everywhere.

However, by the late 1970s annual GNP growth had slowed to be roughly equal to population growth over much of eastern and southern Africa, and during the 1980s it fell to around zero in many countries, leading to sustained widespread material impoverishment (O'Connor 1991). The situation was, of course, even worse in Somalia, Ethiopia and southern Sudan than in most of our region. The 1990s brought a return of macro-economic growth in most countries, but in few did it exceed population growth throughout the decade, whereas in South Asia typical economic growth rates far exceeded population growth. The trends in per capita GDP in the countries of the region in the latter two decades of the twentieth century are shown in Table 4.2.

As can be seen, the trends were not consistent throughout the region and in some cases national variations have reduced earlier disparities. South Africa, Namibia and Botswana remain distinctive in terms of their relative affluence, but South Africa has had many years of very slow growth, and Botswana's exceptional growth slowed sharply in the 1990s (see Chapter 3). Meanwhile, after partial economic collapse in earlier decades, both Uganda and Mozambique have experienced some macro-economic recovery. In addition, there is firm evidence that within many countries the average income gap between urban and rural areas has decreased, as the negative impact of harsh SAPs has hit hardest in the cities, especially through job losses. Nationally, it is not the very poorest (i.e. the rural poor) who have lost most, as some critics of SAPs claim, although all strata of the poor have generally suffered.

The absence of economic development is apparent not only in economic growth no faster than population growth but also in lack of structural change (see, for example, Boxes 4.4 and 4.5 for discussion of the lack of foreign direct investment and the limitations of the much vaunted tourism

Table 4.2   Growth or shrinkage of GDP and per capita GDP, 1980–99

Country	Annual % GDP change		Annual % GDP change per capita	
	1980–90	*1990–99*	*1980–90*	*1990–99*
Angola	+2	−1	−1	−4
Botswana	+10	+4	+7	+2
Burundi	+4	−2	+1	−5
DRC	+1	−5	−2	−8
Kenya	+4	+2	+1	0
Lesotho	+4	+4	+2	+2
Malawi	+3	+4	0	+1
Mozambique	−1	+6	−3	+3
Namibia	+1	+4	−1	+2
Rwanda	+3	0	0	−1
South Africa	+2	+2	0	0
Swaziland	+5	+2	+3	0
Tanzania	+3	+3	0	0
Uganda	+3	+7	0	+4
Zambia	+1	0	−2	−2
Zimbabwe	+4	+2	+2	0

Sources: Diverse – in most cases these figures are the average of disparate figures from UN, World Bank and African Development Bank sources.

Box 4.4   What about big business and multinational corporations?

Discussions of global development and poverty often include strong arguments either advocating or attacking private foreign investment (e.g. Madeley 1999), but the relevance of these arguments for most countries in our region is potential rather than actual, for in few respects is their current 'marginalisation' more evident. The current image of Africa elsewhere in the world is portrayed in a sweeping generalisation by Castells (2000: 82–3), writing of 'the collapse of Africa's economies, disintegration of many of its states, and breakdown of most of its societies'. Poverty is a key element in an image that is hardly likely to attract multinational corporations.

In the late 1990s inflow of foreign investment to Brazil was US$40bn annually, while to Argentina it was US$20bn: in comparison, it was only US$500m. to South Africa, and more typical of our region were 1998 inflows of US$10m. to Kenya and just US$1m. to Malawi. The DRC was providing attractive economic opportunities to military men from Zimbabwe and elsewhere, but not to giant corporations. Clearly, with the exception of oil drilling in Angola, foreign investment at present has little bearing on poverty in eastern and southern Africa.

Sources: Madeley (1999); Castells (2000).

Box 4.5 Does tourism provide significant relief from poverty?

While Africa suffers in many ways from a strongly negative image in other parts of the world, its image contains some positive elements – sufficient to make certain countries attractive tourist destinations. Tourism is of greatest importance in Kenya, contributing over 5 per cent of GNP and over 20 per cent of foreign exchange. However, the equivalent figures of just under 1 and 5 per cent respectively for both Malawi and South Africa are more typical of our region. Furthermore, most academic analyses undertaken so far in Kenya, as elsewhere in tropical Africa (e.g. The Gambia), indicate that even in pure economic terms (setting aside cultural impact) overall benefits and costs are fairly evenly matched.

World Tourist Organisation data suggest that 'Africa' accounts for only 2 per cent of annual international tourism receipts, and that Tunisia and Morocco together account for almost half of this. South Africa accounts for a further 20 per cent and Kenya for 6–7 per cent. While 4 million foreign tourists a year visit South Africa, the total for Kenya is only about 700,000, while Uganda, Tanzania, Zambia and Malawi each receive only 150,000–250,000 – tiny in comparison to the inflows to most countries in Europe. (The 10,000 'tourists' a year recorded for Angola and the DRC no doubt include journalists and diamond smugglers!) It appears that in most of eastern and southern Africa tourism has done almost nothing at national level, and little even at local level, to relieve poverty.

Source: Sindiga (1999).

---

sector). In most countries of our region the share of industry in GDP was no higher in 1999 than in 1980, and in some, such as Zambia, Zimbabwe and especially the DRC, there was clear evidence of de-industrialisation in the 1990s. The most common situation, as in Kenya, Tanzania, Zimbabwe and South Africa, is for large-scale industrial output to have remained almost static while small-scale enterprise has grown roughly in line with population growth. There have been almost no new opportunities for people to break out of poverty, except perhaps in unrecorded illegal activities such as drug trafficking.

The extremely low level of recorded imports in most countries was noted above as an indicator of low material consumption. Here we might note that while the absolute value of recorded imports rose between 1980 and 1999 from US$600m. to US$2,000m. in Botswana, the growth even there was all in the 1980s; and that the total value of imports actually fell in Burundi, the DRC and Zambia. Even in Kenya, Tanzania, Malawi and Zimbabwe imports per capita fell over this period, although in Uganda it rose as economic recovery took place after the years of political turmoil.

World Bank figures for per capita consumption of commercially produced energy also show a fall in most countries over the period from 1980 to 1997, for instance from 793 to 634 kg oil equivalent in Zambia, from 589 to 494 in Kenya and from 553 to 455 in Tanzania. At the individual household level, these decades have often brought a shift back from electricity or kerosene to fuelwood. On overall trends regarding 'consumption', see Box 4.6.

Trends in social welfare have not been quite so discouraging, except in respect of the increasing impact of AIDS. For example, child mortality for sub-Saharan Africa as a whole fell from 260 per 1,000 in 1960 to 200 in 1980 and to 170 in 1999. However, it is noteworthy that this reduction has not matched that in South Asia, where the rate fell from 240 in 1960 to 180 in 1980 and 100 in 1999. Most individual countries in eastern and southern Africa have had first rapid then much slower improvement in both infant and child mortality, but UNDP and UNICEF data suggest some exceptions as indicated in Table 2.2. Thus in Mozambique there has been more improvement since 1990 than in earlier decades, reflecting the improved political situation since the end of the civil war. Conversely, the data suggest that after the normal rate of improvement in the 1960s and 1970s in Zambia, most of the gains there have been lost (see also the discussion in Chapter 2). There is no ready explanation for Zambia's distinctiveness in this respect, though the effects of structural adjustment in a relatively highly urbanised society and of AIDS impact earlier than further south have both contributed.

Changing national figures for life expectancy at birth (see Tables 1.1 and 2.2) have been much affected by the timing of adjustments for the impact of AIDS, but the broad picture has been remarkably similar across the region. Life expectancy increased by 10–12 years in most countries between 1960 and the late 1980s, but has since fallen back by 5–8 years. The setback has been greatest in Botswana, contrasting sharply with the

---

Box 4.6 Private consumption: an indicator of non-development

Tucked away in the *World Development Report 2000/2001* (pp. 276–7) is a data set for 1980–98 annual average growth (or decline) in private household consumption per capita, which demonstrates very clearly the lack of development over this period in most of eastern and southern Africa, and conversely the persistence of poverty at household level. Whereas China and India registered average annual growth of 7 and 3 per cent respectively, the figure for sub-Saharan Africa was −1 per cent. Private consumption per capita grew by 3 per cent a year in Botswana and by 2 per cent in Uganda, but in most countries in our region, from Kenya to South Africa, it remained almost static, while in Zambia, the DRC and especially Angola it fell substantially. The figures are too shaky to merit a table here, but the general picture is very clear and is consistent with other indicators for this period spanning the 1980s and 1990s.

Source of data: World Bank (2000).

economic advance there, and least in the DRC, Angola and Mozambique, in each of which AIDS has either advanced more slowly or been less fully documented (see also Chapter 2). In Angola the situation is not known with any accuracy because of the ravages of war, now hopefully finally ended, but it could be worse than existing fragmentary data suggest.

Food and hunger

Agriculture is the topic of Chapter 6, but the subject cannot be totally excluded here since it is commonly stated that the great majority of the population in countries such as Tanzania, Malawi and Mozambique gain their livelihood from it. This would imply that gross inadequacies in agricultural production must be at the heart of the material poverty of these countries. Low agricultural productivity is indeed a critically important and relevant issue, and therefore this section should be read in conjunction with Chapter 6. However, the statement above should be modified, for most rural households do not depend wholly on farming for their livelihood (Francis 2000; Bryceson and Jamal 1997). They must also allocate time and energy to fetching water, gathering fuel, building their own houses, transporting their own produce, nursing their own children (always bearing in mind that half the people *are* still children), administering themselves, maintaining law and order, and a host of other activities (including other income-generating activities).

To complement Chapter 6, the emphasis here is on food *consumption* rather than production, and especially on undernourishment and malnutrition. Even when imports and food aid are included, several countries in the region do not have sufficient food available year by year to meet even the basic needs of their people (Devereux and Maxwell 2001). The supply might be just adequate if distribution were completely equitable and efficient but some inequity is unavoidable, while spatial and temporal patterns of supply can never closely match those of demand. In reality, inequity is far greater than might be considered inevitable (Sen 1981), the means of linking surplus and deficit areas are desperately lacking, and highly seasonal climates result in much food going to waste whatever efforts at storage are made. FAO data suggest that average daily calorie consumption worldwide is around 2,800, while in rich countries it is around 3,500; but their estimate for sub-Saharan Africa is only 2,200. One cannot say what is really essential, especially in view of differing climates, lifestyles and proportions of children; but there is some consensus in FAO and WHO that this African average represents less than 90 per cent of the requirements for a fully active life. Furthermore, since this is an average, many people must be consuming far less than they need. Indeed, in the cases of Angola, Burundi, the DRC and Mozambique, even the national average is thought to be under 2,000 calories per day.

The FAO data do not support the common claim that average levels of food consumption, and of nutrition, have been declining over the region as a whole for decades; but it is evident that they have failed to increase through the 1970s, 1980s and 1990s, during which time they have advanced substantially elsewhere. Per capita daily calorie consumption increased between 1970 and 1998 from 2,100 to 2,500 in India, and from 2,000 to 2,800 in China. This is clearly another aspect of the persistence of poverty and absence of 'development' in our region.

Both in the early 1970s and in the early 1980s most of the region was spared the dire famines which afflicted both the Horn of Africa, notably Ethiopia, and the West African Sahel; but famine conditions have arisen in certain places at certain times, generally due to a combination of drought and armed conflict. Areas of Mozambique and Angola have suffered most, desperate food shortage forming part of a complex picture in which massive displacements of people have also been prominent; and in 2002 famine conditions built up across much of south-central Africa, with both Malawi and Zimbabwe profoundly affected.

While famine has been experienced by only a tiny proportion of the region's people, undernourishment in the sense of a shortage even of basic staples, alternatively expressed as hunger, affects vast numbers at least periodically (Table 4.3). In most savanna areas the wet season is the busiest time of year, as land is prepared and crops are planted, but it is also the 'hungry season' as the new harvest is awaited. Many urban dwellers are also undernourished, as are some of the military, and Table 4.3 indicates that hunger has become more widespread since 1980, the data being even more disturbing than those for calorie consumption. Malnutrition, in the sense of a diet lacking in some elements essential for health

Table 4.3 Undernourishment

| | Undernourished as % of total population | |
	1980	1996/98
Burundi	38	64
Congo (DRC)	37	60
Mozambique	54	58
Zambia	30	45
Angola	30	44
Kenya	25	41
Tanzania	25	40
Rwanda	24	38
Zimbabwe	30	37
Malawi	26	32
Uganda	31	30
Namibia	25	30
Lesotho	26	28
Botswana	28	25
Swaziland	14	14
India (for comparison)	38	22

Note: no data for South Africa.
Sources: FAO (1999); UNDP, *Human Development Report 2002*.

(Pacey and Payne 1985; Howard and Millard 1997; FAO 1999), is equally widespread especially among children, but it does not exhibit the same spatial pattern. For example, it is less common than hunger among pastoral peoples in semi-arid areas, but more common than hunger in the cities. Various surveys indicate that malnutrition serious enough to cause specific diseases, including kwashiorkor, afflicts a quarter of all children in sub-Saharan Africa, and that it is extensive even in South Africa. Many children recover fully from such diseases, but others suffer permanent physical or mental damage.

The FAO estimates of daily per capita protein supply for the late 1990s are 100–110 g for the rich countries, 75 g for China, 60 g for India, and 50–55 g for sub-Saharan Africa. Within our region the range is from 75 g in South Africa and Botswana to just 35 g in Mozambique – the world's lowest figure by a large margin and perhaps open to question. What is not in doubt is that in countries such as Tanzania and Zambia the level of protein consumption is much lower than the average in India or Pakistan, and has fallen over the past 20 years.

An indicator of the severity of nutrition problems from another angle is provided by UNICEF in their *State of the World's Children* for 2001. Moderate or severe stunting is found in 2 per cent of children aged up to five in the USA, as also in Chile, but the rate is 23 per cent in South Africa, 42 per cent in both Tanzania and Zambia, 48 per cent in Malawi and 53 per cent in Angola. The figures for children 'underweight' are somewhat lower, but the rank order is much the same. With respect to this dimension of poverty, the situation appears to be just as bad in South Asia, but that is no consolation to the people of eastern and south-central Africa.

Healthcare

Disease has already been stressed as a critical element in the environmental context of poverty, and high levels of infant and child mortality have been noted. In this section the focus is on grossly inadequate healthcare as another dimension of poverty. Processes of circular causation are clearly at work in eastern and southern Africa, with low income contributing to ill health and this in turn contributing to low income. Throughout the 1980s and 1990s 'health for all by 2000' was part of the global anti-poverty agenda.

For every family there are spending choices between healthcare and all types of goods. In many countries where basic care was once free, charges are now levied, especially for drugs, while funds may also be needed for bus fares and for 'tips' demanded by medical staff. Similar spending choices must be made by governments, as well as by aid agencies and the religious organisations active in this field. In each country government spending on healthcare is much lower than on education; and in some it is in per capita terms less than 1 per cent of that in western Europe. Total public, private and household annual healthcare expenditure is thought to be just PPP$30–50 per head in Malawi, Uganda and Zambia, compared with PPP$500 in much of Latin America and over PPP$2,000 in western Europe.

For most parts of our region it is unrealistic to think only in terms of healthcare delivery based on Western medicine. Another very different system of largely indigenous practice carried forward from precolonial times remains vitally important, although data on it are scarce. Traditional healers and herbalists are widely consulted and appreciated, though poverty means that resources available to them, and often their effectiveness, are very limited. As Good (1987) has demonstrated for both rural and urban Kenya, their role is not decreasing through any all-pervasive Westernisation. While direct links between indigenous and Western healthcare systems are slight, and should perhaps be strengthened (Last and Chavunduka 1986), they often serve the same people, including both the poorest and the less poor. Ill people may have a choice between them, influenced by relative cost: they may also alternate between them, sometimes with disastrous results.

It would be wrong to dichotomise too sharply between just two healthcare systems. Traditional healers often include some imported drugs in their repertoire; while government medical staff are increasingly setting up in private practice, legally or otherwise, to provide a third strand of healthcare. Partly as a direct response to low or even unpaid salaries, staff time as well as equipment and drugs may thus be siphoned off from the government health system, leaving it further impoverished.

All this means that we can only paint a partial picture of national variations in healthcare, provision and deprivation, especially from sources such as WHO's annual *World Health Report*. Even for the large-scale Western system alone, problems arise due to varying definitions, but some broad orders of magnitude can be indicated, often roughly corresponding to poverty rankings on other criteria. Data from UNDP, UNICEF and WHO sources indicate 50–60 doctors per 100,000 people in South Africa, 15 in Kenya, 4 in Tanzania and only 2 in Malawi. The equivalent figures for nurses are around 175, 25, 45 and 5, but perhaps the definition of nurse is broader in Tanzania than elsewhere. *An Urbanising World* (UNCHS 1996) indicated the proportion of the national population 'with access to health care' as 85 per cent in Zimbabwe, 77 per cent in Kenya, 50 per cent in Uganda and 40 per cent in Mozambique, but gave a high figure of 80 per cent for Malawi (conflicting with the data above). In both Uganda and Mozambique there has been some recovery from the breakdown of healthcare during civil war, but in Angola the collapse in some provinces has been almost total.

Immunisation of infants against measles, polio and tuberculosis made great strides in the 1960s and 1970s, and even in countries such as Malawi and Tanzania the great majority continue to be immunised. In Mozambique there was rapid recovery in the 1990s from the breakdown brought about by civil war. In the DRC, however, the proportion of infants immunised fell during the 1990s, probably to below 20 per cent. In that country regional and urban/rural variations in all aspects of healthcare are particularly extreme, many communities having nothing to supplement their own customary forms of healthcare.

Any discussion of healthcare must make further reference to AIDS, which has hugely added to the burden on all elements in the system in recent years, first in eastern Africa and now even more intensely in southern Africa. No other part of the world is so seriously afflicted, yet the resources available to respond are more scarce than in any other continent, and are used at the expense of other desperate health needs. This means that the overall healthcare situation in countries such as Zambia and Zimbabwe was far worse by 2000 than it had been 30 years earlier.

Education

Low levels of education are clearly part of the persisting poverty scenario (Gould 1993; UNICEF 1999; Watkins 2000). They result both from governments' lack of resources which limits educational provision and from the poverty which prevents many households from fully taking up even what is available. Lack of education then in turn contributes to ongoing poverty at both family and national scales. Education is a basic need and a vital aspect of well-being in the same way as food or shelter, and it incurs costs for families even where basic provision is free. Some cannot afford payments for books or uniforms, or the bribes that head teachers may demand; while sending children to school may be costly in terms of loss of their labour as water-fetchers, firewood-gatherers or infant-minders. In many ways the high priority which many parents give to education throughout our region is remarkable, especially in view of the low quality of much of what is on offer.

Similarly most post-colonial African governments cannot properly be accused of neglecting education, for it has taken up a larger share of government expenditure than in most other parts of the world – in some cases over 20 per cent of total current (as against capital) expenditure. Yet in a country like Uganda total government annual spending on everything (not just education), for a population of over 20 million, is similar to that of one medium-sized British university. Furthermore, while the situation is in many ways grim, there has been much improvement since the 1960s; and it can be argued that educational development was still taking place in some countries in the 1980s when the word 'development' could no longer be applied to most aspects of national life (Bogonko 1992; Ishumi 1994). Sadly deterioration had become the norm almost everywhere by the 1990s, especially in respect of quality.

Better data exist for school enrolments than for most topics considered in this chapter: they can be documented much more easily than incomes or nutrition, though of course there is always scope for falsifying figures. Primary enrolment rates pose a problem because of the variable ages at which children participate; so they may exceed 100 per cent where many 'over-age' children are still in primary schools. However, the broad picture is fairly clear (Table 4.4). Forty years ago primary enrolment rates in most African countries were the lowest in the world, but throughout our region they then increased, generally overtaking those for South Asia. In many countries 70–90 per cent of children aged up to 12 now attend primary school, compared with only 30–40 per cent in the early 1960s. Girls' enrolment has risen faster than that of boys, so gender disparity has narrowed

Table 4.4 Years of schooling and school enrolments (countries ranked by mean years of schooling of the over-14 population in 2000)

Country	Mean years of schooling		Late 1990s % enrolment	
	1980	2000	Secondary	Primary
Mozambique	1.0	1.5	7	70
Rwanda	1.7	2.5	10	85
DRC	2.0	3.0	25	58
Tanzania	2.7	3.1	6	76
Malawi	2.7	3.2	16	135
Uganda	1.8	3.5	12	121
Kenya	3.4	4.2	24	88
Lesotho	3.8	4.4	31	94
Zambia	3.9	5.3	27	94
Zimbabwe	2.5	5.4	48	108
Swaziland	3.9	6.0	54	116
South Africa	3.8	6.1	84	92
Botswana	3.1	6.3	65	118
World	5.2	*c.* 7	58	95

Notes: Data not available for Angola, Burundi and Namibia.
Some gross primary enrolment rates exceed 100% due to enrolment of children above the standard age rate.
Some figures are averages from differing data sets.
Sources: UNDP, *Human Development Report 2002*; UNICEF, *State of the World's Children 2001*.

and is no longer the vital issue that it is, for example, in Sahelian countries from Mauritania to Chad. In many countries something approaching universal enrolment has been achieved for ages seven, eight and nine, though some drop-out then follows.

National variations in poverty are now more clearly demonstrated by secondary enrolment rates, which ranged in the mid-1990s from over 80 per cent in South Africa and almost 50 per cent in Zimbabwe, through 25 per cent in Zambia and Kenya, to under 10 per cent in Mozambique and Tanzania, as shown in Table 4.4. Perhaps surprisingly, the ratio of boys to girls in secondary school is consistent at about 4 : 3 across this range, apart from southern Africa where girls outnumber boys.

Geographical variations in school enrolment within countries are less well documented. Even where district figures are published, as in Kenya, these can mislead in terms of access to education since there is much inter-district movement to secondary boarding schools, and since many city children are sent to live with their rural relatives for their primary school years, both to reduce costs and to avoid the perceived moral hazards of growing up in town. To complicate matters further, movement can also occur in the opposite direction since better quality schools may be available in urban areas. However, regional variations are clearly far greater in some countries, such as Kenya and Zimbabwe, than in others, such as Tanzania and Zambia. In the 1970s the Tanzanian government distributed regional investment in new schools in inverse proportion to existing provision, whereas in Kenya the Harambee movement led to fastest school growth wherever both funds and enthusiasm were most abundant. Tanzania's 1988 census indicated that, in each region, between one-quarter and one-half of the over-10 population had ever attended school; in Kenya regional contrasts in this respect too are much greater. Although based on sample surveys, the data in Table 4.4 showing the length of schooling of the adult population probably give the clearest picture of contrasts over space and time in the educational level of the labour force, reflecting patterns of increased provision in earlier decades rather than the current school situation, and also highlighting Africa's lowly position in global terms.

In terms of school enrolments (and consequent adult literacy – see Box 4.7), education has constituted the most successful aspect of development in post-colonial Africa. Most governments have given it high priority, partly because it is both politically popular and non-threatening to powerful vested interests (except at university level). School building and teacher training are easier to accomplish than increasing productivity in agriculture or setting up viable industries. However, real success involves more than numbers in school, and quantity has often been achieved at the expense of quality. Increased enrolment sometimes just means 50 in each primary class instead of 40. There are many poorly qualified teachers, and many so poorly or irregularly paid that they must devote much time to earning supplementary income. Few classrooms are well equipped, while funds for items such as books are now very scarce in most parts of our region.

Questions also arise about what is taught. What was appropriate when only the future elite attended school may need drastic alteration when the majority do so. Deteriorating formal job opportunities for school-leavers have not always been sufficiently considered. Yet efforts to introduce a curriculum more relevant to a predominantly rural self-employed population often meet strong resistance from teachers, parents and the children themselves.

For individual families almost any education can help to relieve poverty: children who have been to school generally have better economic prospects than those who have not, and can compete better in the job market. However, more education for all may just increase the level of qualifications required to obtain a given income, unless it produces higher productivity. One case of benefit to individuals but not to local society is the 'brain drain' by which the most successful pupils from rural schools go off to the city and even off to work overseas. Fortunately at both national and international scales in Africa remittances commonly provide some compensation for such losses of trained people, bringing some benefits both to their families and to their home areas.

Partly as a direct result of poverty, especially malnutrition, the need for 'special education' for children with profound physical or mental disabilities is particularly great in Africa; yet its high cost largely rules out its provision (Ross 1988). A few institutions do remarkable work, but they

Box 4.7 Adult literacy

Illiteracy is an important facet of poverty, and one which reflects educational deficiency in past decades. South Africa is less exceptional within our region in this respect than in most others, with an adult literacy level only a little higher than Zimbabwe, Zambia and Kenya. Even in Tanzania the level was boosted to 70 per cent by a massive programme of literacy classes for adults associated with villagization in the 1970s, forging ahead of both Malawi and Uganda. In contrast, even today only a minority of adults in Burundi and in Mozambique can read and write. The gender dimension is significant, as the table below shows, with huge disparities in some countries: however, literacy levels have in each of these risen faster for women than for men in recent decades.

	Adult literacy rate (%)					
	1980			*1998/99*		
	All	*m*	*f*	*All*	*m*	*f*
Mozambique	28	44	12	41	57	26
Burundi	29	41	16	44	53	35
Malawi	45	64	27	56	70	42
Uganda	45	60	31	64	76	53
Zambia	61	73	50	75	82	68
Tanzania	51	65	36	76	84	68
Kenya	57	71	43	79	87	71
South Africa	76	78	75	85	85	84
Zimbabwe	74	80	68	87	91	83

Sources: These figures are mainly averages from differing data sets in: UNDP, *Human Development Report 2001*; World Bank, *African Development Indicators 2001*.

represent drops in an ocean which is probably now expanding even faster than the general rate of population growth. Even within the region the provision is least where the need is greatest, for instance across much of Angola.

Despite all these queries about quality and relevance, and despite the certainty that the spread of Western education has increased inequalities in most African societies (especially education at secondary level for a few and tertiary level for a tiny elite), it has certainly also brought economic benefit by increasing many people's productivity and imparting much-needed skills. Without it, material poverty would be even more widespread, and in some ways even more intense. There is thus little doubt that the cutbacks in government spending on education, the introduction of school fees, and the consequent increased drop-out rates among the poorest sectors of the population in the 1990s are now intensifying poverty in much of eastern and southern Africa.

The debt issue

In discussions of poverty in South Asia 'the problem of debt' normally concerns debt at the individual or family level; and this is of some importance in our region also, though with a much smaller role for money-lenders, and most people incurring only small debts to neighbours or local traders. In eastern and southern Africa debt is more clearly a major issue at national level. The poverty of African governments led them to borrow heavily in the 1970s, and it has prevented their repaying loans which were often at variable rates that later rose steeply. Total debt increased by over 25 per cent annually between 1972 and 1982, and while the rate then slowed the total burden has continued to rise, making debt a more intractable problem in Africa than anywhere else (Parfitt and Riley 1989). The absolute sums owed are much smaller than for some Latin American and South-East Asian countries, but they represent a higher proportion of GNP and of annual export earnings (Box 4.8), and for some countries are far more debilitating. In contrast to Latin America, most debt is not to private banks in the rich world, but to governments and to international institutions.

The largest debt is the US$25bn owed by South Africa, but this represents only 18 per cent of its annual GNP. In sharp contrast, the US$13bn owed by the DRC represents 200 per cent of annual GNP, and is most unlikely ever to be repaid. Zambia borrowed heavily when copper prices fell in the 1970s (see Chapter 3), hoping to repay when they rose again: sadly no such price revival occurred, and the US$7bn debt there equals 180 per cent of annual GNP. In Zambia, as also in Mozambique, Tanzania

---

Box 4.8   Poverty intensified by external debt

The critical aspect of external debt in many African countries is not its absolute size but its size in relation to the national economies. While Brazil's total debt is US$230bn, this equals only 30 per cent of annual GNP; and India's debt of US$100bn equals only 20 per cent of GNP. South Africa is in a similar position, but in most countries of eastern and south-central Africa external debt equals more than 70 per cent of annual GNP. It also equals more than 500 per cent of annual export earnings. In many of these countries annual debt-service payments are equivalent to between 10 and 25 per cent of export earnings, while in both Burundi and Zambia this ratio exceeded 45 per cent in 1999. By 2001, the attention of the richest countries was at last becoming focused on the situation of the heavily indebted poor countries (HIPCs) – the first time 'poor' has been used officially for country classification.

and elsewhere, allocation of government funds to pay off some of the debt has dug deeply into the budget available for social welfare spending, thus intensifying poverty at household level.

In Zimbabwe, somewhat cut off from the world during UDI in the 1970s, debt at independence in 1980 was very low; but it rose sharply in the early 1980s, forcing cutbacks in government spending there too (see Chapter 3). In Uganda, likewise, most debt arose later as it tried to recover from the political turmoil which had discouraged any lending to it in the 1970s. Among the smaller countries of the region, Swaziland has the least intense and Burundi the most intense debt problem.

Conclusions

This chapter has painted a gloomy picture – inevitably so, given its topic and in view of African realities at the start of the new century. Whatever their value as rallying cries, there was never any real prospect of 'health for all' or 'clean water for all' 'by the year 2000'. Without invoking imaginary and arbitrary 'poverty lines', it can be asserted with confidence that the extent of poverty has increased in eastern and southern Africa over the past 20 years, in terms of absolute numbers either below any given income or suffering any particular form of deprivation. This change is largely a function of overall population growth. Whether there has also been a general impoverishment, involving a reduction in average levels of income and welfare, is much harder to determine. In several countries, such as Kenya and Mozambique, gains and losses in different periods and for different sectors seem to have roughly balanced out. It is clear, however, that there has not been widespread poverty alleviation, and therefore that speaking of 'development' in this region during the late twentieth century is really not appropriate.

There are important variations in the extent and intensity of poverty from one country to another, demonstrated in the text and tables above, and there have also been substantial differences with regard to recent trends. However, there are also great regional variations within most countries; and these entities inherited from colonial rule, with which many people do not yet identify at all strongly, are far from ideal units for discussing poverty. Unfortunately, we are generally tied to them in respect of quantitative data, while any alternative framework would be highly subjective.

It has not been possible in one brief chapter to analyse the causes of the region's persisting poverty, as White and Killick (2001) have done in respect of economic and policy causes; but one can assert that there is no single primary cause, even in particular areas. The causes lie both in the past and in factors still operating today: they also lie both within Africa and beyond it in the global economic and political system. Among the internal factors, drought is not the cause of poverty, but has most certainly contributed to it. Disease is not the explanation, but its huge importance is often not fully recognised. Rapid population growth is not the culprit, but it is far from irrelevant, especially where pressure on resources is intense.

The colonial experience is also clearly relevant, perhaps less through economic exploitation than through the legacy of a political framework which has proved highly unsatisfactory. Specific government policies have often been inappropriate, as institutions such as the World Bank have stressed, but so has much of the advice of those institutions. External factors include deteriorating commodity prices and terms of trade, and also rising interest rates on borrowed money. Perhaps the reality is that some parts of the world must stay materially poor so that others can stay rich, since there is no prospect of the whole world's fast-expanding population living at a North American or West European level of consumption.

When all these matters are considered together, Africa's present condition ceases to be surprising. Indeed, some might argue that the greater mystery is why there was quite so much optimism, and also so much real development, in the 1960s and early 1970s. The fact that many forces working together contribute to persisting poverty across most of eastern and southern Africa means that many things must change before there is hope of substantial and widespread improvement. However, it also means that there are many possible changes each of which might benefit at least some people in some areas, and each of which is therefore worth striving for. The prospects for such change in this new century, both in Africa's own circumstances and in attitudes and policies elsewhere, are left for the reader to judge.

References

Adams, W.M. (1992) *Wasting the Rain: Rivers, people and planning in Africa.* Earthscan, London.

African Development Bank (annual) *African Development Report.* Oxford, Oxford University Press.

Ahmed, A. and **Mlay, W.** (eds) (1998) *Environment and Sustainable Development in Eastern and Southern Africa.* Macmillan, London.

Anderson, D. and **Broch-Due, V.** (eds) (1999) *The Poor Are Not Us: Poverty and pastoralism in Eastern Africa.* James Currey, Oxford.

Barnett, T. and **Blaikie, P.** (1992) *AIDS in Africa: its present and future impact.* Wiley, London.

Baylies, C. and **Bujra, J.** (eds) (2000) *AIDS, Sexuality and Gender in Africa: Collective strategies and struggles in Tanzania and Zambia.* Routledge, London.

Benson, C. and **Clay, E.** (1998) *The Impact of Drought on Sub-Saharan African Economies.* World Bank, Washington.

Blavo, E.Q. (1999) *The Problem of Refugees in Africa.* Ashgate, Aldershot.

Bogonko, S.N. (1992) *Reflections on Education in East Africa.* Oxford University Press, Nairobi.

Bryceson, D. and **Jamal, V.** (eds) (1997) *Farewell to Farms: De-agrarianization and employment in Africa.* Ashgate, Aldershot; African Studies Centre, Leiden.

Cairncross, F. (2001) *The Death of Distance 2.0.* Texere, London.

Castells, M. (2000) *The Information Age.* Vol. III, *End of Millennium*, 2nd edn, Blackwell, Oxford.

Chabal, P. and **Duloz, J.-P.** (1999) *Africa Works: Disorder as political instrument.* James Currey, Oxford.

Chege, M. (1995) Sub-Saharan Africa: underdevelopment's last stand, in Stallings, B. (ed.) *Global Change, Regional Response.* Cambridge University Press, Cambridge: 309–45.

Christie, F. and **Hanlon, J.** (2001) *Mozambique and the Great Flood of 2001.* James Currey, Oxford.

Devereux, S. and **Maxwell, S.** (eds) (2001) *Food Security in Sub-Saharan Africa.* Intermediate Technology, London.

FAO (1999) *The State of Food Insecurity in the World 1999.* FAO, Rome.

Francis, E. (2000) *Making a Living: Rural existence in Africa.* Routledge, London.

Global Coalition for Africa (annual) *African Social and Economic Trends.* Washington.

Good, C. (1987) *Ethnomedical Systems in Africa.* Guilford, New York.

Gould, W.T.S. (1993) *People and Education in the Third World.* Longman, London.

Grove, A.T. (1991) The African environment, in Rimmer, D. (ed.) *Africa 30 Years On.* James Currey, London: 39–55.

Hope, K.R. (ed.) (1999) *AIDS and Development in Africa.* Haworth, New York.

Hope, K.R. and **Chikulo, B.C.** (eds) (2000) *Corruption and Development in Africa.* Macmillan, London.

Howard, M. and **Millard, A.** (1997) *Hunger and Shame: Child malnutrition and poverty on Mount Kilimanjaro.* Routledge, New York.

Iliffe, J. (1987) *The African Poor: A history.* Cambridge University Press, Cambridge.

Ishumi, A. (1994) *30 Years of Learning: Educational development in Eastern and Southern Africa from independence to 1990.* IDRC, Ottawa.

Jamal, V. (1998) Changes in poverty patterns in Uganda, in Hansen, H. and Twaddle, M. (eds) *Developing Uganda.* James Currey, Oxford: 73–97.

Jamal, V. and **Weeks, J.** (1993) *Africa Misunderstood: Or whatever happened to the rural–urban gap?* Macmillan, London.

Jones, S. and **Nelson, N.** (eds) (1999) *Urban Poverty in Africa.* Intermediate Technology, London.

Landes, D. (1998) *The Wealth and Poverty of Nations.* Little, Brown and Co, London and Boston.

Last, M. and **Chavunduka, G.** (eds) (1986) *The Professionalization of African Medicine.* Manchester University Press, Manchester.

Madeley, J. (1999) *Big Business, Poor Peoples.* Zed, London.

Nafziger, E.W., **Stewart, F.** and **Väyrynen, R.** (eds) (2000) *War, Hunger and Displacement*, Vol. II, *Case Studies.* Oxford University Press, Oxford.

O'Connor, A.M. (1978) *The Geography of Tropical African Development*, 2nd edn. Pergamon, Oxford.

O'Connor, A.M. (1991) *Poverty in Africa.* Belhaven, London.

O'Connor, A.M. (2001) Information needs for urban policy making in Africa, in Livingstone, I. and Belshaw, D. (eds) *Renewing Development in Sub-Saharan Africa.* Routledge, London: 328–35.

Oxfam (1993) *Africa: Make or break.* Oxford.

Pacey, A. and **Payne, P.** (eds) (1985) *Agricultural Development and Nutrition.* Hutchinson, London.

Parfitt, T. and **Riley, S.** (1989) *The African Debt Crisis.* Routledge, London.

Poku, N. (2001) *The Political Economy of AIDS in Southern Africa.* Ashgate, Aldershot.

Potter, R., **Binns, T.**, **Elliott, J.** and **Smith, O.** (1999) *Geographies of Development.* Longman, London.

Potts, D. (1997) Urban lives: adopting new strategies and adapting rural links, in Rakodi, C. (ed.) *The Urban Challenge in Africa.* UN University Press, Tokyo.

Ross, D.H. (1988) *Educating Handicapped Young People in Eastern and Southern Africa.* UNESCO, Paris.

Sai, F.T. (1988) Changing perspectives of population in Africa, *African Affairs* 87(347): 267–76.

Sen, A.K. (1981) *Poverty and Famines.* Clarendon Press, Oxford.

Sindiga, I. (1999) *Tourism and African Development: Change and challenge of tourism in Kenya.* Ashgate, Aldershot.

Spielman, A. and **D'Antonio, M.** (2001) *Mosquito: Man's deadliest foe.* Faber and Faber, London.

UNCHS (1996) *An Urbanizing World: Global report on human settlements.* Oxford University Press, Oxford.

UNCHS (2001) *Cities in a Globalizing World.* Earthscan, London.

UNDP (annual) *Human Development Report*. Oxford University Press, New York.

UNICEF (annual) *State of the World's Children*. New York.

UNICEF (1999) *State of the World's Children 1999*. Special edition on education. New York.

Wallman, S. (ed.) (1996) *Kampala Women Getting By: Wellbeing in the time of AIDS*. James Currey, Oxford.

Watkins, K. (ed.) (2000) *The Oxfam Education Report*. Oxfam, Oxford.

Webb, D. (1997) *HIV and AIDS in Africa*. Pluto, London.

White, H. and **Killick, T.** (2001) *African Poverty at the Millennium*. World Bank, Washington.

WHO (annual) *World Health Report*. Geneva.

World Bank (annual) *African Development Indicators*. Washington.

World Bank (annual) *World Development Indicators*. Washington.

World Bank (annual) *World Development Report*. Oxford University Press, New York.

World Bank (2000) *World Development Report 2000/2001*. Special edition: *Attacking poverty*. Oxford University Press, New York.

Yeager, R. (1989) *Tanzania: An African experiment*, 2nd edn. Westview, Boulder.

CHAPTER 5

Natural resources: use, access, tenure and management

Sian Sullivan and Katherine Homewood

The significance of gathered or hunted resources to African livelihoods has received increasing attention in recent years. These are here defined as plant and animal resources, generally indigenous as opposed to introduced or 'alien' species, that are hunted, gathered or otherwise procured from the wider landscape rather than cultivated or husbanded close to home-steads and settlements. Although called 'natural' or 'wild', most have been influenced over millennia by African peoples utilising and inhabiting the continent's diverse landscapes.

Such resources are now recognised as conferring important benefits to their users and, for some, may be the primary sources of subsistence and welfare. Food security, for example, may be enhanced in several ways: through direct consumption of accessible 'wild' foods which, even in small quantities, may provide essential nutrients and diversify otherwise monotonous diets; through the sale or exchange of gathered products which increases purchasing power and the ability to obtain alternative foods (de Merode *et al.* forthcoming, a); and through holding trees as a form of 'savings bank', to be converted into income in response to unexpected contingencies (e.g. Chambers and Leach 1989; Barrow 1990: 168). Gendered dimensions of resource use mean that gathering may provide a source of independence and extra income for women who are often the primary collectors and processors of specific products. On the other hand, gendered associations between animal wildlife and men as hunters mean that the current plethora of schemes to increase local access to wildlife resources may focus on men as the recipients and obscure women's knowledge about the wider environment (e.g. Sullivan 2000). Infusing these utilitarian dimensions of resource-gathering are less tangible aspects of cultural iden-tity and symbolism bound up with enacting resource-use practice, and through which culture, tradition and identity are renewed and revisited (cf. Bourdieu 1990; Posey 1999).

This chapter discusses current issues pertinent to the use and management of indigenous biotic (i.e. biological or living) resources in East and southern Africa. We do not cover soil, water and mineral resources, or debates over farming systems and soil mining, rangeland management, or environmental 'degradation': for these see Chapters 6, 7 and 8 of this volume.

'Wild' plant and animal resources: biophysical determinants of availability

Land and habitat types in southern and eastern Africa are dominated by arid, semi-arid and subhumid vegetation types with varying degrees of woody vegetation cover. They include grasslands and savannas, wetland and riverine habitats, and forests ranging from dry deciduous woodlands to tropical montane, coastal and riverine forests. In contrast to the rest of the region covered by this volume, the Democratic Republic of Congo (DRC) is largely rain, riverine or swamp forest.

Patterns of occurrence and changes in these habitat or vegetation types are best predicted through three interlinked biophysical factors: soil nutrient and water availability; climate and seasonality; and biogeographical influences on the distribution of species. Fire and herbivory are also important in moulding and maintaining species assemblages and structural formations, particularly those with significant grass cover (i.e. grasslands and savannas).

Plant available moisture and nutrients (PAM and PAN)

The availability of soil moisture and nutrients for uptake by plants constrains the possibilities for productivity. In the drier areas water is the main determinant (Solbrig 1991). High moisture and nutrient availability (due to rapid nutrient cycling) make possible the forests of the DRC. At the other extreme, limited soil moisture and nutrients permit the dominance of ephemeral grasslands and shrublands, the species of which display a disjunct (i.e. discontinuous) distribution between the drylands of northeast and south-west Africa. Table 5.1 illustrates the interacting effects of these two variables.

Climate: effects of seasonality and aridity

Climate characteristics in eastern and southern Africa are discussed in Chapters 6 and 8 where the extreme significance of the amount of rainfall – and its seasonality and unpredictability – as a determinant of vegetation, agricultural and hydrological patterns is emphasised. Use of natural resources is one means of coping with this variability, whether as the basis

Table 5.1 Soil and water availability, and influence on vegetation types

	High water availability	*Low water availability*
High soil nutrient availability	High potential land. Mosaic of forest and other vegetation depending on land-use. Examples occur in the East African Rift Valley (fertile volcanic soils with orographic rainfall)	Seasonal pulse of very high primary production during rains allows seasonal influx of temporarily very high densities of migratory or transhumant grazers. For example the Serengeti short grass plains (on fertile volcanic soils in rainshadow area)
Low soil nutrient availability	Abundant water but poor soil nutrient supply means dense growth but low quality forage, hence low herbivore density. Dry season fires sweep through the mass of dry matter left standing, resulting in a fire-dominated ecosystem with fire-resistant grasses and trees. Examples include the *miombo* and *mopane* woodlands of Tanzania, Zambia, Malawi, Zimbabwe and Botswana, and the wooded 'elephant' grass stands of deforested interfluvial parts of the Congo Basin	Limited soil nutrients allow limited but nutritious forage growth during brief rainy season. Ephemeral and seasonal primary productivity allows limited seasonal influx of wild and domestic herds. Examples include the arid and semi-arid lands of the East African Sahel (e.g. north Kenya, southern Sudan, southern Ethiopia), Namibian *thornveld*, and northern Cape and Namibian succulent-dominated shrublands

Source: After Bell (1982).

of major production systems, or as irregular but proactive practices which capitalise on variable productivity to promote livelihood reliability (Roe *et al*. 1998). As such, natural resources constitute complementary elements in flexible and resilient networks of livelihood strategies. Understanding the ways that seasonality and the unpredictability of rainfall affect productivity and therefore resource availability in different parts of East and southern Africa is important for understanding patterns of human use of wild plants and animals.

Biogeography and biodiversity

Vegetation structure may be dictated largely by local soil and water conditions together with land-use patterns. Species composition of local vegetation, however, also depends on biogeographic factors. Following White (1983) and Davis *et al*. (1994), East and southern Africa incorporate seven major plant biogeographic zones or regional centres of endemism (RCE) and a further seven transition zones between these RCEs (see Table 5.2 and Figure 5.1). These affect the availability of resources useful to people. The species composition at particular locations is further affected by: species–area relationships, with larger areas normally having a richer species complement, all other things being equal) (Rodgers *et al*. 1982;

Table 5.2 Eastern and southern African phytochorological regions

Phytochorological region (RCE) Regional Centre of Endemism	Dominant habitat	Total area ('000 km²)	Total no. of plant species	No. of endemic species
I Guineo-Congolian RCE	Evergreen and semi-evergreen rainforest	2,800	8,000	6,400
II Zambezian RCE	>95% savanna	3,770	8,500	4,590 (54%)
III Sudanian RCE	>95% savanna	3,731	2,750	910
IV Somalia–Masai RCE	90% savanna	1,873	4,500*	1,250 (31%, incl. 2 families and 50 genera)
V Cape RCE	Sclerophyllous thicket (fynbos)	71	7,000	1,250
VI Karoo–Namib RCE	Desert; the most extensive and distinctive shrubland assemblage	661	>7,000*	35–50% (incl. 1 family and 160 genera)
VIII and IX Afromontane and Afroalpine archipelago-like RCE	Montane grassland interspersed with forest patches	715	4,000	3,000 (75%, incl. 2 families and 200 genera)
X Guinea–Congolia/Zambezia regional transition zone	Forest with 20% savanna	705	2,000	Few
XI Guinea–Congolia/Sudania regional transition zone	Forest with 30% savanna	1,165	2,000	Few
XII Lake Victoria regional mosaic	Forest with 30% savanna	224	3,000	Few
XIII Zanzibar–Inhambane regional (coastal) mosaic	50% savanna	2,482	1,200	40
XIV Kalahari–Highveld regional transition zone	75% savanna	1,223	3,000	200
XV Tongaland–Pondoland regional mosaic	50% savanna	148	3,000	200
XVI Sahel regional transition zone	50% savanna	2,482	1,200	40

Note: Numbers of species and endemics are continually being revised as more research is undertaken, particularly in remote areas or in areas that have experienced protracted periods of conflict. Also, note that an alternative system is used in the IUCN *Directory of Afrotropical Protected Areas* (1987b): we have chosen to use the system established by White as this is probably more generic and widely acknowledged. *Source:* Derived from information on physiognomy or form, floristics or species assemblages, and physical environment, to designate major biogeographic zones (after White 1978, 1983; Scholes and Walker 1993: 12; Davis et al. 1994). Where figures in Davis et al. (1994) differ from White (1978, 1983) the former, as the more recent publication, are used and marked with an asterisk.

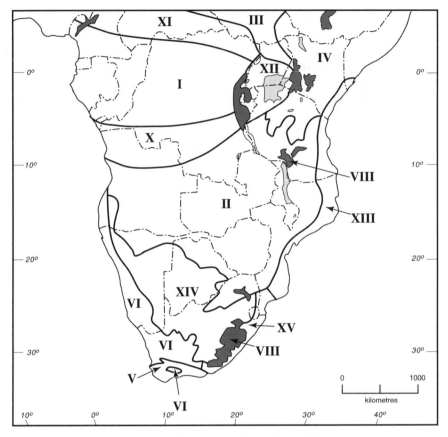

Figure 5.1 Main phytochoria of eastern and southern Africa
Source: White (1983: 38). For key refer to Table 5.2

Western and Ssemakula 1981); topographic and substrate diversity (e.g. Sullivan 1999a); and climate. In southern Africa, for example, the richness of woody edible plant species is strongly correlated with higher rainfall and lower evapotranspiration (O'Brien 1988).

As can be seen from Table 5.2, RCE zones II, III and IV with, to a lesser extent, zone VI, are dominated by a variety of savanna vegetation types. Savanna species of all taxa are generally widely distributed in contrast to their forest counterparts (Davis *et al.* 1994; Stattersfield *et al.* 1998). This means that, despite their considerable regional endemism, savannas are less likely to host site-endemic animal species than are forests. For example, of 23 African endemic bird areas, only the Juba-Shebelle valley, Somalia, is savanna (Stattersfield *et al.* 1998). For their individual and cumulative areas, therefore, forests represent sites of outstanding species richness and endemism and thus are of high conservation value. This is particularly true of the long-established forests associated with the Congo refuge in the DRC[1] and of the block-faulted mountains of East Africa formed over 20 million years BP, including the Eastern Arc forests of the Usambaras, Ulugurus and Udzungwas ranges in northern Tanzania (Rodgers and Homewood 1982; Rodgers *et al.* 1982; Lovett and Wasser 1993; Myers *et al.* 2000). Forests on recent volcanic mountains in savanna areas, e.g. Mts Kilimanjaro, Kenya and Meru, tend to be less rich in terms of both species numbers and endemics. These patterns apply to a wide range of plant, vertebrate and invertebrate taxa and exist despite the considerable species richness of East African savannas.

Common uses of 'natural resources' and relevance for livelihoods

Several peoples conventionally classified as 'hunter-gatherers', including the 'Bushmen' of southern Africa, the Hadza, Dorobo and Ik of East Africa, and the Twa and Mbuti 'pygmies' of the Zaire Basin, retain a high dependence on, and knowledge of, natural resources, but also rely to varying extents on alternative sources of livelihood. 'Hunter-gatherers' have generally fared rather badly from movement into 'their' territories by cultivators and pastoralists, the imposition of colonial rule and the setting aside of conservation areas (e.g. Turnbull 1972; Wilmsen 1992; Lewis and Knight 1995; Hitchcock 1996; Simpson 1997; Gordon and Sholto Douglas 2000). It is important not to perpetuate popular representations which may romanticise their 'hunter-gatherer' lifestyle and present these people as 'in harmony' with, and dependent on, their immediate environment, when in many (if not all) cases their natural resource use practices have been extremely circumscribed. Furthermore, automatic presumptions that social and economic 'development' for these peoples should revolve around 'hunting and gathering' can be problematic – such decisions preferably should lie within communities themselves.

More generally, the hunting and gathering of wild resources, including specialist resource extraction such as honey-harvesting and charcoal production, are elements of many major production systems in the region which are normally classified as agriculture or pastoralism and/or agropastoralism. Commercial use of natural resources, particularly wild-life hunting and/or wildlife viewing based enterprises, are also significant in the African context. The uses of different natural resources are often intertwined, either as they are procured and/or consumed, or because the resource can fulfil many functions (e.g. nutritional, curative, cosmetic, symbolic). This makes it difficult to construct an effective typology of resources or their uses without excluding the complexities of the roles

---

[1.] 'Refugia' are areas that act as core and continuous habitat 'islands' for species when environmental change causes extinctions to occur through the contraction of surrounding habitats.

they play in people's livelihoods (see Box 5.1). Below we introduce some resources and their uses under the broad categories of 'non-timber plant products', 'timber' and 'animals', our logic being that these categories are broadly different ecologically, affecting both the type and impacts of harvest practices, as well as embodying some coherence regarding uses within these categories.

Box 5.1 Categories and complexities in the uses of gathered plants by Damara herders in north-west Namibia

The following examples of plant use by Damara herders in north-west Namibia (see Figure 5.2) indicate how simple classifications may overlook much that is significant concerning natural resources.

1. Many items apparently consumed as food or beverages are considered to have other beneficial values, making their categorisation problematic and frequently misleading. For example, the flowers, leaves and stems of the herb *Thamnosma africana* (*khanab*) may be consumed as a herbal tea, but a stronger brew (or decoction) is used to treat a variety of complaints, from coughs to menstrual problems. This herb also is a component of perfume or *sâi*, a fine powder of aromatic plants which is made, used and traded by women and which has complex symbolic as well as cosmetic values. Similarly, stems of the succulent plant *Hoodia* spp. are consumed as food but also are considered to lower blood pressure and to prevent mosquito bites, while *Stipagrostis* spp. grass seeds are consumed as a nutritious porridge-like food but are also used in the production of beer and liquor, which provide an important source of cash income for many women.

2. Complications also arise over the categorisation of gathered items as 'wild', a label that frequently obscures the investment that people make in controlling or otherwise ensuring the future productivity of such resources. For example, many leafy species are left growing as weeds in cultivated fields because their leaves provide a nutritious source of 'relish' to add to starchy staples. In the settlement of Sesfontein, north-west Namibia, seeds of the wild spinach *Amaranthus* sp. have been planted in people's gardens, having been brought to the settlement by Owambo men from the wetter areas of north-central Namibia who have married into Damara families. *Amaranthus* spp. also grow 'in the wild' in the environs around Sesfontein. Both sources of spinach are harvested and consumed by people in the settlement. Similarly, although the consumption of stems of the spectacular and near-endemic succulent *Hoodia* spp. (see above) is the cause of some concern among conservationists, Damara people also frequently plant this species around their homesteads and in their gardens. They can be seen, therefore, as propagating, as well as harvesting, this valued species.

Source: Sullivan (2000).

Non-timber plant products

Of the range of natural resources utilised, non-timber plant products, i.e. fruits, nuts, seeds, leaves, flowers, stems, gums, and underground plant parts such as bulbs, corms and tubers, are frequently the dominant sources of 'wild' foods and medicines, both in terms of quantity of product and diversity of species consumed.[2] While it may be rare today to find people relying solely on gathered plants for food and/or medicine, non-timber plant products are frequently important complements to foods and medicines procured from elsewhere (i.e. cultivated crops, husbanded animals or shop-bought).

Gathered foods, for example, are not often consumed as staples.[3] Instead they contribute essential nutrients and diversity to frequently starchy and monotonous diets. They may also confer nutritional success under circumstances where food provision is unpredictable (Grivetti 1978). This is the case for many indigenous fruit species, which often contain greater quantities of vitamin C than domesticated species (e.g. Wehmeyer 1986: 8). These are often consumed when encountered on common land in rural areas (e.g. Maundu 1987), which means that their dietary contribution may be missed by household diet surveys. Similarly, wild spinaches (e.g. *Amaranthus* spp.), which grow as weeds in cultivated plots, are high in proteins, minerals and vitamin A and are consumed throughout East and southern Africa (e.g. Fleuret 1979; Zmarlicki *et al.* 1984; Ogle and Grivetti 1985; Kiwasila and Homewood 1998). Underground plant parts, on the other hand, may have high energy quantities and are particularly important dry season foods for people inhabiting drier areas of the subcontinent (e.g. Heinz and Maguire 1974; Grivetti 1979). Herbs consumed as teas also may contain important minerals (e.g. Wehmeyer 1986: 29).

Plant products frequently form the basis for indigenous medical treatments, particularly for household remedies for common illnesses but also for medical practices based on the sympathetic or magical properties of plant items (Maundu 1987; Maundu *et al.* 2001). The use of plant products in the treatment of livestock diseases is particularly important among pastoralist societies (e.g. Malan and Owen-Smith 1974; Fratkin 1996). Many medicinal plants contain active chemical compounds which also make these plants aromatic and popular for use as perfumes and cosmetics (e.g. Sullivan 2000).

Finally, as rainfed cultivation is unreliable in some two-thirds of southern and East Africa there is a strong dependence across extensive areas

---

[2.] For general references documenting uses of non-timber plant products in various regions of East and southern Africa see, for example, Getahun (1974); Brokensha and Riley (1980, 1986); Benefice *et al.* (1984: 241–2); le Floc *et al.* (1985); Storey (1985); Becker (1986: 61); Campbell (1986); FAO (1986); Gura (1986); Malaisse and Parent (1985); Stiles and Kassam (1991); Sullivan (1998, 2000, forthcoming 2003).

[3.] Although see Lee (1973, 1979), Biesele *et al.* (1979), Peters (1987) and Widlok (1999) for documentation of the continuing importance of the staple food *mangetti* – nuts from the tree *Schinziophyton* (formerly *Ricinodendron*) *rautanenii* – among 'Bushmen' populations in Botswana and Namibia.

on animal production (domestic livestock and/or wildlife) using natural forage. Unlike commercial ranches, where beef production under private ownership on the western European model tends to be the goal, African pastoralists employ a number of strategies in order to capitalise on the vagaries of rainfall-driven primary productivity and which make use of a wide range of indigenous plant resources as forage for domestic livestock. These strategies include mobility (trans-humance and nomadism), in order to utilise pastures when and where they become available and to avoid seasonal disease outbreaks, and the utilisation of browse (e.g. leaves and pods from woody plants) as well as grassy pastures (e.g. Sandford 1983; Homewood and Rodgers 1991; Scoones 1994; Niamir-Fuller 1999).

Timber

The majority of people in rural Africa rely on timber for fuelwood and building materials: for example, over 90 per cent of Malawi's energy needs are met by fuelwood (Moyo *et al.* 1993: 98; also see Bradley 1991). The utilisation of indigenous woody species for these purposes has evoked concern among foresters and the conservation fraternity, although in some areas of West Africa it appears that local uses of woody plants have actually promoted secondary growth and the expansion of forested areas (e.g. Fairhead and Leach 1996, 1998). Frequently, negative impacts on woody species in rural areas are associated with requirements for fuel and charcoal for commercial purposes (e.g. Box 5.2; see also Ribot 1998) or in urban areas. In rural areas, and particularly drylands where woody biomass is low, measured fuelwood and building timber consumption tends to be rather conservative and is often lower than estimated rates of use (e.g. Barnes *et al.* 1984). For example, a small survey of Damara households in arid north-west Namibia in 1996 found that only 0.105 m^3 of wood per capita per year was used (Sullivan 1998), well below the average rate of 0.5 m^3 per capita per year for 'arid and sub-arid' areas of Africa (FAO 1981 cited in deLucia 1983: 9). Timber is also important for the production of household utensils such as mortars and pestles. In recent decades this has generated concern regarding the sustainability of an increasing use of high-quality and slow-growing hardwoods for the production of carved curios for tourists.

Animals

The consumption of 'bushmeat' – the meat from wild animals – has commonly been viewed as driven by a nutritional need for protein (e.g. Eltringham 1984), although this may be somewhat misleading as human daily protein requirements are surprisingly low and relatively easily met from alternative sources (e.g. plant proteins, invertebrates, fish). Where the tsetse fly challenge is endemic (causing trypanosomiasis in cattle and sleeping sickness in people), domestic livestock (and thus the protein from their milk and meat) have been limited until recently.[4] The consumption of animal protein from hunted wildlife has thus been particularly

Box 5.2   Fuelwood extraction and canopy loss in Lake Malawi National Park

Tropical dry woodlands throughout the world are thought to be declining due to human activity, particularly domestic requirements for fuelwood and building poles. In a recent study of vegetation changes in Lake Malawi National Park during 1982–90, aerial photograph analysis showed measurable conversion of closed canopy *miombo* to sparse woodland. This study used a multidisciplinary approach to investigate possible contributions to these changes by the domestic use of fuelwood and construction poles, and by the requirements of fuelwood in fish-smoking for commercial sale.

Domestic fuelwood use was measured in 30 households in each of 2 enclave villages over 12 and 5 months respectively. Domestic fuelwood use consumes a large biomass of mainly dead wood and small branches with a wide species range. Mean total annual domestic fuelwood consumption by the total enclave population was less than half the mean annual deadwood biomass production in the park, estimated from three quadrats harvested monthly over a year. In other words, the domestic consumption of fuelwood was low compared to the availability of this resource.

Construction poles were mostly small, have extended durability and come from a broad species range. Fencing poles commonly take root to form live hedges and *Eucalyptus* trees are grown for poles. Construction pole use thus appeared sustainable and also showed signs of substitution for indigenous species.

The 305 commercial fish-smoking stations in the enclaves used a relatively lower mean annual fuelwood biomass than domestic fuelwood consumption, but targeted large branches and logs from a narrow species range and therefore involve destructive felling of canopy species. Unlike domestic fuelwood collectors (normally women), 95 per cent of men collecting fuel for fish-smoking used cutting tools and three-quarters transport the wood by boat or bicycle.

The scale, size classes and species involved in commercial fish-smoking suggest that it is this activity which is driving vegetation changes from closed canopy to sparse woodland and *not* domestic uses of fuelwood as is generally assumed. Traditional local fishing previously focused on small fish species sundried for preservation. Commercial fish-smoking, introduced relatively recently by in-migrants and utilising gill nets which harvest larger fish, requires smoking for preservation. Seventy per cent of commercial fish-smoking stations were owned by northern in-migrants. At the same time the increasing demand for fish by a growing urban population underpins the continuing growth of the fish-smoking industry.

The disaggregation of different wood-use practices should allow informed management policy for the park. At the time of the study management targets and penalises domestic fuelwood collectors. While seeking to reduce demand and provide alternative fuelwood sources, law enforcement and forestry extension should be reoriented to address the extraction of fuelwood for fish-smoking.

Source: Abbot and Homewood (1999).

important in these areas (e.g. Child 1970; Butyinski and Von Richter 1974; de Merode *et al.*, forthcoming a and b; see Box 5.3). Under recent and contemporary circumstances, however, it has been rather difficult to ascertain the continuing significance of animal wildlife products for local livelihoods. This is because local people's hunting of wildlife was largely criminalised under colonial regimes throughout East and southern Africa through the institution of game laws and the establishment of protected areas from which local people were excluded.

Despite the criminalisation of wildlife hunting, studies of bushmeat consumption indicate that bushmeat can constitute an extremely high-value commodity, underpinning local livelihoods but also moving through far-reaching commodity chains to be sold to urban elites, sometimes across international borders (see Box 5.3). The monetary value of bushmeat tends to vary according to species (e.g. generally small-bodied species are of low value and vice versa), cultural preferences and whether the meat has been procured legally or not. Put simply, there are two extremes of illegal hunting practice (commonly known as poaching). First, there is hunting carried out by local people for their own consumption or local trade and exchange, using low-tech methods (e.g. snares) which target smaller-bodied species, and producing relatively low returns with low impacts on the animal resource (e.g. Butyinski and Von Richter 1974). Second, is hunting for commercial profit by highly organised operators, frequently outsiders using high-tech weapons (e.g. automatic rifles), who pursue large-bodied species with high commercial and conservation values (e.g. elephant), and transport the products of their hunt to locations far from the source of species. One of the main challenges for conservation practice in the wildlife-rich areas of Africa is to find a way of policing and preventing the latter, while accommodating the legitimate needs of the former (e.g. Child 1970, 1984; Chabwela 1990).

Wildlife-based enterprises have also remained attractive to European settlers in East and southern Africa for both economic and cultural reasons. Economic, because in circumstances where settlers have inalienable rights to large tracts of land (i.e. freehold tenure) and to many of the wildlife resources on that land, the possibilities exist for profit-making through game cropping, trophy hunting and/or non-consumptive tourism based on game viewing and the provision of accommodation in game lodges. Cultural, because constructions of an expatriate and usually masculine identity linked economically and psychologically to hunting tend to be inextricably bound up with the spectacular wildlife of large mammals for which East and southern Africa are famed (e.g. MacKenzie 1987; Ellis 1994; Carruthers 1995; Skidmore-Hess 1999; Wels 1999).

[4.] Although note that pastoralism – the herding of domestic livestock – has been significant in Africa for millennia. At the onset of colonial rule towards the end of the nineteenth century, livestock herds were generally much smaller than they had been previously, and the tsetse-infested areas larger. This was due to the rinderpest epidemic of the 1890s (among other factors) which decimated livestock herds by 90 per cent in many areas, and also reduced wildlife populations substantially. In other words, the limited extent of livestock herding and the extended areas of tsetse observed during the twentieth century until recently may be partly an artefact of this disease event (e.g. Bell 1987).

Box 5.3 Wild resource use among the Zande in Kiliwa, Democratic Republic of Congo

In 1996, a 24-hour dietary recall survey of 128 households giving 1,245 'diet-days' over a 5-month period showed that gathered plants, fish and wild meat or 'bushmeat' made up 15–20 per cent of the total market value of Zande household diets in Kiliwa. Analysis of wild food use with respect to wealth indicated the following patterns:

- while plant foods are important in the diet of poor households, these families eat little bushmeat;
- wealthy families consume more wild meat as well as most of the fish recorded during the survey;
- the bushmeat eaten by the rich generally is purchased or received as gifts from poorer households, although in some cases it is derived from recreational hunting.

In the collapsing DRC economy, therefore, wild meat seems to represent a source of income for the poor and a source of prestige protein for the wealthy.

Bushmeat is a high-value commodity, extracted by both local farmers and hunters and by commercial hunters who supply a strong urban and even international demand. The commodity chain of which it is a part is regulated by various networks of power and control, i.e. through: local chiefs; local, regional and national government officials; the military; and, importantly, wealthy patrons who bankroll hunters' firearms and ammunition, as well as giving middlemen and women traders a degree of protection and/or 'official' licence to trade.

Some wild meat consists of legally hunted prey (e.g. cane-rats or duikers) apparently harvested sustainably from *domaines de chasse* (i.e. buffer zones surrounding protected areas where hunting of non-restricted species is allowed), or snared on farmers' plots. Large quantities of commercially marketed meat, however, come from officially protected species (e.g. elephant) which are harvested illegally from protected areas (such as Garamba National Park in the north-east of the country, see Figure 5.2). Severe penalties, including shoot-to-kill conservation policies, have little impact because people's livelihoods are so precarious and income-earning possibilities so few. Enforcement is hampered by the lack of a legal framework of any relevance to rural resource users. Given both the prestige attached to various wild foods, and the role that these can play in supporting the livelihoods of the poor, this study suggests that the consumption of wild food and particularly of bushmeat is unlikely to decline in the DRC, whether future economic change is positive or negative, both nationally and for local people.

Source: de Merode (1998).

Trophy hunting and wildlife viewing by tourists have become major sources of foreign exchange for governments and entrepreneurs in eastern and southern Africa (Cumming 1990; Hulme and Murphree 1999). Although demand fluctuates because of political, security and health scares, the market for wildlife viewing and hunting is surprisingly lucrative (e.g. Leader-Williams *et al.* 1996). Returns to local communities, however, tend to remain rather low, as discussed below. Wildlife cropping, ranching and domestication have been less successful enterprises (Eltringham 1984). Commercial cropping, which involves culling wild animals from 'natural' environments, appears attractive because it seems as though one can make money from an essentially 'free' good. In fact, because there is frequently little infrastructure in areas suitable for wildlife (e.g. roads, water provision, storage facilities, abattoirs), there are serious logistical problems in producing meat of commercially acceptable quality. The use of fixed facilities is also problematic due to the fact that wildlife tends to be mobile and seasonally migratory. Experience shows such commercial enterprises are unprofitable unless: they take trophy animals; operate in national parks or on ranches where access is easy; hygiene restrictions are waived; and/or hunting is subsidised. If the motivation driving commercial hunting is to provide local people with more animal protein, then arguably it makes more economic sense to allow them to hunt for themselves (e.g. Eltringham 1984; also Box 5.3). Game ranching and the domestication of wild animals (e.g. Carles *et al.* 1981) involve substantial inputs for fencing, supplying forage, water, mineral licks, veterinary care and for removing predators. Recent animal health problems have also meant products face major restrictions on export to European Union and other countries. In other words, the associated costs mean that often there are few economic advantages to keeping wildlife as opposed to livestock.

Invertebrates

Invertebrates are usually little mentioned when the consumption of animals is discussed in the literature. This reflects not only cultural preferences but, frequently, also distaste on the part of donors (who often dictate the direction of research) (Speight *et al.* 1999: iii) and other 'outsiders'. Insects, however, often comprise an important nutritional contribution to local diets in terms of energy, minerals and vitamins. Caterpillars, for example, primarily the larval stages of various emperor moth species (Saturniidae), are 'the most commonly utilised food insects in Southern Africa' (Marais 1996: 1–2; also Grivetti 1979; Silbauer 1981: 217; Mkanda and Munthali 1994; Sullivan, forthcoming 2003). Nutritionally they are extremely valuable: for example, 100 g of dried mopane worm (*Imbrasia belina* Lepidoptera: Saturniidae) provides 76 per cent of an average person's daily protein requirement and 100 per cent of the daily requirements for many vitamins and minerals (Speight *et al.* 1999: 21). As such significant food items they form the basis for a lucrative informal trade.[5] The *products* of insects also may be significant, as with the harvesting of honey from wild

---

[5.] Analysis of data in Marais (1996: 8) suggests a mean return of US$4.19/kg for primarily *Imbrasia belina* (Lepidoptera: Saturniidae) caterpillars traded in Windhoek, Namibia, in May 1996.

and/or set hives of the honey-bee (*Apis mellifera*) (Brokensha *et al.* 1972; Grivetti 1979; Ntenga and Mugongo 1991; Cunningham 1996; Kiwasila and Homewood 1998; Sullivan 1999b). To give some idea of the potential scale of this resource, it has been estimated that in 1997 there were 52 hives per hectare in Ethiopia, 43 in Kenya and 28 in Tanzania (Speight *et al.* 1999: 15). Income derived from honey is also significant in Malawi for both commercial and subsistence farmers (Mkanda and Munthali 1994).

Fish

As suggested by the livelihoods explored in Boxes 5.2 and 5.3, fish are also important for both direct consumption and income generation. For example, although landlocked, about 6 per cent of Zambia's surface area is under water with more than 150 species of fish supporting both commercial and subsistence fisheries (IUCN 1987a). Frequently, fisheries interact with uses of other natural resources, as illustrated for Malawi in Box 5.2 where fuelwood is important for fish smoking. In some places (e.g. Zambia and Malawi) the introduction of alien fish species to increase productivity and support trophy fishing unfortunately had serious adverse effects on native fish populations (Stuart *et al.* 1990: 134, 232). Coastal marine fish resources similarly are known to have constituted significant food resources for millennia. Coastal archaeological sites focused around shellfish middens suggest that shellfish was a staple food, at least for particular periods (e.g. Kinahan 1991). These resources remain important today (e.g. Hockey *et al.* 1988).

Tenure and access rights

Tenurial arrangements in the region include both indigenous and intro-duced forms. Indigenous, or customary, tenure and resource access rights in eastern and southern Africa encompass a wide range of rights. These include strict controls approximating private and individual ownership; collectively managed resources known as 'common property'; 'open access' resources with little or no control over access or use; and state-owned property where individuals are liable to prosecution should they trans-gress legal guidelines determining access and use. These categories are not static, either geographically or historically, but they comprise a useful typology for discussion of the different types of tenure influencing the way in which African natural resources are used (Toulmin and Quan 2000).

Private ownership

In general, the higher and more predictable the productive potential of land and/or specific natural resources, the more likely they are to come under strict control over access and use. These sorts of resources and access rights would include:

- Land that is vested in, maintained and inherited by individual families, households or lineages, usually delineated as fields under cultivation or fallow. This type of tenure is characteristic of farmland under forms of rainfed arable cultivation in much of southern and East Africa. Examples are delineated fields cultivated by farmers in the mountainous areas of Meru and Arusha in Tanzania (Spear 1997) and delineated and inherited fields on Swazi Nation Land in Swaziland (Dlamini 1989). In these contexts, ultimate customary ownership of land and resources rests with the collective group associated with an area, under the control of the traditional leadership: thus the individual farmer or household is not in a position to alienate land from the group in the long term, even where land may be inherited over more than one generation (e.g. Dlamini 1989).

- Individual ownership of trees with both commercial and subsistence values. Examples include gum arabic acacias (*Acacia senegal*) in northeast Africa or fruit trees left standing on land otherwise cleared for cultivation (Wilson 1989).
- Personally constructed wells for the provision of water (Lewis 1961).
- Beehives: these may be constructed and set by harvesters – for example, Pare farmers living around Mkomazi Game Reserve in northern Tanzania (Kiwasila and Homewood 1998) – or may occur in the 'wild' but be considered 'owned' by individual harvesters, generally men (Sullivan 1999b).
- Other 'patches' of valued resources. In north-west Namibia, for example, harvester ant nests, from which large quantities of grass seeds can be collected, may be thought of as the property of individual harvesters, usually women (Sullivan 1999b).

The idea of resources as completely private property is generally associated with the imposition of European concepts of ownership and tenure as recognised under Roman or formal law. This was instituted under colonial, or white settler, rule in most countries. This frequently led to the expropriation of large areas of high potential land for private ownership, particularly where settler populations were significant, as in South Africa, Namibia, Zimbabwe and Kenya. Such expropriation was based on denial of the proprietary character of 'common land' under African management and ownership (Okoth-Ogendo 2000). In the remaining areas where residence and use by Africans were permitted, usually the land was, and largely remains, legally owned by the state but is utilised and allocated under communal (i.e. locally administered) forms of tenure. In higher rainfall regions, for example KwaZulu-Natal in South Africa, many of these areas are under smallholder arable farming with farm plots being passed down through the family. Under these circumstances, resources occurring outside of farm plots (such as grazing land and other plant and animal resources) tend to be used and managed as common property resources (see below). Following independence and transition to African rule, some of the land expropriated by Europeans has reverted through land reform to land-poor Africans (for example, in Zimbabwe and Namibia). More commonly, however, private ownership of such estates has been either retained by settler farmers or passed to a new generation of elite, African landowners (e.g. Galaty 1999a; see Box 5.4).

Box 5.4   Land tenure and subdivision on Maasai group ranches, Kenya

Lemek group ranch near the Maasai Mara (see Figure 5.2) in Kenya (745 km²) was established in 1969. The group ranch chairman and land adjudication committee allocated land to educated or influential Maasai in a belt along the western portion of the group ranch boundary bordering the Mara River. These allocations were cemented under private ownership with the issuing of title deeds, the process being facilitated by the local administrative chief and land registry staff. Ostensibly to guard against the continued westward movement of non-Maasai cultivating groups onto Maasai lands, beneficiaries included Maasai administration chiefs, MPs, councillors, county council officials and a police inspector. Ironically, many of these new landowners rapidly sold land on a piecemeal basis to the same in-migrant cultivating groups apparently causing concern to Maasai pastoralists.

On the northern portions of Lemek, outside entrepreneurs, since 1984, have been approaching the administration chief and group ranch chairman to culti- vate wheat on leases of upwards from 2,000 to 4,000 acres per contractor. In addition to arranging these leases for their own benefit, the administration chiefs and chairmen have been giving responsibility to other group ranch com- mittee members, councillors and associates to arrange leases with contractors. On subdivided land on Lemek, each registered member was supposed to be entitled to receive 100 acres of land (in fertile places) or 128 acres on steeply sloping or marshy areas. The process of registering involves all circumcised men deemed to have been resident on the group ranch by the land adjudica- tion committee prior to the closing of the register in 1993. According to the Narok County Council there were 1,021 registered members on Lemek. Initial attempts by local elites to allocate larger shares to themselves were thwarted in 1995 when, under the supervision of the district commissioner, a revised survey was undertaken to ensure plots were of equal size.

Despite this, locally influential people (with access to the register and map providing the location of the plots) have still been able to exercise control for personal benefit of the land subdivision process. Examples include:

- Those previously involved in leasing land for wheat cultivation using the considerable sums generated to buy the permanent/modern houses con- structed by contractors. Once owners of the permanent housing, their stake to the land on which the house is located is secure, thus ensuring a position in the lucrative wheat-leasing belt.
- Those involved in leasing out the land for wheat farming use the money accrued to buy out poorer neighbours' shares in land. Once agreement has been reached (usually a handwritten confirmation signed or marked with a fingerprint) the position of the selling party's land is changed to ensure it is located on the wheat belt.
- Influential people registering their younger (uncircumcised) sons and ensuring that the shares are located adjacent to each other in the wheat belt. In this way, farms of up to 1,000 acres in extent are established.

All of these facilitate the further consolidation of land in the hands of the wealthy, while excluding poorer land-users whom the subdivision process is ostensibly intended to benefit.

Source: Thompson and Homewood (2002).

Common property resources (CPRs)

Indigenous or customary land and resource tenure throughout Africa often involves various forms of collective ownership. In these circumstances a clearly defined group collectively own, manage and have access to a specific resource, with the group establishing and enforcing rules of access and use. In CPR systems, a leader, chief or elite group such as a committee of elders, commonly acts as custodian(s) of the land or other CPRs, in the sense of presiding over allocation, regulating access and resolving disputes that may arise. Sometimes these allocative powers are exercised, at least in theory, by elected 'committees' of some sort (e.g. in Zimbabwe although here the powers of the chiefs were strengthened again in the 1990s). Local hotspots of productive potential (for example, wetlands in drylands, or highland drought refuges), while being common property, may fall under the control of a dominant group with the power to exclude others or exact payment for use of the resource.

As mentioned above, areas of uncultivated land between villages or fields are often held and used on a common property basis. For example, seasonally waterlogged grasslands, (known as *vleis* in some parts of southern Africa and as *dambos* in, for example, Malawi and parts of Zambia) are areas that sustain grasses collected for thatching, and which often play an important role as dry season grazing resources (Scoones 1991). In contexts where land is relatively abundant, fields and plots also may be allocated as a CPR such that plots are designated to be worked by particular individuals or households for one or more farming seasons, or until the household head has died, after which it reverts to the pool of common land for reallocation (Birley 1982).

As a general rule, the more arid and infertile the land, and the more seasonally and annually variable its productivity and ensuing use, the more likely it is that the area and its resources will be under communal control rather than individual tenure. This makes common property regimes typical of the tenure system of indigenous pastoralists in East and southern Africa, for whom movement with livestock herds is essential in order to access forage and other resources. Common components of CPR management by pastoralists include:

- Management of a dry season grazing area, often with a committee of elders who decide when and where to reserve, or allow access to, dry season grazing. This system has been well documented for the Tanzanian Maasai (e.g. see Potkanski 1994; Brockington and Homewood 1998).
- Sophisticated collaborative management, of both the timing of herd access and the coordination of labour, to enable group access to shared water sources. A good example can be found among the Borana pastoralists of southern Ethiopia (see Cossins and Upton 1987).
- Negotiation of group access to other 'key resources' – local 'hotspots' of productive potential. For example, access to and inheritance of riverine tree resources for dry season forage is managed by Turkana pastoralists in north Kenya (Barrow 1988, 1990, 1992).

The mobility practices of both pastoralist and hunter-gatherer peoples, and the consequent apparent 'vacancy' of 'their' land when they are elsewhere, leaves them open to land dispossession due to pressures from elsewhere. This was a common mode of land loss to European settlers during the colonial period. A more recent case occurred when Barabaig pastoralists in Tanzania were displaced from communally held pastures and ancestral grave sites to accommodate a vast wheat-growing programme sponsored by Canada. The assumption on the part of donors was that the land had been standing 'idle' (Lane and Pretty 1990: 7). For so-called 'hunter-gatherers', and despite conventional stereotypes of their relentless mobility and their inability to recognise land and natural resources as belonging to any individual or group, a number of anthropological studies indicate complex conceptualisations of access and tenure rights. Like cultivators and pastoralists, these are mediated via kin relations and rules guiding inheritance (see Box 5.5). Relating this point back to the issue of how their use of natural resources in their livelihoods has been circumscribed by encroachment on their lands, a priority for 'hunter-gatherer' livelihood and security today is the development of very clear, and legally enforceable, rights over their land and resources. Often this means the reinstatement of theoretically existing rights conveniently ignored by other, encroaching groups. Other policies relating to natural resources involving indigenous knowledge or sustainable offtakes, while evidently fashionable, are actually less important. In any case, they are unlikely to be effective if hunter-gatherer rights are generally being sidelined.

A second and related source of pressure for those requiring access to spatially and temporally dispersed resources is the imposition of private forms of land tenure, usually accompanied by the delineation of land areas using fencing. In extremely broad terms, this may occur in two contexts. First, due to formal land tenure reform at the level of national policy, based on assumptions guiding farming practices for commercial markets (e.g. Birley 1982; Rohde *et al.* 2001). In this case, land enclosure is associated with capital-intensive, commercialised production for export markets and, as such, has usually been associated with European settler farmers producing items for single product markets. In the livestock sector, for example, the production of meat for export markets becomes a primary objective and is based on low stocking rates per unit area of land and the regular harvest of a surplus 'crop' of young cattle for meat. An assumption here is that inalienable title to land will increase investment in agriculture and thereby increase commercial productivity, although this is not necessarily what ensues (cf. Haugerud 1989; Platteau 2000). Second, land enclosure may occur as a result of the fencing off and *de facto* 'privatisation' of land *in situ* by wealthy herders (Graham 1988; Behnke 1988; Hitchcock 1990; Prior 1994). Importantly, as capitalist relations of production and the demands of a global 'free' market increasingly penetrate African farming sectors, a land-privatising trajectory becomes ever more likely, even in contexts where land redistribution to poorer farmers on communal land is a stated objective (as, for example, in the post-apartheid contexts of Zimbabwe, South Africa and Namibia).

Box 5.5 Traditional concepts of landownership among Ju'/hoansi 'Bushmen'

Although conventionally thought to have little concept of land tenure or resource ownership, a consideration which has undermined their formal claims to land throughout southern and East Africa, 'hunter-gatherer' populations conceptualise land and natural resources in terms of socially defined access rights determined through kin relatedness and inheritance. Here we review categories of land among the Ju'/hoansi, speakers of a central !Kung language who inhabit the Nyae Nyae area of western Botswana and eastern Namibia (see Figure 5.2). Ju'/hoansi recognise two types of communal land; the broad category of *gxa/kxo* and the named places of *n!oresi*. These are discussed separately below.

1. Gxa/kxo

This term translates literally as 'face of the earth' and refers to all the land and its resources in Nyae Nyae, to which all Ju'/hoansi have use and habitation rights as individual members by descent. The *gxa/kxo* thus is not the property of any corporate body within the Ju'/hoansi. The rights of individuals within the *gxa/kho* include the following:

- the right to use major plant-food resources such as the *tsi* or *morama* bean (*Tylosema esculentum*) and *g/kaa* or *mangetti* nuts (*Schinziophyton rautanenii*, formerly *Ricinodendron rautanenii*);
- the right to hunt and track animal wildlife, such that a hunted animal belongs to the hunter who strikes it, and not to the owners of the recognised territory or *n!ore* (see below) in which it was hit or in which it dies from the effects of arrow poison;
- the freedom to travel;
- the right to live at a permanent source of water during drought periods.

2. *N!oresi*

The *n!oresi* are named territories without fixed boundaries, usually with important focal resources such as permanent or semi-permanent waterholes and concentrations of valued plant-food species. Individual rights to residence within a *n!ore*, and to use its resources, are inherited directly from both parents and ownership of a *n!ore* is only recognised if this traceable descent can be demonstrated. As such, 'ownership' of a *n!ore* is exclusive to a group related through kin alliances who manage its resources communally. 'Ownership' cannot be conferred on outsiders, even though they may reside within a *n!ore* for a prolonged period of time with permission of its recognised owners. An individual chooses in adulthood which of their parents' *n!ore* they wish to claim as their own and, through marriage to someone from outside that *n!ore*, gain rights of access and resource use to a second *n!ore*. In this sense, kinship networks underpin in a fundamental way an individual's rights to land and resources.

Sources: Ritchie (1987); Botelle and Rohde (1995).

Privatisation of previously communal lands is thus occurring incrementally in many countries of the region. Even when this is through legal, official channels it is often highly controversial, because cash-strapped African governments are ceding huge tracts of land to wealthy individuals or corporations, sometimes expatriate, without much care being taken to ensure that the land is unoccupied or unused. There is great potential for such land transfers to involve corruption in high places. Any people who thereby are displaced are rarely consulted, let alone compensated properly, which is manifestly unjust. In Tanzania, for example, acquisition of private title to formerly communal lands has proceeded far from the location of the land itself in a way that is neither transparent nor equitable: little or no warning has been given to inhabitants that they can be made, and indeed are becoming, landless and effectively squatters on 'their own' land (Igoe and Brockington 1999). Privatisation of communal land *can* enhance security and political representation for some, as illustrated in Box 5.4 by the case of landed Maasai in Kenya. The other side of the coin is that those already vulnerable tend to be impoverished by this process. In many cases this has a gender component in that land title is usually vested in male family heads (e.g. Talle 1988; Hodgson 1999, 2000). Ethnic and class dimensions to land acquisition processes also complicate matters, sometimes conferring a tacitly violent edge to dealings regarding land. For example, the acquisition of legal title to land in Maasai areas by Kikuyu and Kipsigis people has frequently been accompanied by violence (Galaty 1999a; Thompson and Homewood 2002; Homewood *et al.* 2003).

Further, a number of studies indicate that when privatisation of land occurs in this region, the actual use and management of land and resources may still allow flexible and reciprocal access to geographically dispersed land areas with, for example, the movement of livestock between ranches which are long distances apart. In other words, a compromise between new systems of private tenure and ecologically suitable CPR management practices tends to emerge in ways which may be unanticipated by policymakers. Movement remains essential during drought periods (Niamir-Fuller 1999) and continues in Kenya, even where pastoralists are settled on delineated group ranches (e.g. Grandin and Lembuya 1987). Even settler livestock farmers in Namibia and South Africa, who have exclusive use of huge ranches under freehold tenure, are documented as needing to move their herds across ranch boundaries, and sometimes over large distances, in order to maintain herd numbers in the face of variable forage productivity (Sullivan 1996a). In South Africa, from about 1910 to the 1930s, the government even subsidised transhumance by rail in response to pressure from white livestock farmers (Beinart 2003, forthcoming).[6] The evidence

[6.] There was, however, considerable pressure from government agricultural officials for such practices to end. The favoured policy was for farmers to adopt rotational grazing on their farms, to avoid transhumance which, *inter alia*, was thought to spread livestock diseases. Some farmers managed in the short term to reduce the need for transhumance by, for example, investing in fodder. One plant used was the spineless cactus but when this was destroyed by cochineal in 1946, recourse had again to be made to transhumance (see Beinart 2003, forthcoming).

thus suggests that where access to extensively distributed resources is important, as is the case for dryland environments in southern and East Africa, it might be inappropriate to assume that individualised land tenure holdings are essential for increased economic productivity (Behnke and Scoones 1993; Homewood 1995; Sullivan 1996b; Niamir-Fuller 1999; Platteau 2000; Sullivan and Rohde 2002). At the same time, herders who are unable to qualify for, or otherwise maintain access to, privatised pastures and the other natural resources occurring on these lands, tend to experience disproportionately adverse effects due to privatisation and the application of monetarist macro-economic policy. The increase in wealth differentials between rich and poor is a common outcome of such agrarian reforms, in some cases inviting protest over land policy and other changes (Graham 1988: 7; Rohde *et al.* 2001).

There has been much misunderstanding over the implications of common property tenure regimes for natural resource management. Its detractors often confuse the system with open access tenure (see below) and erroneously assume that environmental degradation is an inevitable by-product. Its merits, however, can also be over-romanticised, for example by overestimating the egalitarian and environmentally sound characteristics of the system. They work best where small, long-established user groups cooperate closely and are tied into reciprocal arrangements so that trust is optimised.

CPR systems take time to evolve and are not easily established *de novo*, although current development interventions are often predicated on this. Their prevalence and effectiveness have recently been elevated in development and resource management discourse. Many current 'integrated conservation and development projects' (ICDPs) and 'community-based' development and conservation initiatives (discussed in detail in a later section) are based on strengthening and/or creating new common property management institutions and practices. Ironically, in some cases today, environment and development conceptualisations that favour CPRs are driving donor-funded policies at the same time as customary tenure practices are being dismantled wholesale by the state. This, for example, is the case in Tanzania where the 1994 Land Act attempts to extinguish customary tenure based on common property.

Open access

Land labelled 'open access' is that for which neither access nor use comes under any form of management or governing body. Examples include 'frontier' or 'no man's land' zones, where access is continually contested and control has ultimately depended on *force majeure* (Lewis 1961; Kurimoto and Simonse 1998; Fukui and Markakis 1994), and areas where the costs of developing and maintaining a system of territorial control outweigh the limited benefits of low or sporadic production. During the twentieth century, such areas frequently expanded due to the tumultuous effects of colonialism, apartheid and contemporary conflict.

In a number of African states that aspired to some form of socialism in the initial post-colonial era (e.g. Tanzania and Mozambique), all land

reverted to the state at independence or, as in Namibia, formerly communal land was thus transferred. This resulted, to some extent, in a *de facto*
open access situation replacing a formerly CPR system (Moris 1981; Bromley
and Cernea 1989; Homewood 1995). In these contexts, local rural communities often nevertheless maintain their CPR management systems, even
if the allocative powers now rest with new, and perhaps political party
based, local institutions. However, given the limited resources and power
of many emerging African states to regulate or enforce access to land and
natural resources, these situations can be ripe for well-placed individuals
to take advantage of the tenure vacuum and profit at the expense of
a wider group of customary users (Galaty 1999a; also see Box 5.4). In
Namibia, for example, the post-independence constitution allows all
citizens to move to wherever they choose on communal (i.e. state) land,
with the *proviso* that the customary rights of others be observed. Problems
have arisen because no procedures or resources exist to monitor this process, with the result that some groups have been marginalised in the face
of incoming, and frequently wealthy, herders (Botelle and Rohde 1995;
Sullivan, 2002c).

Early analyses of African land-use tended to represent what were
complex common property regimes as situations of 'open access' – with
resources used on an ad hoc and 'free-for-all' basis until 'degradation'
occurred and people were forced to move or turn to alternative resources.
The most famous exposition of this scenario is Hardin's (1968) 'Tragedy
of the Commons'. In relation to African pastoralism, this model alleges
that environmental degradation is inevitable since pastoralists 'free-ride'
– benefiting from the profits of individual herd accumulation while
bearing none of the costs of communal range use and possible degradation. Although still often invoked, this is deeply misleading (Platteau 2000).
As explained in the preceding section, CPRs are *communally* managed. Individual profit-maximising behaviour, the conceptual basis of much economic theorising under capitalist modes of production, is thus constrained.

State land

Apart from land used communally but under state ownership, much of
the rest of state-owned land in southern and East Africa is set aside for the
'conservation estate', as national parks, forest and game reserves, etc. In
these areas access and resource use by local people are either prohibited,
or there are strict limits on the nature of access and use. These restrictions
are often deeply resented by local communities, especially if they were
summarily evicted to make way for the conservation areas, as has all too
frequently occurred in the region. Restricted access to state land may also
occur in order to protect significant economic resources, as, for example,
with the so-called diamond areas of Namibia.

***

The above categories of resource tenure by no means constitute a strict
or static typology. Any one geographical area may encompass multiple

and changing types of tenure, making up a diverse mosaic of different, site-specific ways of managing natural resources. On top of this, the spatial and social distribution of land may change through time, together with the resources (labour, legal and enforcement) available to maintain territorial control.

In summary, most African states face conflicts between customary tenure regimes and imposed national (formal) legislation defining both state and privately owned land (individual or corporate), at both the level of national legal frameworks and at individual sites (Shivji 1998; Toulmin and Quan 2000). This is made worse where successive administrations have brought in conflicting systems, and where alternative sociopolitical hierarchies exist through which disputes are contested (e.g. 'traditional' leaders versus formal district and regional government). Land tenure and natural resource access in many parts of Africa thus comprise what Mortimore (1998) has called a palimpsest of systems, evolved by accretion and displacement with each new wave of migration or conquest or change of policy. Each new event has left a new layer in the hierarchy of tenure relations.

Indigenous knowledge, resources and trade-related intellectual property rights (TRIPs)

In the last two decades of the twentieth century there was much wider recognition of the depth and importance of the knowledge held by many rural Africans of their local environment (e.g. Brokensha and Riley 1986; Riley and Brokensha 1988; Juma 1989). However, it is also noted that such knowledge has frequently been eroded through the alienation of people from their land; through the institution of internationally led economic policies which contribute to the replacement of indigenous agricultural practice; through greater access to alternatives; and through the devastating effects of conflict. Nevertheless local environmental knowledge remains a highly significant resource, and its resilience and dynamism are often greater than anticipated (Sullivan 1999b, 2000; Redhead 1985; Maundu *et al.* 2001). Working with local people to develop resource management not only fits with the contemporary emphasis on participation in development but also, by tapping into this knowledge, may avoid much wasted or misdirected effort.

This issue also relates to the accelerating global search for potentially commercial 'natural products' – particularly of pharmaceuticals, botanical medicines and cropseeds (e.g. Moss 1988). This has engendered increasing concern regarding the protection of Africa's biological and/or genetic resources, of the local knowledge surrounding their use, and of the economic status of the source community (Ten Kate and Laird 1999). A particular danger is that local knowledge and practice regarding biodiversity are exploited in the development of indigenous genetic resources (including synthesis and patenting of isolated components), without recompense to people whose pre-existing biological knowledge is rendered invisible by being communally held and part of an oral, unpublished tradition. In response to these concerns, countries in southern Africa are reviewing and creating new legislation to guide 'biotrade'. Such legislation is also in line with the international framework of the 1992 Convention on Biological Diversity. This convention's signatories assert a commitment to protect biological diversity and to institute benefit-sharing (e.g. payments, training, royalties, technology transfer). Namibia's proposed biotrade legislation, for example, tries to ensure that 'bioprospecting' and the commercial exploitation of indigenous species are accompanied by protection of local-user and intellectual property rights, as well as the contractual return of economic benefits (see Craven and Sullivan 2002).

Conservation of natural resources: from criminals to community

Fortress conservation: the separation of people from 'nature'

Since the turn of the century, conservation in Africa has been dominated by ideals transported and imposed from a recently industrialised Europe, which saw the continent's widlife-rich savannas as a seemingly recovered Eden, and emphasised the preservation of 'wilderness' landscapes containing animals and not people (e.g. Abel and Blaikie 1986; Anderson and Grove 1987). Following the model of Yellowstone National Park established in North America in the 1800s, this approach led to the delineation of protected areas from which local people were excluded.[7] Ironically, national parks were usually conceived of, and run by, 'penitent butchers': European hunters who had decimated the continent's wildlife through hunting for trophies to provide an overseas market with items procured from large animals (e.g. elephant ivory, rhino horn and ostrich feathers) (MacKenzie 1987; Carruthers 1995). Parks were set aside for the pursuit of a wilderness aesthetic by European elites from the aristocracy, the colonial administration and growing numbers of natural historians. African uses of natural resources and the links between wildlife and human welfare were invariably severely compromised.

The perceived incompatibility between wilderness and human occupation necessitated the construction of various 'supporting narratives', or justifications, for the conservation policies that were detrimental to indigenous people (Galaty 1999b). Thus local consumptive use of wildlife was portrayed as ignoble and non-sporting, and as unethically and unsustainably destructive, justifying the further prohibition of activities such as through-passage, resource-gathering and livestock grazing in protected areas. At the same time, even utilising wildlife *outside* protected areas was frequently prohibited through legislation (e.g. game laws), as was the use of vegetation resources which were protected in forest

---

[7] It is pertinent to note that it was the genocide of indigenous populations that accompanied European settlement of North America (e.g. Brown 1970) which allowed Yellowstone to be established on the principle of excluding people.

reserves and under Forestry Acts (e.g. Juma 1989). In other words, much that was formerly part of 'normal' subsistence in rural Africa was reconstructed as illegal and criminal (Marks 1984; Bell 1987). People were (and still are) seen as potential if not actual poachers, and the parks/people relationship was (and is) based to a large extent on policing and law enforcement. The whole process of resource conservation and protection, therefore, acted to impoverish African populations who were already suffering the combined effects of the slave trade, imported disease, land alienation, taxation and requirements by the colonial state for labour to support new industry and settler agriculture, or to produce cash crops.

Typically eviction and exclusion have taken place with little or no compensation (Brockington 2001). In some cases, compensation was negotiated for the loss of livestock, personal injury and the trampling of crops by wildlife, and some 'problem animals' were shot following attacks on livestock (Ansell 1989; Galaty 1999b). Africans have not been passive recipients of such exclusionary environmental policies, however. In many instances they have actively resisted and protested against restrictions in circumstances where the balance of power has been stacked against them (e.g. Gordon and Sholto Douglas 2000). The exclusion of people from resources and decision-making in protected areas, however, has meant the removal of very significant areas from African use and habitation in some countries (see Table 5.3). Over 40 per cent of Tanzania, for example, comprises some form of conservation estate, and 27 per cent of its total

Table 5.3 Protected areas in selected countries of southern and eastern Africa (% land area)

Country	Protected areas (national parks and game reserves) (%)	Other
Angola	6	+ large controlled hunting area
Botswana	17	18% (controlled hunting areas – licences issued for subsistence and recreational hunting)
Malawi	11	10% (forest reserves and protected hill slopes)
Mozambique	13	+ hunting reserves and fauna utilisation zones
Namibia	12	+ 8% private game farms
Tanzania	27 (incl. forest reserves)	+ other game reserves
Zambia	32 = national parks and game management areas	9.8% = protected forest
Zimbabwe	12.5	2.3% = state forests; also CAMPFIRE project areas on communal land and wildlife conservancies on private land

Note: The IUCN recommends that some 10% of a country's land surface area is set aside for conservation purposes (Musters *et al.* (2000)).
Source: After CDC (1984: 15, 18); du Plessis (1992: 132); Moyo *et al.* (1993); Wildlife Sector Review Task Force (1995); Nhira *et al.* (1998).

land surface area allows no habitation or use by local people (Wildlife Sector Review Task Force 1995). In some parts of the region, the policing and enforcement of fortress conservation in national parks and game reserves have often meant drastic measures (including 'shoot-to-kill' policies in Zimbabwe, Zambia and the DRC). These tend to be portrayed as being directed against armed and organised international poaching gangs, for whom there is, perhaps understandably, little sympathy. In reality, however, these measures may also be used against historically marginalised people (Peluso 1993). A very different set of conservation issues is faced in other countries, where conflict situations and an effective breakdown in civil society mean that protected areas are poorly staffed and little policed. Moyo *et al.* (1993: 23) observe for early 1990s Angola, for example, that most conservation areas have been 'completely abandoned by the government, with absolutely no control being exercised over the hunting of animals, the burning of forests, and human settlements in prohibited areas'. The destabilisation era of the late 1970s and early 1980s, when white minority regimes struggled to maintain their supremacy in southern Africa, also saw dramatic destruction of wildlife, encouraged or orchestrated by the South African Defence Force (SADF) in order to help finance this process (and line private pockets; see Ellis 1994).

These colonial paradigms of separation between wildlife and people, and of the erosion of indigenous rights to natural resources, together with their associated narratives, have come under attack from various quarters in recent years. These include:

- development and human rights groups who assert the need for improvement of local livelihoods, and often affirm indigenous peoples as 'natural' conservationists;
- conservation groups alarmed by the costs of enforcement and the implications of structural adjustment which, through 'rolling back the state', reduces state resources for policing and funding conservation areas. This has prompted recognition of the pragmatic need to enlist the support of people living in communal areas, particularly those adjacent to conservation areas, to ensure the long-term sustainability of conservation, particularly given the new climate of 'participatory development' guiding funding from donors (e.g. Lindsay 1987; Western 1982, 1984; Lewis *et al.* 1991; Wily and Mbaya 2001);
- resource economists who argue that the 'sustainable use' of resources, including slow reproducers such as elephant and rhino, is necessary for the conservation of their habitats, *provided that significant benefits accrue to local people* (e.g. Barbier *et al.* 1990);[8]

[8.] The late ecologist Graham Caughley (1993) maintained that, for slow-reproducing species such as elephants, money may accumulate interest in the bank more rapidly than resource stocks can reproduce, thereby making the conversion of the resource into cash (e.g. through hunting for ivory or through felling hardwoods) the most economically rational act. He used this to explain the economic incentives operating against sustainable use of such species. However, due to low and even negative interest rates in many southern and East African countries, it may indeed make economic sense to retain animals, whether cattle or wildlife, as cash 'on the hoof'.

- local populations who may primarily experience wildlife on 'their' land as crop pests, a source of danger to life and limb, and as resources which could be utilised as food and a means of income. Given historical circumstances, 'illegal' uses of, or attacks on, wildlife can also be viewed as a form of local resistance against a system of restrictions viewed correctly as illegitimate and discriminatory.

'Community-based conservation': a radically new approach?

Current political and economic realities, and awareness of the rights, needs and aspirations of rural African populations, make it clear that exclusion from natural resources may not be the optimal policy for their conservation, however attractive this option may remain for environmentalists and conservation biologists (e.g. Ansell 1989; Kramer *et al.* 1997; Struhsaker 1998; Robinson and Bennett 2000). The outcome has been the construction of a new conservation paradigm influenced by the World Conservation Strategy (IUCN/UNEP/WWF 1980) that is reflected in the national conservation strategies of many African countries (e.g. CDC 1984; Duffy 2000). This paradigm is based on ideals of 'participation', 'benefit-sharing' and, ultimately, of 'community-based natural resources management' (CBNRM) for both conservation and local income generation. In other words, it has been increasingly acknowledged that a 'win–win solution' for both conservation and rural development will require the support of local communities (e.g. Leader-Williams and Albon 1988; Kiss 1990; Davis *et al.* 1994; IIED 1994; Western and Wright 1994; Brockington and Homewood 1998; Wily and Mbaya 2001).

'Community-based conservation' however, has emerged from a palette of by no means mutually compatible ideas and ideologies. Conservationists argue that, if communities are to benefit, they will need to share the conservation vision. Rousseauists[9] aver that if local indigenous people are in control they will, by definition, manage resources sustainably. Development workers and human rights activists believe that local residents should have control *whether or not* they eventually choose to pursue a conservation outcome. 'Community conservation' (CC) can thus be a catch-all term to cover many different possible arrangements (e.g. Western and Wright 1994), and some authors only use the term to refer to initiatives in which resource ownership and decision-making devolve to local people (e.g. Hartley and Hunter 1997). Table 5.4 considers some of the variations in conceptual approach which have come under the broad label of 'community-based conservation'. Reading from top to bottom, these approaches can be seen to represent a sliding scale of participation in four different factors:

1. The direction of information flow (e.g. one-way and top-down; two-way; bottom-up);

---

[9] Jean-Jacques Rousseau was an eighteenth-century French writer and philosopher who argued that 'man' in his 'natural state' was a creature of noble instincts who lived in harmony with his environment, expressed in the recent populist view of the 'ecologically noble savage'.

Table 5.4 Some typologies of types of participation in 'community conservation'

Typology by publication: Kiss (1990)	Wells et al. (1992)	Pimbert and Pretty (1996)	Kiwasila and Brockington editorial (1996)
• Participation in benefits	• Information-gathering	• Passive participation	• Passive
• Participation in planning and design	• Consultation	• Participation by consultation	• Interactive (benefit sharing)
• Participation in implementation and management	• Decision-making	• Participation for material incentives	• Dynamic (agendas determined by local communities)
	• Initiating action	• Functional participation	
	• Evaluation	• Interactive participation	
		• Self-mobilisation/active participation	

Sources: Compiled from information in Kiss (1990); Wells *et al.* (1992); Pimbert and Pretty (1996); Kiwasila and Brockington (1996).

2. The degree of involvement in setting the agenda (e.g. goals imposed; co-operation achieved through coercion; goals negotiated and co-operation won; goals set by 'community'; local people initiate and are prime-movers);
3. The nature of benefits accruing to local people (e.g. opportunistic handouts; regular proportional revenue-sharing; 'community' ownership of resources with the right to issue leases or offtake licences and set quotas for offtake);
4. The degree of contribution to decision-making (e.g. from nil to total control).

In practice a country may combine many of these approaches in an overlapping mosaic of varying access to different resources, and of multiple interactions with local people to gain support for the conservation endeavour.

Indirect subsidies versus direct payments?

Community-based conservation has been celebrated as a radical departure from the exclusive, centralised and alienating 'fortress' conservation practices of the past (Hulme and Murphree 1999; Jones 1999). Recent critique, however, suggests that the label 'CC' often obscures circumstances and practices which are not qualitatively different from earlier approaches (e.g. Alexander and McGregor 2000; Murombedzi 1999; Gillingham and Lee 1999; Roe *et al.* 2000). Identified problems are detailed in Box 5.6.

In theory, CC initiatives attempt to bring benefits from conservation to local people, and thus make conservation more successful and sustainable. However, an overriding desire to obtain a conservation outcome means that policies may be introduced which are not, on critical inspection, really viable. The following constraints or problems may feature in such policies:

- Setting a conservation goal which is not rooted in local priorities and may conflict with them;
- Establishing a CPR system where none currently exists;
- Compensating for *major* livelihood losses and/or opportunity costs with *minor* benefits;
- Subsidising commercial ventures which may not be more lucrative than existing or alternative activities and which may not be commercially viable in the absence of subsidies;
- Working within, rather than reforming, existing inequalities in land and resource distribution.

CC can be meaningful to the rural poor but only when it genuinely improves their livelihoods. If the costs are greater than the benefits then evidently no outreach or education programme can help them 'own' the enterprise. Clearly, it is preferable that local people benefit from the animal wildlife with which they live instead of remaining alienated from these resources. But beneath the rhetoric, the reality of CC is that, often, it may not be the radically and qualitatively different approach to conservation that is frequently claimed. Escobar (1996) has argued that 'community-based

---

Box 5.6 Community-based conservation: a critique

Material benefits for community

- Reviews of CC suggest that initiatives tend to deliver *negligible* amounts at the household level, despite some specific exceptions.
- Many problems identified of local corruption in use of money generated.
- Costs of running CC often heavily subsidised by conservationist donors – sustainability of benefits to local people is thus questionable.

Continuity with past conservation/preservationist policies

- Run by same people/organisations with little retraining.
- Primarily promotes continued access of private safari/hunting operators to animal wildlife for profit.
- Promotion of armed community game guards increases policing aspect of conservation, and access to small arms in often politically unstable areas.
- Focus on large (and dangerous) mammals and macho identity of related conservationists tends to confine local participation in decisions to men.

Community identity

- As with other community-based development initiatives, problem of heterogeneity within communities and thus conflict over 'desired' resource management.
- Long-term community liaison required to mediate for fair and sustainable participation usually not funded.

Sources: Marindo-Ranganai and Zaba (1994), Bergin (1995), Emerton (1996, 1998), Norton-Griffiths (1996), Simpson and Sedjo (1996), Patel (1998), Gillingham and Lee (1999), Matenga (1999), Taylor (1999), Sullivan (1999b, 2000), Wøien and Lama (1999).

---

conservation' is inseparable from a Northern modernising development discourse which asserts conformity and control through donor-funding to the countries of 'the South'. In the case of conservation in Africa, this means that support is available only to 'communities' to the extent that they agree to construct themselves as 'suitable' custodians of internationally valued biodiversity, particularly animal wildlife (see Box 5.7). While now stressing that local people should benefit from wildlife, a number of

Box 5.7   CAMPFIRE in Zimbabwe: tensions between the philosophy and practice of CBNRM

The Communal Area Management Programme for Indigenous Resources – or CAMPFIRE – in Zimbabwe involves 36 of Zimbabwe's 57 districts and 85 local communities representing 200,000 households. The scale of the programme is thus huge, reflecting both the dynamism of those involved in wildlife conservation – particularly in seeking new means of bringing previously excluded local people 'on board' – as well as the high degree of donor funding available for 'community-based conservation' initiatives in southern African countries in the last couple of decades.

CBNRM initiatives like CAMPFIRE are, in theory, meant to combine wildlife conservation with rural development aims, such that local inhabitants benefit materially from wildlife on their land *and* are empowered regarding decision-making processes. However, a range of critical studies has emerged in recent years which identify flaws in the assumptions guiding this programme, as well as significant problems in implementation 'on the ground'. In Nkayi and Lupane districts of Matabeleland North in Zimbabwe, for example, a CAMPFIRE initiative in the 1990s met with outright resistance from most of the local people, yet the state went to great lengths to try and impose the project, disregarding their reasons.

In the Gwampa valley in southern Nkayi and Lupane (see Figure 5.2), the population was made up of a small number of early settlers, mainly of Nyai origin, and a much greater number of people who had been evicted from 'white' farms. While the former had some tradition of hunting, the relationship between wildlife and the evictees was deeply antagonistic, for the remote and very difficult physical environment in the valley was alien to them, and they had struggled long and hard to establish farming livelihoods and 'tame' or 'civilise' their surroundings. To complicate matters further, people who lived on the Gwampa river's southern banks and watershed were on Forestry Commission land and their tenure was insecure. Between 1970 and 1972 many were evicted without compensation. Others living in the communal areas on the other side of the valley were also subject to removals in line with colonial state conservation policy, which was a matter of huge controversy. Many people had had to move several times during the colonial period. During the liberation war in the late 1970s some people moved back to the forests with the support of the liberation army, only to find themselves being evicted again in the late 1980s as the Forestry Commission reasserted its exclusion policies. This was referred to as 'the forest war' as it was often accompanied by state violence.

In these circumstances it is evident that any CBNRM initiative which further affected people's rights to access resources and land would have to be introduced with great sensitivity and would have to benefit them in ways which they deemed to be significant. Unfortunately the attempt to implement the CAMPFIRE proposal did not meet these criteria. The project had funding pledged from the Canadian International Development Agency (CIDA) and the central government as well as personnel and vehicles from a Danish NGO, MS-Zimbabwe. There was little big game in the Gwampa valley and the plan was to stock it with wildlife that would attract photo safaris which would bring revenues to the local area. However, this would involve moving people out of a strip of land running the length of the valley, measuring 1.5–2 km from the river itself, which would then be fenced off allowing wildlife to move between the Forestry Commission's lands and the river. For some of these people their eventual destination was not even specified. The donors were promising large sums of money and the local councils were keen on the project as they believed it would raise their revenue. Yet the local communities were not fully consulted, which is meant to be a cornerstone of CBNRM principles.

In these circumstances it is scarcely surprising, perhaps, that most of the local people in Gwampa valley steadfastly refused to agree to the project. They were subsequently treated with astonishing contempt. For example, the provincial governor asserted they 'were backward and had nothing to lose because they lived only in "grass huts"'. Again, in complete contrast to the idea that CAMPFIRE should be empowering, 'executive officers and district administrators argued that development was necessarily coercive'. Committees that resisted CAMPFIRE were dissolved and new ones handpicked; minutes of meetings were falsified; CAMPFIRE opponents were arrested by the police. Experience with local councils over timber exploitation had also made people cynical about the likelihood of CAMPFIRE revenues being used for their real benefit and, in any case, a World Wildlife Fund report of 1994 felt the project was probably not very viable. The proposal also ignored the local history of rounds of evictions and people's antipathy towards wildlife. Their main resource grievance was about land. Since the proposal would reduce land availability even further it 'was met with horror'. Talking about the liberation war residents argued, 'We didn't fight to stay in this sandy area, we fought for rich land. We are 16 years independent, but nothing has been done – people are still piled up like these melons here' and, 'We're going to be grouped together like buffaloes while land is given to animals. This makes us think of war, this is terrible. . . . There's so much empty land – Forestry, commercial farms – and they come here to where people are living.'

This example illustrates a worrying gap between the philosophy of CBNRM and the practice of some conservation projects which try to assume a spurious legitimacy (and donor funding) by labelling themselves as 'community-based'. There are also many other complex issues that identify CAMPFIRE as an intensely politicised and contested area of activities. These include conflicts regarding ethnicity and gender in constraining access to CAMPFIRE benefits, and accusations that CAMPFIRE simply maintains the privileged access to wildlife enjoyed by foreign business interests and tourists. For example, Tembo-Mvura hunter-gatherers in the Zambezi valley have argued, 'What CAMP-FIRE does is to stop us from hunting so that white people can come from far away to kill animals for fun. We have heard that these people pay money but we have never seen any of it.' A noticeable trend for fencing to facilitate safari hunting has led to local perceptions elsewhere in Zimbabwe 'that the safari operator wanted to create a private farm out of their land, . . . to prevent people from accessing . . . resources . . . [and] to reintroduce white colonialism'. Furthermore, it is clear that distant interests in African wildlife play an important role. Thus the views and desires of foreign tourists and trophy hunters, and of development and conservation 'experts' working within a neo-liberal framework which assumes wildlife to be an economic 'resource' characterised by its use as opposed to intrinsic value, are all being played out on the Zimbabwean stage.

Sources: Alexander and McGregor (2000), Dzingirai (1995), Hasler (1999), Marindo-Ranganai and Zaba (1994), Wels (1999).

perhaps unrealistic, and generally unvoiced, expectations remain (Sullivan 2002b). First, there is the expectation that some African communal area residents should continue to live with dangerous wildlife on 'their' land. Second, that efforts should be made to foster the increase of populations of these same dangerous, but threatened, species. Third, that this should occur over and above investment in alternative sources of livelihood. Fourth, that, as donor-funding is phased out, revenue received from conservation efforts should be used to finance new communal area wildlife management institutions. Fifth, that a primary responsibility of these institutions should be the negotiation of business agreements which allow private safari operators continued access to the wildlife resources on which their profits depend.

A view is currently emerging, therefore, that suggests that, in practice, 'community-based conservation' is the fine-tuning of an existing status quo of inequality in the global and national distribution of capital, rather

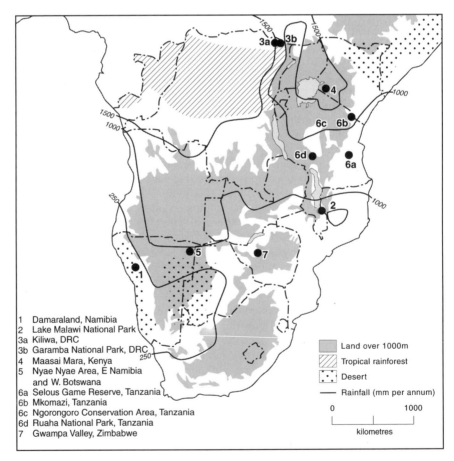

1	Damaraland, Namibia
2	Lake Malawi National Park
3a	Kiliwa, DRC
3b	Garamba National Park, DRC
4	Maasai Mara, Kenya
5	Nyae Nyae Area, E Namibia and W. Botswana
6a	Selous Game Reserve, Tanzania
6b	Mkomazi, Tanzania
6c	Ngorongoro Conservation Area, Tanzania
6d	Ruaha National Park, Tanzania
7	Gwampa Valley, Zimbabwe

Land over 1000m

Tropical rainforest

Desert

Rainfall (mm per annum)

0   1000

kilometres

Figure 5.2 Location of places discussed in boxes in Chapter 5

than a radical way of devolving power and improving livelihoods. This view asks if it is reasonable to expect a structurally entrenched rural poor to continue to service the fantasies of African wilderness projected by predominantly expatriate environmentalists, conservationists, tourists and trophy hunters? Or that a communalising discourse equating rural development and 'empowerment' with wildlife preservation and foreign tourism will be 'sustainable', given both the constraints it imposes on individual aspirations and the dissatisfaction it may produce in people who feel excluded? An alternative suggestion is that wildlife conservation will be 'sustainable' only if accompanied by a 'consumer pays' approach, amounting to economically realistic and long-term subsidies directly to African land-users in recognition of the ways in which their land-use and livelihoods are being manipulated to meet national *and international* conservation objectives (see, for example, Simpson and Sedjo 1996; Kiss 1999; Sullivan 2002b). This implies nothing short of a secure commitment to substantial and long-term (upwards of several decades) international subsidies, directly to local land-users, of amounts realistic enough to compensate for the opportunity costs of not converting either land to alternative uses or large mammals to cash (e.g. Norton-Griffiths 1996). Failing this, it seems logical that policing and law enforcement, whether by government officials, NGO employees or community game guards, will remain the foundation on which preservation of internationally valued African wildlife and 'wilderness' depends.

Some thoughts for the future . . .

It probably goes without saying that the countries and people of East and southern Africa face severe challenges which will affect both the availability of natural resources and people's uses and perceptions of them. Factors such as the tension between population increase and HIV–AIDS and the demographic implications of these (Chapter 2); the international economic policy arena and its effects on local land-use decisions (Chapters 3, 6 and 10); and the volatile and unpredictable incidence of violence and conflict (Chapter 9), are all significant for natural resources and livelihoods based on them. This is not the place to review or make projections regarding the complex interplay of these factors and their effects on natural resources and their uses. What we would like to emphasise, however, are a few issues which we consider important in the arena of natural resources use, management and conservation.

- Within conservation policy and implementation a greater and more explicit recognition is needed of the effects of gross structural inequality which protects wildlife and wild areas for wealthy consumers while expecting African land-users to shoulder the costs. In agreement with Simpson and Sedjo (1996), we consider that direct payments to land-users in return for conservation practice, instead of subsidised and constrained enterprise developments, should be explored as an approach to conservation which is more honest about the distribution of costs and benefits.

- Initiatives based on natural resources, whether oriented primarily towards conservation or economic returns, should be accompanied by the collection of biological field data and a monitoring programme regarding the population and recruitment status of the resource in relation to its uses by people (see, for example, Konstant *et al.* 1995; Sullivan *et al.* 1995; Homewood forthcoming). It is noticeable that few such initiatives currently collect and/or make public such data, despite being rooted in ideas of 'sustainability' of both resource offtake and local livelihoods.
- In a globalising but diverse world it is crucial to give attention to divergences that may emerge between local contexts and the more general policy *vis-à-vis* environment that is developed at national and global levels. This relates particularly to issues of equity in what remains an extremely unequal global distribution of wealth, opportunity and resources; to a lack of sensitivity to this inequity in approaches to conservation; and to the related potential for conflict and local resistance to emerge in the arena of 'natural resources' conservation.

References

Abbot, J. and **Homewood, K.** (1999) A history of change: causes of miombo woodland decline in a protected area in Malawi, *Journal of Applied Ecology* **36**(3): 422–33.

Abel, N. and **Blaikie, P.** (1986) Elephants, people, parks and development: the case of the Luangwa Valley, Zambia, *Environmental Management* **10**: 735–51.

Alexander, J. and **McGregor, J.** (2000) Wildlife and politics: CAMPFIRE in Zimbabwe, *Development and Change* **31**: 605–27.

Anderson, D. and **Grove, R.** (eds) (1987) *Conservation in Africa: People, policies and practice.* Cambridge University Press, Cambridge.

Ansell, W.F.H. (1989) Book review: The imperial lion: human dimensions of wildlife management in Central Africa, *South African Tydskr. Natuurnav.* **19**(3): 126–7.

Barbier, E., Burgess, J., Swanson, T. and **Pearce, D.** (1990) *Elephants, Economics and Ivory.* Earthscan, London.

Barnes, C., Ensminger, J. and **O'Keefe, P.** (1984) *Wood Energy and Households: Perspectives on rural Kenya.* Beijer Institute and Scandinavian Institute of African Studies, Uppsala.

Barrow, E. (1988) Trees and pastoralists: the case of the Pokot and Turkana, *ODI Social Forestry Network Paper 6b*, London.

Barrow, E. (1990) Usufruct rights to trees: the role of Ekwar in dryland central Turkana, Kenya, *Human Ecology* **18**(2): 163–76.

Barrow, E. (1992) Tree rights in Kenya: the case of the Turkana, *ACTS Biopolicy International* Series No. 8. Nairobi.

Becker, B. (1986) Wild plants for human nutrition in the Sahelian zone, *Journal of Arid Environments* **11**: 61–4.

Behnke, R. (1988) Range enclosure in central Somalia. ODI Pastoral Development Network Paper 25b, London.

Behnke, R.H. and **Scoones, I.** (1993) Rethinking range ecology: implications for rangeland management in Africa, in Behnke, R.H., Scoones, I. and Kerven, C. (eds) *Range Ecology at Disequilibrium: New models of natural variability and pastoral adaptation in African savannas*. ODI, IIED and Commonwealth Secretariat, London: 1–30.

Beinart, W. (forthcoming 2003) *The Rise of Conservation in South Africa: Settlers, livestock, and the environment 1770–1950*. Oxford University Press, Oxford.

Bell, R. (1982) Effect of soil nutrient availability on community structure in African ecosystems. *Ecological Studies* **42**: 193–216.

Bell, R. (1987) Conservation with a human face: conflict and reconciliation in African land use planning, in Anderson, D. and Grove, R. (eds) *Conservation in Africa: People, polices and practice*. Cambridge University Press, Cambridge.

Benefice, E., Chevassus-Agnes, S. and **Barral, H.** (1984) Nutritional situation and seasonal variations for pastoralist populations of the Sahel (Senegalese Ferlo), *Ecology of Food and Nutrition* **14**: 229–47.

Bergin, P. (1995) Conservation and development: the institutionalisation of community conservation in Tanzania National Parks. PhD thesis, University of East Anglia.

Biesele, M., Bousquet, J. and **Stanford, G.** (1979) A Kalahari food staple: *Ricinodendron rautanenii*, in Goodin, J.R. and Northington, D.K. (eds) *Arid Land Plant Resources: Proceedings of the International Arid Lands Conference on Plant Resources*. ICASALS: Lubbock, Texas.

Birley, M.H. (1982) Resource management in Sukumaland, Tanzania, *Africa* **52**(2): 1–29.

Botelle, A. and **Rohde, R.** (1995) *Those Who Live on the Land: A socio-economic baseline survey for land use planning in the communal areas of eastern Otjozondjupa*. Ministry of Land, Resettlement and Rehabilitation, Land Use Planning Series, Report 1, Windhoek, Namibia.

Bourdieu, P. (1990) *The Logic of Practice*. Polity Press, Cambridge.

Bradley, P. (1991) *Woodfuel, Women and Woodlots*. Vol. 1. *The Foundations of a Woodfuel Development Strategy for East Africa*. Macmillan, London.

Brockington, D. (2001) *Fortress Conservation: The preservation of the Mkomazi Game Reserve*. African Issues Series, James Currey, Oxford.

Brockington, D. and **Homewood, K.** (1998) Pastoralism around Mkomazi Game Reserve: the interaction of conservation and development, in

Coe, M., McWilliam, N., Stone, G. and Packer, M. (eds) *Mkomazi: The ecology, biodiversity and conservation of a Tanzanian savanna.* Royal Geographical Society (with the Institute of British Geographers), London.

Brokensha, D., Mwaniki, H.S.K. and **Riley, B.W.** (1972) Beekeeping in Embu District, Kenya, *Beeworld* **53**: 114–23.

Brokensha, D. and **Riley, B.W.** (1980) Mbeere knowledge of their vegetation and its relevance for development: a case study from Kenya, in Brokensha, D.W., Warren, D.M. and Werner, O. (eds) *Indigenous Knowledge Systems and Development.* University Press of America, Lanham, Md.

Brokensha, D. and **Riley, B.W.** (1986) Changes in uses of plants in Mbeere, Kenya, *Journal of Arid Environments* **11**: 75–80.

Bromley, D.W. and **Cernea, M.M.** (1989) *The Management of Common Property Natural Resources: Some conceptual and operational fallacies.* World Bank Discussion Papers 57, Washington, DC.

Brown, D. (1970) *Bury My Heart at Wounded Knee: An Indian history of the American West.* Rinehart and Winston, New York.

Butyinski, T. (1973) Life history and economic value of the spring hare (*Pedetes carpernis*) in Botswana, *Botswana Notes and Records* **5**: 209–13.

Butyinski, T.M. and **Von Richter, W.** (1974) In Botswana most of the meat is wild, *Unasylva* **26**(106): 24–9.

Campbell, A. (1986) The use of wild food plants, and drought in Botswana, *Journal of Arid Environments* **11**: 81–91.

Carles, A., King, J. and **Heath, B.** (1981) Game domestication for animal production in Kenya: an analysis of growth of oryx, eland and zebu cattle, *Journal of Agricultural Science, Cambridge* **97**: 453–63.

Carruthers, J. (1995) *The Kruger National Park: A social and political history.* University of Natal Press, Pietermaritzburg.

Caughley, G. (1993) Elephants and economics, *Conservation Biology* **7**: 943–5.

Chabwela, H.N. (1990) The exploitation of wildlife resources in Zambia: a critique, in Lungwangwa, D. and Sinyandwe, I. (eds) *Utilizing Local Resources for Development. Proceedings of the 9th PWPA conference, Eastern, Central and Southern Region,* Musungwa Lodge, Zambia, July 1988.

Chambers, R. and **Leach, M.** (1989) Trees as savings and security for the rural poor, *World Development* **17**(3): 329–42.

Child, G. (1970) Wildlife utilization and management in Botswana, *Biological Conservation* **3**: 18–32.

Child, G. (1984) Managing wildlife for people in Zimbabwe, in McNeely, J. and Miller, K. (eds) *National Parks, Conservation and Development.* Smithsonian Institute, Washington.

CDC (Conservation for Development Centre) (1984) *Towards Sustainable Development: A national conservation strategy for Zambia.* CDC, IUCN, Gland, Switzerland.

Cossins, N. and **Upton, M.** (1987) The Borana pastoral system of Southern Ethiopia, *Agricultural Systems* **25**: 199–218.

Craven, P. and **Sullivan, S.** (2002) *Inventory and Review of Ethnobotanical Research in Namibia: First steps towards a central 'register' of indigenous plant knowledge.* National Botanical Research Institute, Occasional Contributions 3, Windhoek.

Cumming, D. (1990) *Developments in Game Ranching and Wildlife Utilisation in East and Southern Africa.* WWF, Multispecies Animal Production Systems Project, Project Paper 13, Harare.

Cunningham, A.B. (1996) *People, Park and Plant Use: Recommendations for multiple-use zones and development alternatives around Bwindi Impenetrable National Park, Uganda.* UNESCO Division of Ecological Sciences, People and Plants Initiative, People and Plants Working Paper 4, Paris.

Davis, S.D., Heywood, V.H. and **Hamilton, A.C.** (1994) *Centres of Plant Diversity: A guide and strategy for their conservation.* Vol. 1. *Europe, Africa, Southwest Asia and the Middle East.* WWF, IUCN Publications, Cambridge.

de Lucia, R. (1983) Defining the scope of wood fuel surveys, in FAO (ed.) *Wood Fuel Surveys.* FAO Forestry for Local Community Development Programme, Rome.

de Merode, E. (1998) Protected areas and local livelihoods: contrasting systems of wildlife management in the Democratic Republic of Congo. PhD Thesis, University of London.

de Merode E., Hillman-Smith K., Homewood K., Pettifor R., Rowcliffe, M. and **Cowlishaw G.** (forthcoming, a) Wildlife conservation and armed conflict.

de Merode, E., Homewood K., Cowlishaw G. (forthcoming, b) The value of bushmeat and other wild foods to rural households living in extreme poverty in the eastern Democratic Republic of Congo.

Dlamini, P.M. (1989) Land rights and reforms and their impact on food production in rural Swaziland, *Royal Swaziland Journal of Science and Technology* **10**(2): 1–10.

Duffy, R. (2000) *Killing for Conservation: Wildlife policy in Africa.* International African Institute, James Currey, Oxford; Indiana University Press, Indiana.

Du Plessis, W. (1992) In situ conservation in Namibia: the role of national parks and nature reserves, *Dinteria* **23**: 132–41.

Dzingirai, V. (1995) *'Take Back Your CAMPFIRE'. A Study of Local Level Perceptions to Electric Fencing in the Framework of Binga's CAMPFIRE Programme.* Centre for Applied Social Sciences (CASS), Harare.

Ellis, S. (1994) Of elephants and men: politics and nature conservation in South Africa, *Journal of Southern African Studies* **20**(1): 53–69.

Eltringham, S. (1984) *Wildlife Resources and Economic Development.* John Wiley, New York.

Emerton, L. (1996) *Valuing Domestic Forest Use: Communities and conservation in Kenya.* ODI Rural Development Network Paper 19e, London.

Emerton, L. (1998) Why wildlife conservation has not economically benefited communities in Africa. Paper No. 5 in *Community Conservation Research in Africa, Principles, and Comparative Practice.* Institute of Development Policy and Management (IDPM), University of Manchester.

Escobar, A. (1996) Constructing nature: elements for a post-structural political ecology, in Peet, R. and Watts, M. (eds) *Liberation Ecologies: Environment, development, social movements.* Routledge, London.

Fairhead, J. and **Leach, M.** (1996) *Misreading the African Landscape: Society and ecology in a forest–savanna mosaic.* Cambridge University Press, Cambridge.

Fairhead, J. and **Leach, M.** (1998) *Reframing Deforestation: Global analysis and local realities: Studies in West Africa.* Routledge, London.

FAO (1981) *Map of the Fuelwood Situation in the Developing Countries, with Explanatory Note.* FAO, Rome.

FAO (1986) Focus on traditional food plants, *Food and Nutrition* **12**: 2–57.

Fleuret, A. (1979) Methods for evaluation of the role of fruits and wild greens in Shambaa diet: a case study, *Medical Anthropology* **3**, 249–69.

Fratkin, E. (1996) Traditional medicine and concepts of healing among Samburu pastoralists of Kenya, *Journal of Ethnobiology* **16**(1): 63–97.

Fukui, K. and **Markakis, J.** (1994) *Ethnicity and Conflict in the Horn of Africa.* James Currey, London.

Galaty, J.G. (1999a) Double-voiced violence in Kenya, paper presented at workshop on 'Conflict's fruit: poverty, violence and the politics of identity in African arenas', 21–24 October 1999, Sophienburg Castle, Denmark.

Galaty, J.G. (1999b) Unsettling realities: pastoral land rights and conservation in East Africa, paper presented at conference on 'Displacement, Forced Settlement and Conservation', 9–11 September 1999, Queen Elizabeth House, University of Oxford, Oxford.

Getahun, A. (1974) The role of wild plants in the native diet in Ethiopia, *Agro-Ecosystems* **1**: 45–56.

Gillingham, S. and **Lee, P.** (1999) The impact of wildlife-related benefits on the conservation attitudes of local people around the Selous Game Reserve, Tanzania, *Environmental Conservation* **26**: 218–28.

Gordon, R.J. and **Sholto Douglas, S.** (2000) *The Bushman Myth: The making of a Namibian underclass*, 2nd edn. Westview Press, Boulder, Colo.

Graham, O. (1988) Enclosure of the East African rangelands: recent trends and their impact, ODI Pastoral Development Network Paper 25a, London.

Grandin, B. and **Lembuya, P.** (1986) The impact of the 1984 drought at Olkarkar Group Ranch, Kajiado, Kenya. Pastoral Development Network Paper 23e. ODI, London.

Graziani, M. and **Burnham, P.** (2003) Legal pluralism in the rain forests of Southeastern Cameroon, in Homewood, K. (ed.) *Rural Resources and Local Livelihoods in Africa*. James Currey, Oxford.

Grivetti, L.E. (1978) Nutritional success in a semi-arid land: examination of Tswana agro-pastoralists of the eastern Kalahari, Botswana, *American Journal of Clinical Nutrition* **31**: 1204–20.

Grivetti, L.E. (1979) Kalahari agro-pastoral-hunter-gatherers: the Tswana example, *Ecology of Food and Nutrition* **7**: 235–56.

Gura, S. (1986) A note on traditional food plants in East Africa: their value for nutrition and agriculture, *Food and Nutrition* **12**(1): 18–26.

Hardin, G. (1968) The tragedy of the commons, *Science* **162**: 1234–48.

Hartley, D. and **Hunter, N.** (1997) Community wildlife management: turning theory into practice, paper prepared for the DFID Natural Resources Advisers' Conference, Sparsholt College, Winchester, 6–10 July.

Hasler, R. (1999) *An Overview of the Social, Ecological and Economic Achievements and Challenges of Zimbabwe's CAMPFIRE Programme*. Evaluating Eden Discussion Paper No. 3. IIED, London.

Haugerud, A. (1989) Land tenure and agrarian change in Kenya, *Africa* **59**(1): 61–90.

Heinz, H.J. and **Maguire, B.** (1974) Ethnobotany of the !Ko Bushman: Their ethnobotanical knowledge and plant lore, *Occasional Papers of the Botswana Society* **1**: 1–53.

Hitchcock, R. (1990) Water, land and livestock: The evolution of tenure and administrative patterns in the grazing areas of Botswana, in Galaty, J. and Johnson, D. (eds) *The World of Pastoralism*. Guilford Press, New York: 216–54.

Hitchcock, R. (1996) *Kalahari Communities: Bushmen and the politics of the environment in southern Africa*. IWGIA document No. 79, Copenhagen.

Hockey, P., Bosman, A. and **Siegfried, W.** (1988) Patterns and correlates of shellfish exploitation by coastal people in Transkei: an enigma of protein production, *Journal of Applied Ecology* **25**: 353–63.

Hodgson, D. (1999) Pastoralism, patriarchy and history: changing gender relations among Maasai in Tanganyika, 1890–1940, *Journal of African History* **40**: 41–65.

Hodgson, D. (ed.) (2000) *Rethinking Pastoralism in Africa*. James Currey, Oxford.

Homewood, K. (1995) Development, demarcation and ecological outcomes in Maasailand, *Africa* **65**: 331–50.

Homewood, K. (ed.) (forthcoming 2003) *Rural Resources and Local Livelihoods in sub-Saharan Africa*. James Currey, Oxford.

Homewood, K., Coast, E. and **Thompson, M.** (forthcoming) In-migrants and exclusion in East African rangeland buffer zones.

Homewood, K.M. and **Rodgers, W.A.** (1991) *Maasailand Ecology: Pastoralist development and wildlife conservation in Ngorongoro, Tanzania*. Cambridge University Press, Cambridge.

Hulme, D. and **Murphree, M.** (1999) Communities, wildlife and the 'new conservation' in Africa, *Journal of International Development* **11**: 277–85.

Igoe, J. and **Brockington, D.** (1999) *Pastoral Land Tenure and Community Conservation in East African Rangelands: A case study from northeastern Tanzania*, Pastoral Land Tenure Series No. 11, IIED, London.

IIED (1994) *Whose Eden? An overview of community approaches to wildlife management*. Report to the Overseas Development Administration of the British Government, IIED, London.

IUCN (1987a) *The Nature of Zambia: Conservation and development issues*. IUCN 2, Geneva.

IUCN (1987b) *Directory of Afrotropical Protected Areas*. International Union for Conservation of Nature, Gland, Switzerland.

IUCN/WWF/UNEP (1980) *World Conservation Strategy*. International Union for Conservation of Nature, Gland, Switzerland.

Jones, B.T.B. (1999) Policy lessons from the evolution of a community-based approach to wildlife management, Kunene Region, Namibia, *Journal of International Development* **11**: 295–304.

Juma, C. (1989) *Biological Resources and Innovation: Conserving and utilizing genetic resources in Kenya*. African Centre for Technology Studies (ACTS), Nairobi, Kenya.

Kinahan, J. (1991) *Pastoral Nomads of the Central Namib Desert: The people history forgot*. Namibian Archaeological Trust and New Namibian Books, Windhoek, Namibia.

Kiss, A. (1990) *Living with Wildlife: Wildlife resource management with local particpation in Africa*. World Bank Technical Paper No. 130, Africa Technical Department Series, World Bank, Washington DC.

Kiss, A. (1999) Making community-based conservation work, paper presented at a Society for Conservation Biology annual meeting, College Park, Md, 18 June.

Kiwasila, H. and **Brockington, D.** (1996) Combining conservation with community developments, *Miombo Technical Supplement* **1**: 8–9.

Kiwasila, H. and **Homewood, K.** (1998) Natural resource use by reserve-adjacent farming communities, in Coe, M., McWilliam, N., Stone, G. and Packer, M. (eds) *Mkomazi: The ecology, biodiversity and conservation of a Tanzanian savanna*. Royal Geographical Society (with the Institute of British Geographers), London.

Konstant, T.L., Sullivan, S. and **Cunningham, A.B.** (1995) The effects of utilization by people and livestock on *Hyphaene petersiana* basketry resources in the palm savanna of north-central Namibia, *Economic Botany* **49**(4): 345–56.

Kramer R., von Schaik, C. and **Johnson, J.** (1997) *Last Stand. Protected areas and the defence of tropical biodiversity*. Oxford University Press, Oxford, 242 pp.

Kurimoto, E. and **Simonse, S.** (eds) (1998) *Conflict, Age and Power in North East Africa*. James Currey, Oxford; Ohio University Press, Athens.

Lane, C. and **Pretty, C.** (1990) *Displaced Pastoralists and Transferred Wheat Technology in Tanzania*, Sustainable Agriculture Programme, Gatekeeper Series 20, IIED, London.

Leader-Williams, N. and **Albon, S.** (1988) Allocation of resources for conservation, *Nature* **336**: 533–5.

Leader-Williams, N., Kayera, J. and **Overton, G.** (1996) *Community-based Conservation in Tanzania*. IUCN Species Survival Commission No. 15, Cambridge.

Lee, R.B. (1973) Mongongo: the ethnography of a major wild food resource, *Ecology of Food and Nutrition* **2**: 307–21.

Lee, R.B. (1979) *The !Kung San: Men, women and work in a foraging society*. Cambridge University Press, Cambridge.

le-floc, H.E., Lemordant, D., Lignon, A. and **Rezkallah, N.** (1985) Ethnobotanical practices of the populations of the Middle Awash Valley, Ethiopia, *Journal of Ethnopharmacology* **14**(2–3): 283–314.

Lewis, D., Kaweche, G.B. and **Mwenya, A.** (1991) Wildlife conservation outside protected areas – lessons from an experiment in Zambia, *Conservation Biology* **4**(2): 171–80.

Lewis, I.M. (1961) *A Pastoral Democracy – A study of pastoralism and politics among the northern Somali of the Horn of Africa*. Oxford University Press, London.

Lewis, J. and **Knight, J.** (1995) *The Twa of Rwanda*. World Rainforest Movement, Chadlington and International Work Group for Indigenous Affairs, Copenhagen.

Lindsay, W.K. (1987) Integrating parks and pastoralists: some lessons from Amboseli, in Anderson, D. and Grove, R. (eds) *Conservation in Africa: People, policies and practice*. Cambridge University Press, Cambridge.

Lovett, J.C. and **Wasser, S.K.** (eds) (1993) *Biogeography and Ecology of the Rainforest of Eastern Africa*. Cambridge University Press, Cambridge.

MacKenzie, J.M. (1987) Chivalry, social Darwinism and ritualised killing: the hunting ethos in Central Africa up to 1914, in Anderson, D. and Grove, R. (eds) *Conservation in Africa: People, policies and practice*, Cambridge University Press, Cambridge.

Malaisse, F. and **Parent, G.** (1985) Edible vegetable products in the Zambezian woodland area: a nutritional and ecological approach, *Ecology Food and Nutrition* **18**: 43–82.

Malan, J.S. and **Owen-Smith, G.L.** (1974) The ethnobotany of Kaokoland, *Cimbebasia* (B) **2**(5): 131–78.

Marais, E. (1996) Omaungu in Namibia: Imbrasia belina (Saturniidae: Lepidoptera) as a commercial resource, paper presented at the First Multidisciplinary Symposium on Phane (mopane caterpillars), 18 June, Department of Biological Sciences and Kalahari Conservation Society, Botswana.

Marindo-Ranganai, R. and **Zaba, B.** (1994) Animal conservation and human survival: a case study of the Tembo-Mvura people of Chapoto Ward in the Zambezi Valley, Zimbabwe. Harare University Research Paper, Harare.

Marks, S. (1984) *The Imperial Lion: Human dimensions of wildlife management in Central Africa*. Westview Press, Boulder, Colo.

Matenga, C.R. (1999) Community-based wildlife management schemes in Zambia: empowering or disempowering local communities? paper presented at a conference on African Environments – Past and Present, St. Anthony's College, University of Oxford, 5–8 July.

Maundu, P. (1987) The importance of gathered foods and medicinal plants in Kakuyuni and Kathama areas of Machakos, in Wachiira, K.K. (ed.) *Women's Use of Off-Farm and Boundary Lands: Agroforestry potentials*, Final Report, ICRAF, Nairobi; Indiana University Press, Bloomington.

Maundu, P. *et al.* (2001) *Ethnobotany of the Loita Maasai*. People and Plants Working Paper No. 8, UNESCO, Paris.

Mkanda, F.X. and **Munthali, S.M.** (1994) Public attitudes and needs around Kasungu national park, Malawi, *Biodiversity and Conservation* **3**: 29–44.

Moris, J. (1981) A case in rural development: the Maasai Range Development Project, in Moris, J. (ed.) *Managing Induced Rural Development*, Indiana University Press, Bloomington: 99–113.

Mortimore, M. (1998) *Roots in the African Dust.* Cambridge University Press, Cambridge.

Moss, H. (1988) *Under-Exploited Food Plants in Botswana.* FAO, Rome.

Moyo, S., O'Keefe, P. and **Sill, M.** (1993) *The Southern African Environment: Profiles of the SADC countries.* Earthscan, London.

Murombedzi, J. (1999) Devolution and stewardship in Zimbabwe's CAMP-FIRE programme? *Journal of International Development* **11**: 287–93.

Musters, C., de Graaf, H. and **Ter Keurs, W.** (2000) Can protected areas be expanded in Africa? *Science* **287**: 17–18.

Myers, N., Mittermeier R.A., Mittermeier C.G., da Fonseca G. and **Kent, J.** (2000) Biodiversity hotspots for conservation priorities, *Nature* **403**: 853–8.

Nhira, C., Baker, S., Gondo, P., Mangono, J.J. and **Marunda, C.** (1998) *Contesting Inequality in Access to Forests,* Policy That Works for Forests and People Series, No. 5, Zimbabwe Country Study, IIED, London.

Niamir-Fuller, M. (ed.) (1999) *Managing Mobility in African Rangelands: The legitimization of transhumance.* IT Publications, London; FAO, Rome.

Norton-Griffiths, M. (1996) Property rights and the marginal wildebeest: an economic analysis of wildlife conservation options in Kenya, *Biodiversity and Conservation* **5**: 1557–77.

Ntenga, G.M. and **Mugongo, B.T.** (1991) *Honey Hunters and Beekeepers: A study of traditional beekeeping in Babati District, Tanzania.* Swedish University of Agricultural Sciences, International Rural Development Centre, Working Paper 161, Uppsala.

O'Brien, E.M. (1988) Climatic correlates for species richness for woody 'edible' plants across southern Africa, *Monograph of Systematic Botany, Missouri Botanical Garden* **25**: 385–401.

Ogle, B.M. and **Grivetti, L.E.** (1985) Legacy of the chameleon: edible wild plants in the Kingdom of Swaziland, southern Africa, a cultural, ecological, nutritional study. Part II – demographics, species availability and dietary use, analysis by ecological zone, *Ecology of Food and Nutrition* **17**: 1–30.

Okoth-Ogendo, H.W.O. (2000) The tragic African commons: a century of expropriation, suppression and subversion, Keynote Address delivered at a workshop on Public Interest Law and Community-Based Property Rights, organised by the Lawyers Environmental Action Team, Tanzania, and the Centre for Environmental Law, USA, held at the MS-TSC DC Danish Volunteer Centre.

Patel, H. (1998) *Sustainable Utilization and African Wildlife Policy: The case of Zimbabwe's Communal Areas Management Programme for Indigenous Resources (CAMPFIRE).* Report for Indigenous Environmental Policy Centre, Cambridge, Mass.

Peluso, N. (1993) Coercing conservation? The politics of state resource control, *Global Environmental Change* **3**: 199–217.

Peters, C.R. (1987) *Ricinodendron rautanenii (Euphorbiaceae):* Zambezian wild food plant for all seasons, *Economic Botany* **41**(4): 494–502.

Pimbert, M. and **Pretty, J.** (1996) Parks, people and professionals: putting 'participation' into protected area management, in Ghimire, B. and Pimbert, M. (eds) *Social Change and Conservation.* Earthscan, London.

Platteau, J-P. (2000) Does Africa need land reform? in Toulmin, C. and Quan, J. (eds) *Evolving Land Rights, Policy and Tenure in Africa.* DFID/IIED/NRI, London: 51–74.

Posey, D. (ed.) (1999) *Cultural and Spiritual Values of Biodiversity. A complementary contribution to the Global Biodiversity assessment.* United Nations Environment Programme, IT Publications, London.

Potkanski, T. (1994) *Property Concepts, Herding Patterns and Management of Natural Resources among the Ngorongoro and Salei Maasai of Tanzania.* Pastoral Land Tenure Series No. 6, IIED, London.

Prior, J. (1994) *Pastoral Development Planning.* Oxfam Development Guidelines 9, Oxfam, Oxford.

Redhead, J. (1985) Decline and revival of traditional food plants in East Africa, *Food and Nutrition* **11**(2): 17–22.

Ribot, J. (1998) Theorising access: forest profits along Senegal's charcoal commodity chain, *Development and Change* **29**, 301–41.

Riley, B.W. and **Brokensha, D.** (1988) *The Mbeere in Kenya.* Vol. 1. *Changing Rural Ecology.* USA Institute for Development Anthropology (IDA) and University Press of America, Lanham, Md.

Ritchie, C. (1987) The political economy of resource tenure: San survival in Namibia and Botswana. Unpublished MA thesis, Boston University, Boston.

Robinson, J. and **Bennett, E.** (eds) (2000) *Hunting for Sustainability in Tropical Forests.* New York, Columbia University Press.

Rodgers, W.A. and **Homewood, K.M.** (1982) Species richness and endemism in the Usambara Mountain forests, Tanzania, *Biol. J. Linn. Soc.* **18**: 197–242.

Rodgers, W., Owen, C. and **Homewood, K.M.** (1982) Biogeography of East African forest mammals, *Journal of Biogeography* **9**: 41–54.

Roe, D., Mayers, J., Grieg-Gran, M., Kothari, A. and **Fabricius, C.** (2000) *Evaluating Eden: Exploring the myths and realities of community-based wildlife management.* Evaluating Eden Series No. 8, IIED, London.

Roe, E., Huntsinger, L. and **Labnow, K.** (1998) High reliability pastoralism, *Journal of Arid Environments* **39**: 39–55.

Rohde, R.F., Benjaminsen, T.A. and **Hoffman, M.T.** (2001) *Land Reform in Namaqualand: Poverty Alleviation, Stepping Stones and 'Economic Units'.* PLAAS Land Reform and Agrarian Change in Southern Africa Occasional Paper No. 16, University of the Western Cape, Cape Town.

Sandford, S. (1983) *Management of Pastoral Development in the Third World.* John Wiley, Chichester.

Scholes, R.J. and **Walker, B.H.** (1993) *An African Savanna: Synthesis of the Nylsvley study.* Cambridge University Press, Cambridge.

Scoones, I. (1991) Wetlands in drylands, *Ambio* **20**(8): 366–71.

Scoones, I. (ed.) (1994) *Living with Uncertainty: New directions in pastoral development in Africa.* IT Publications, London.

Shivji, I. (1998) *Not Yet Democracy: Reforming land tenure in Tanzania.* IIED, London.

Silbauer, G.B. (1981) *Hunter and Habitat in the Central Kalahari Desert.* Cambridge University Press, Cambridge.

Simpson, R.D. and **Sedjo, R.A.** (1996) Paying for the conservation of endangered ecosystems: a comparison of direct and indirect approaches, *Environment and Development Economics* **1**: 241–57.

Simpson, T. (1997) *Indigenous Heritage and Self Determination.* IWGIA 86, Copenhagen.

Skidmore-Hess, C. (1999) Flora and flood-plains: technology, gender and ethnicity in northern Botswana 1900–1990, paper presented at conference on 'African Environments – Past and Present', St. Anthony's College, University of Oxford, 5–8 July.

Solbrig, O. (1991) Savanna modelling for global change, *Biology International*, special issue **24**: 1–47.

Spear, T. (1997) *Mountain Farmers.* James Currey, Oxford.

Speight, M.R., Blench, R. and **Bourn, D.** (1999) Insect diversity and rural livelihoods. Unpublished report, ODI, London.

Stattersfield, A.J., Crosby, M.J., Long, A.J. and **Wedge, D.C.** (1998) *Endemic Bird Areas of the World – Priorities for biodiversity conservation.* Birdlife Conservation Series No. 7. Birdlife International, Cambridge.

Stiles, D. and **Kassam, A.** (1991) An ethno-botanical study of Gabra plant use in Marsabit District, Kenya, *Journal of East African Natural History Society and National Museum* **81**(198): 14–37.

Storey, R. (1985) *Some Plants Used by Bushmen in Obtaining Food and Water,* Botanical Survey of South Africa Memoir, No. 30.

Struhsaker, T. (1998) A biologist's perspective on sustainable harvest in conservation, *Conservation Biology* **12**: 930–2.

Stuart, S.N., Adams, R.J. and **Jenkins, M.D.** (1990) *Biodiversity in Sub-Saharan Africa and its Islands: Conservation, management and sustainable use.* Occasional Papers of the IUCN Species Survival Commission, IUCN, Geneva.

Sullivan, S. (1996a) The 'Communalization' of Former Commercial Farmland: Perspectives from Damaraland and implications for land reform. Research Report 25, Social Sciences Division of the Multidisciplinary Research Centre, University of Namibia.

Sullivan, S. (1996b) Towards a non-equilibrium ecology: perspectives from an arid land, *Journal of Biogeography* **23**: 1–5.

Sullivan, S. (1998) People, plants and practice: socio-political and ecological dimensions of resource-use by Damara farmers in north-west Namibia. PhD thesis, University College London.

Sullivan, S. (1999a) The impacts of people and livestock on topographically diverse open wood- and shrub-lands in arid north-west Namibia, *Global Ecology and Biogeography*, Tropical Open Woodlands Special Issue **8**: 257–77.

Sullivan, S. (1999b) Folk and formal, local and national: Damara knowledge and community-based conservation in southern Kunene, Namibia, *Cimbebasia* **15**: 1–28.

Sullivan, S. (2000) Gender, ethnographic myths and community-based conservation in a former Namibian 'homeland', in Hodgson, D. (ed.) *Rethinking Pastoralism in Africa: Gender, culture and the myth of the patriarchal pastoralist.* James Currey, Oxford: 142–64.

Sullivan, S. (2002a) How can the rain fall in this chaos? Myth and metaphor in representations of the north-west Namibian landscape, in LeBeau, D. and Gordon, R.J. (eds) *Challenges for Anthropology in the 'African Renaissance': A Southern African contribution.* University of Namibia Press, Windhoek.

Sullivan, S. (2002b) How sustainable is the communalising discourse of 'new' conservation? The masking of difference, inequality and aspiration in the fledgling 'conservancies' of north-west Namibia, in Chatty, D. and Colchester, M. (eds) *Conservation and Mobile Indigenous People: Displacement, forced settlement and sustainable development.* Berghahn Press, Oxford.

Sullivan, S. (2002c) Protest, conflict and litigation: dissent or libel in resistance to a conservancy in north-west Namibia, in Berglund, E. and Anderson, D. (eds) *Ethnographies of Environmental Under-privilege: Anthropological encounters with conservation.* Berghahn Press, Oxford.

Sullivan, S. (forthcoming 2003) Detail and dogma, data and discourse: food-gathering by Damara herders and conservation in arid north-west Namibia, in Homewood, K. (ed.) *Rural Resources and Local Livelihoods in Sub-Saharan Africa.* James Currey, Oxford.

Sullivan, S., Konstant, T.L. and **Cunningham, A.B.** (1995) The impact of utilization of palm products on the population structure of the Vegetable Ivory Palm (*Hyphaene petersiana, Arecaceae*) in north-central Namibia, *Economic Botany*. **49**(4): 357–70.

Sullivan, S. and **Rohde, R.F.** (2002) Guest editorial. On nonequilibrium in arid and semi-arid grazing systems: a critical comment on Illius, A. and O'Connor, T.G. (1999) On the relevance of nonequilibrium concepts to arid and semiarid grazing systems, *Ecological Applications*, **9**, 798–813, *Journal of Biogeography* **29**: 1–26.

Talle, A. (1988) *Women at a Loss. Changes in Maasai Pastoralism and their Effects on Gender Relations.* Stockholm Studies in Social Anthropology, No. 19.

Taylor, M. (1999) 'You cannot put a tie on a buffalo and say that is development': differing priorities in community conservation, Botswana, paper presented at a conference on African Environments – Past and Present, St. Anthony's College, University of Oxford, 5–8 July.

Ten Kate, K. and **Laird, S.A.** (1999) *The Commercial Use of Biodiversity: Access to genetic resources and benefit-sharing.* Earthscan, London.

Thompson, M. and **Homewood K**. (2002) Elites, entrepreneurs and exclusion in Maasailand, *Human Ecology* **30**(1): 107–38.

Toulmin, C. and **Quan, J.** (2000) *Evolving Land Rights, Policy and Tenure in Africa.* DFID/IIED/NRI, London.

Turnbull, C. (1972) *The Mountain People.* Paladin, London.

Wehmeyer, A.S. (1986) Edible wild plants of southern Africa: data on the nutrient contents of over 300 species. Unpublished report, Pretoria.

Wels, H. (1999) The origin and spread of private wildlife conservancies and neighbour relations in South Africa, in a historical context of wildlife utilization in Southern Africa, paper presented at a conference on African Environments – Past and Present, St. Anthony's College, University of Oxford, 5–8 July.

Wells, M., Brandon, K. and **Hannah, L.** (1992) *Parks and People: Linking protected area management with local communities.* World Bank, Washington.

Western, D. (1982) Amboseli national park: enlisting landowners to conserve migratory wildlife, *Ambio* **11**(5): 302–8.

Western, D. (1984) Amboseli National Park: human values and the conservation of a savanna ecosystem, in McNeely, J. and Miller, K. (eds) *National Parks, Conservation and Development?* Smithsonian Institute, Washington.

Western, D. and **Ssemakula, J.** (1981) The future of the savannah ecosystems: ecological islands or faunal enclaves? *African Journal of Ecology* **19**: 7–19.

Western, D. and **Wright, R.M.** (eds) (1994) The background to community conservation, in *Natural Connections: Perspectives in community-based conservation*. Island Press, Washington DC: 1–12.

White, F. (1978) The Afromontane Region, in Werger, M.J.A. (ed.) *Biogeography and Ecology of Southern Africa*. Dr W. Junk, The Hague.

White, F. (1983) *The Vegetation of Africa. A descriptive memoir to accompany the UNESCO/AETFAT/UNSO vegetation map of Africa*. UNESCO, Paris.

Widlok, T. (1999) *Living on Mangetti: 'Bushman' autonomy and Namibian independence*. Oxford University Press, Oxford.

Wildlife Sector Review Task Force (1995) *A Review of the Wildlife Sector in Tanzania*. Vol. 1: *Assessment of the Current Situation*. Ministry of Tourism, Natural Resources and the Environment, Dar es Salaam.

Wilmsen, E. (1992) Pastoro-foragers to 'Bushmen', in Galaty, J.G. and Bonte, P. (eds) *Herders, Warriors and Traders: Pastoralism in Africa*. Westview Press, Boulder and Oxford.

Wilson, K.B. (1989) Trees in fields, Southern Zimbabwe, *Journal of Southern African Studies* **15**(2): 1–15.

Wily, L.A. and **Mbaya, S.** (2001) *Land, People and Forests in Eastern and Southern Africa at the Beginning of the 21st Century: The impact of land relations on the role of communities in the future*. Natural Resources International and IUCN, Nairobi.

Wøien, H. and **Lama, L.** (1999) *Market Commerce as Wildlife Protector? Commercial initiatives in community conservation in Tanzania's northern rangelands*. IIED Pastoral Land Tenure Series No. 12, London.

Zmarlicki, C., Wehmeyer, A.S. and **Rose, E.F.** (1984) Important indigenous plants used in the Transkei as food supplements, *Ecology of Food and Nutrition*.

CHAPTER 6

Agricultural production in eastern and southern Africa: issues and challenges

John Briggs

The current status of agricultural production

For many years, there have been concerns about the ability of Africa's agricultural sector to feed the people of the continent (Binns 1994; McMaster 1992; Obia 1997; Okoth-Ogendo 1993; Raikes 1988). According to the Food and Agriculture Organisation (FAO), while the output of crops and livestock in Africa increased in the 1990s, it did not keep pace with the population so that by 1998 agricultural production per capita had declined to 98 per cent of that in 1989–91 (Table 6.1). Moreover, the performance of the 16 countries of eastern and southern Africa was weaker than for the continent as a whole; the FAO estimate that total agricultural production rose only 6 per cent over the same period, and that per capita production in 1998 had dropped to only 87 per cent of that in 1989–91.

The FAO figures suggest that only four countries in eastern and southern Africa experienced an increase in production per capita, the rest being characterised by decline. However, a word of caution needs to be expressed in interpreting these figures. The problem is illustrated by the fact that one of the countries estimated to have 'enjoyed' relative agricultural success during the 1990s is Angola, a country characterised by extreme political instability. Although its long-standing civil war was broken by two shaky periods of peace during the 1990s, by the end of 1998 all-out war was re-established. Even when there was a ceasefire large parts of this huge country were not under government control. Peace was again established in 2002. Whether agricultural ouptut can be measured with any reliability in such circumstances is extremely questionable, and any suggestion that Angola's farmers or agricultural policy had somehow been relatively successful compared with elsewhere in the region would be patently absurd. Indeed, for the region in general, FAO figures are, at best,

Table 6.1   Index of agricultural production by country in 1998
(base 1989–91 = 100)

	Total production	*Production per capita*
Angola	136.7	105.4
Botswana	90.1	74.0
Burundi	89.2	74.2
DRC	94.8	72.1
Kenya	111.8	90.1
Lesotho	102.1	83.2
Malawi	117.0	104.7
Mozambique	140.9	107.3
Namibia	123.5	100.9
Rwanda	80.5	84.8
South Africa	98.7	82.6
Swaziland	92.5	73.9
Tanzania	101.8	80.6
Uganda	108.8	85.1
Zambia	101.5	84.3
Zimbabwe	109.1	90.1
Region	106.2	87.1
Africa	121.1	97.9

Source: FAO (1999).

only an indication of prevailing agricultural conditions, as they are based on incomplete data sources in some cases, and on only summary data in others. A particular problem is that a large proportion, and often the majority, of food produced in the countries of eastern and southern Africa is consumed on-farm and therefore is not formally recorded for statistical purposes. The FAO makes assumptions about the volume and growth trends of such production but, evidently, if these are wrong, serious distortions can creep into the country's agricultural figures. In other instances, cross-border smuggling of crops may lead to significant underestimation of real production figures for a particular country. For example, coffee produced in Tanzania may find its way onto Kenyan or Ugandan markets, depending on relative prices, or vice versa. Population data may also be collected in different ways in different countries, making per capita production comparisons difficult. During the 1990s, fertility rates have also declined, particularly in southern African and Kenya (Potts and Marks 2001), and therefore production per capita estimates in the later 1990s may be rather better than the FAO figures suggest. It is important to recognise, therefore, that the FAO figures used are only helpful in terms of identifying broad trends and estimating current 'snapshot' circumstances.

Bearing in mind these reservations, there are, nonetheless, some interesting variations with regard to agricultural production within the region and, in this regard, three groups of countries can be identified. The first comprises a group where political instability and warfare have had an important impact on agricultural production. Mozambique, for example, has experienced the biggest increase in both total and per capita production over the time period. Having experienced destabilising and debilitating political conditions and armed conflict in the 1980s (and, indeed, earlier), these conditions eased considerably during the 1990s. As greater political stability has ensued, so have the conditions for increasing agricultural output. Conversely, deteriorating political conditions during the 1990s in the Democratic Republic of Congo (DRC, formerly Zaire), Burundi and especially Rwanda had the effect of reducing not only production per capita but also total production.

A second group of countries comprises those which are estimated to have experienced modest increases in total production, but declines in production per capita as high population growth rates have continued to erode the gains made in overall agricultural production. These include mainly the countries of East and central Africa. The final group comprises the southern African countries of South Africa, Botswana, Lesotho and Swaziland, all of which have experienced a decline in both total and per capita production levels. This may reflect the particularly poor climatic conditions experienced in the south during the 1990s, with the worst drought of the century at the beginning of the decade, but may be more to do with the increasing proportions of their populations becoming economically active outside the agricultural sector. Indeed urbanisation is very often a highly significant factor in explaining agricultural trends and demand for food imports (see, for example, Bryceson 1993 for Tanzania). Declining *national* per capita output, and rising food imports, can occur even if the per capita output of farmers is steady, or even increasing slowly, if farmers do not increase their production sufficiently to keep up with a shift to a more urban population. Even then the food imports may not necessarily reflect an incapacity on the part of the farming population, but may be influenced by changes in urban tastes and the cheapness of (sometimes subsidised) imports (see Raikes 1988). Although eastern and southern Africa is performing less well than the continent as a whole on these production indices, there is rather greater similarity in terms of cultivated land. Both the percentage of cultivated land and the amount of cultivated land per capita are similar for both the region and for Africa as a whole (Table 6.2). Unsurprisingly, given the vast range of ecological and natural resource conditions in the region, there are significant variations in the *amount* of cultivated, and cultivable, land across the region. Rwanda, Burundi and Uganda are the most intensively cultivated, largely because they all have access to high-quality soils and reliable rainfall, typically in excess of 1200 mm annually. It is these very conditions which have led to these countries being the most densely settled in the region. Conversely, countries with limited areas of cultivated land include those with significant dryland areas, such as Botswana, Namibia and Tanzania. Significantly, they are also joined by countries that have experienced politically destabilising conditions over the last couple of decades, notably Angola and Mozambique. In both of these countries, and in Tanzania, Zambia and the DRC, there is still considerable scope for extending the area of cultivated land. More irrigation would allow vast areas throughout the region

Table 6.2 Cultivated land as percentage of all land, and cultivated land per capita, 1997

Country/region	% arable land	Hectares/capita
Angola	2.4	0.26
Botswana	0.6	0.22
Burundi	30.0	0.12
DRC	3.0	0.14
Kenya	7.0	0.14
Lesotho	10.7	0.16
Malawi	16.9	0.16
Mozambique	3.8	0.16
Namibia	1.0	0.50
Rwanda	34.5	0.14
South Africa	12.6	0.40
Swaziland	9.8	0.18
Tanzania	3.5	0.10
Uganda	25.3	0.25
Zambia	7.1	0.61
Zimbabwe	8.0	0.28
Region	5.4	0.21
All of Africa	5.9	0.24

Source: FAO (1999).

which are as yet uncultivated to be brought into production, but as yet financial costs and technical problems generally mitigate against such developments. It is worth noting, however, that there are often significant amounts of smallholder, indigenous irrigation on, for example, local wetlands in eastern and southern Africa, although these are rarely recognised in national statistics on irrigated land (Woodhouse *et al.* 2000; Bell *et al.* 1987; Adams *et al.* 1997).

The salient staple crop of the region remains maize, with sorghum and millet important in some areas (particularly where it is drier) despite the increasing popularity of wheat, mainly imported, and rice, especially among the growing urban populations. The enduring popularity of maize is borne out by Table 6.3, which shows that about 52 per cent of Africa's production comes from the region. According to these FAO figures, maize yields per hectare are lower in eastern and southern Africa than for the continent as a whole. However, production per capita is over 50 per cent higher.

Livestock still constitute an important component of regional agricultural activity, providing a range of products including milk, meat, manure, wool, skins, means of transport and, in some instances, a vital source of draught power. However, increases in livestock numbers since 1989–91 have been modest. Indeed, the number of livestock per capita had declined in most countries in the region by 1998 according to the FAO (Table 6.4). This reflects stock losses during the droughts in the early 1990s in the southern countries (when the worst drought of the twentieth century occurred) and further north in the later years of the decade. In both cases, productivity

Table 6.3 Maize production, yields and production per capita, 1998

	Production (t)	*Yields (kg/ha)*	*Production (kg) per capita*
Angola	504,662	7,369	43.0
Botswana	1,000	1,250	0.7
Burundi	131,830	11,463	20.7
DRC	900,000	7,500	18.8
Kenya	2,450,000	16,225	86.1
Lesotho	94,000	9,400	46.6
Malawi	1,724,755	13,853	171.3
Mozambique	1,124,000	9,006	60.9
Namibia	14,888	7,444	9.1
Rwanda	78,000	10,400	13.1
South Africa	7,574,000	21,511	195.4
Swaziland	107,185	17,864	115.9
Tanzania	2,750,000	13,560	87.5
Uganda	750,000	12,542	37.5
Zambia	649,537	14,434	75.7
Zimbabwe	1,418,000	11,585	126.4
Region	20,271,857	11,588	82.7
All of Africa	38,728,808	15,555	53.0

Source: FAO (1999).

Table 6.4 Index of livestock numbers by country in 1998 (base 1989–91 = 100)

	Total production	*Production per capita*
Angola	110.4	85.2
Botswana	95.0	78.0
Burundi	90.8	75.5
DRC	105.0	79.8
Kenya	108.1	87.4
Lesotho	105.2	85.8
Malawi	120.7	108.0
Mozambique	95.7	72.9
Namibia	128.7	105.2
Rwanda	97.6	102.8
South Africa	103.1	86.4
Swaziland	109.1	87.4
Tanzania	115.9	91.8
Uganda	115.4	90.3
Zambia	125.4	104.2
Zimbabwe	101.1	83.5
Region	108.0	89.0
Africa	119.2	96.4

Source: FAO (1999).

has again been lower than for Africa as a whole. Notably, densely populated and overcrowded Malawi is estimated to have had the second highest per capita increase in livestock although, as Table 6.5 indicates, cattle and sheep numbers fell. The highest increase was in Zambia where there is still much potential grazing.

Cattle numbers in the region as a whole increased only marginally (1.3 per cent) between 1990 and 1998, well below human population increase (Table 6.5). The 'big three' cattle countries, South Africa, Kenya and Tanzania, with over 60 per cent of the region's cattle between them, all experienced increases, although for the first two this was only around the low, regional average. The largest proportionate increases were estimated for Angola, Lesotho, Tanzania, Uganda and Zambia. Significant falls are recorded for Botswana, Rwanda, Swaziland and Zimbabwe and, in particular, for the DRC and Burundi. In countries such as Botswana and Zimbabwe, where cattle are particularly crucial for many rural people's livelihoods, such falls are very significant (although in Botswana a generally positive economic climate provides far more alternative livelihood options than in Zimbabwe). Again a significant cause will have been the terrible drought of the early 1990s which decimated national herds. Sheep fared even less favourably in the region than cattle, decreasing 8 per cent between 1990 and 1998 with most countries experiencing declines – in four cases, the decline was in excess of 20 per cent.

Table 6.5 Numbers of cattle and sheep (thousands) by country in 1990 and 1998

Country/region	Cattle			Sheep		
	1990	*1998*	*% change*	*1990*	*1998*	*% change*
Angola	3,100	3,500	+12.9	240	245	+2.1
Botswana	2,696	2,330	−13.6	317	240	−24.3
Burundi	432	346	−19.9	361	320	−11.4
DRC	1,535	1,000	−34.9	927	1,020	+10.0
Kenya	13,793	14,116	+2.3	6,516	5,700	−12.5
Lesotho	523	580	+10.9	1,378	1,100	−20.2
Malawi	836	800	−4.3	148	120	−18.9
Mozambique	1,380	1,300	−5.8	121	123	+1.6
Namibia	2,087	2,192	+5.0	3,328	2,086	−37.3
Rwanda	582	500	−14.1	389	270	−30.6
South Africa	13,500	13,800	+2.2	32,665	29,980	−8.2
Swaziland	716	650	−9.2	24	25	+4.2
Tanzania	13,047	14,302	+9.6	3,557	3,960	+11.3
Uganda	4,913	5,370	+9.3	1,350	1,960	+45.2
Zambia	2,878	3,100	+7.7	60	65	+8.3
Zimbabwe	6,407	5,450	−14.9	599	520	−13.2
Region	68,425	69,336	+1.3	51,980	47,734	−8.2
Africa	188,575	217,388	+15.3	203,249	233,161	+14.7

Source: FAO (1999).

Since 1970, the proportion of every country's population engaged in agriculture has declined (see Table 6.6). In some cases, especially in southern Africa, the decline has been marked. For example, the FAO reports that the proportion of Botswana's population engaged in agriculture dropped from over 81 per cent in 1970 to only 45 per cent in 1997; and in South Africa from 40 to 15 per cent. These figures reflect significant economic and social changes in these countries which have encouraged the development of the non-agricultural sector, or the relative decline of agriculture. The South African economy has been essentially driven by mining and industrial enterprises for well over 100 years, and in the latter years of apartheid and post-1994 rates of urbanisation of the African population significantly increased (see Chapter 11, this volume). Botswana's economic structure has been utterly transformed since 1970 with the discovery and successful exploitation of diamonds, and again urbanisation and urban employment have developed rapidly, largely supplanting the migrant flows to the South African mines which used to be an important aspect of the Botswana economy. In Swaziland there was significant economic development in the 1970s and particularly the 1980s, although much of its manufacturing industry is in agro-processing (particularly of sugar). However, the 1990s saw relative economic stagnation and the FAO estimate that only about 36 per cent of the population was agricultural by 1997 must be based on outdated projections. In fact, the level of urbanisation did not increase at all over the intercensal period 1986–97, remaining at 23 per cent (see Table 11.1), and monetary employment (including employment in the informal sector) fell slightly in absolute terms. Also, according to the

Table 6.6 Agricultural population as a percentage of the total population, 1970–97

	1970	*1980*	*1990*	*1997*
Angola	78.3	76.4	74.5	72.7
Botswana	81.5	63.9	46.4	45.1
Burundi	93.5	92.8	91.6	90.7
DRC	74.8	71.6	67.8	64.6
Kenya	85.7	82.2	79.5	76.7
Lesotho	42.8	40.2	40.1	38.6
Malawi	88.4	83.0	82.2	78.9
Mozambique	82.8	80.4	78.6	76.8
Namibia	70.7	64.0	56.9	51.2
Rwanda	93.6	92.8	91.7	90.8
South Africa	40.9	25.0	19.9	15.9
Swaziland	64.9	50.0	39.4	35.6
Tanzania	88.8	84.0	82.5	79.4
Uganda	89.3	86.3	83.6	80.4
Zambia	79.1	76.1	74.7	71.1
Zimbabwe	76.6	72.4	68.2	64.4
Region	74.8	70.0	67.2	64.4

Source: FAO (1999).

Swazi Statistical Office (Swaziland 2001), 70 per cent of the economically active population in 1995 were not in monetary employment but working on the Swazi National Lands (i.e. farming) and/or casually, and a further 25 per cent of formal employment was in agriculture. Although there was a 25 per cent unemployment rate in the rural areas, it is evident that the majority of Swaziland's population is still agricultural. The Swazi example gives an indication of how agency data sources, such as the FAO, can sometimes be misleading. Major structural change has not been the norm generally in the region and the majority of the 16 countries still have more than 70 per cent of their populations engaged in agriculture, with both Rwanda and Burundi exceeding 90 per cent.

From the FAO figures, contemporary agriculture in eastern and southern Africa can be summarised as experiencing a decline in output per capita, even though absolute production has increased. This reflects the inability of the sector to grow at a level commensurate with that of population. The performance has been even less impressive with regard to livestock production, and especially of cattle and sheep, both of which have performed below the African average. However, despite the fact that agricultural performance has not been as good as elsewhere in Africa and that there has been a steady decline in the relative numbers involved in agriculture, although probably not in absolute numbers, most countries of the region are still overwhelmingly agricultural in terms of economic activity.

Macro-environmental operating conditions for agriculture

Natural resource base

In this region, the available natural resource base remains a more important factor of production for agriculture than in most of the world's more developed economies. The key reason is limited capital availability, apart from some areas of South Africa, to modify water and soil conditions. Consequently, most farmers in the region have little choice but to make the best use of, and to work within the constraints of, the available natural resource base.

Most agriculture in the region is thus highly dependent on rainfall patterns which are generally highly seasonal. The exceptions are the highland areas of the East African plateau, including western Kenya, Uganda, Rwanda and Burundi, where seasonality is less marked due to the moderating influence of altitude and relief. For most of the rest (apart perhaps from the southern Cape which has a 'Mediterranean'-type climate), there is a distinct pattern of rainfall distribution throughout the year associated with the annual migration northwards and southwards of the intertropical convergence zone (ITCZ) (Griffiths 1972; Jackson 1989; Osei and Aryeetey-Attoh 1997). This is an area of air mass instability which migrates northwards during the early part of the year, being located over the Equator in March–April, the Tropic of Cancer in June–July, back over the Equator again in October–November, and over the southern Tropic of Capricorn in January–February. Consequently, those areas nearest to the Equator, such as parts of Kenya and Tanzania, can expect four distinctive seasons during the year (two wet and two dry). However, this distinction becomes increasingly blurred further northwards and southwards, such that in much of southern Africa there is only one distinctive wet season followed by a long dry season (see also Chapter 8, this volume). It is within this rhythm that African farmers have to set their agricultural production parameters.

This pattern, however, is modified in a number of ways. For example, the cold Benguela current off the west coast of the region contributes to significant aridity in much of western South Africa, Namibia and Botswana. Eastern Kenya and eastern and central Tanzania also receive rather less annual rainfall than might be expected given their latitudes. This is due to the location of arid regions to the north, particularly the Arabian Peninsula and the Horn of Africa, over which air masses from the north flow, and from which there are only limited opportunities for water acquisition, and of Madagascar to the south which operates as a rain shadow on air masses coming from the south and south-east (Hills 1978; Nieuwolt 1979).

Agriculture's dependence on rainfall patterns is problematic in a number of ways. The seasonal pattern is a problem in itself. For example, the Nairobi region in Kenya receives about 900 mm of rainfall annually, on the face of it more than enough to sustain the successful cultivation of maize, which requires about 500 mm during its growth cycle. However, about 500 mm of the annual total is received in the 'long rains' between March and May, a period which is then followed by a long dry period until the 'short rains' in November–December. Given this distribution, maize can in fact be a marginal crop in this area, and particularly so in years of below-average rainfall during the 'long rains'. Seasonality is also important in that agriculture, for most farmers, can take place only during the wet season. Water storage for large-scale irrigation in the dry seasons is not common in the region, as there are only a few large rivers relative to the total land area (such as the Limpopo, Vaal, Orange, Zambesi, Congo and Rufiji) and such schemes are so expensive. Consequently, capital-intensive water-storage schemes are not common outside South Africa (apart from a few exceptions such as Kariba and Cahora Bassa) (see Chapter 7).

Equally important is the annual rainfall variability experienced by farmers. Low rainfall years cannot be predicted but they have devastating impacts on crop production in the absence of irrigation. This is illustrated in Figure 6.1 for commercial and communal (i.e. African smallholder) maize production in Zimbabwe where the extremely high level of correlation between annual rainfall total and output is evident for both sectors, although it is especially marked for the communal sector. Unfortunately, in eastern and southern Africa the areas with lower levels of annual rainfall experience higher levels of annual variability. In Kongwa (central Tanzania), the mean annual rainfall (based on data for 1920–95) is 510 mm, but this masks a variation of extremes between 177 and 711mm. An annual total of 510 mm clearly makes this a marginal area for cultivation, and many crops will fail entirely in those years when rainfall is nearer to the lower end of

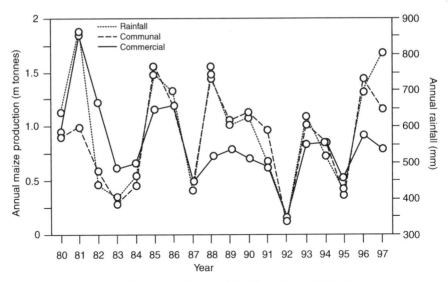

Figure 6.1 Maize production and rainfall in Zimbabwe, 1980–97
Source: Alderson (2001)

the range. Even in areas with higher mean annual rainfall, there can be significant annual variations. In extensive areas of southern Africa, for example, the 1997–98 season was drier than normal, being linked to an El Niño event (Dawson and O'Hare 2000). Conversely, the 1999–2000 season was significantly wetter than normal, culminating in devastating floods in South Africa (mainly in KwaZulu-Natal, Mpumalanga and Gauteng), and especially in Mozambique in February–March 2000. Add this variability to the seasonal pattern of rainfall described above, and it becomes clear that agriculture can be a very precarious experience in these areas. This was tragically exemplified in 2002, when a very wet start to the 2001–2 season in much of southern Africa, followed by drought from January 2002, led to devastating food shortages in parts of Malawi, Zambia and Zimbabwe (compounded in Zimbabwe by disruption in normal growing patterns because of uncertainties caused by fast-track land resettlement).

There is the added problem for farmers of predicting the timing of the onset of the rainy season, and hence optimising the timing of seed planting. To plant too early can mean wasting seed material; an early downpour can provide enough water to initiate germination, but if this is then followed by a 2–3 week dry period before the full onset of the rains, production can be lost. Equally, planting after the rains have become established can be problematic. Saturated soils are much more difficult to work, and valuable water is not being optimised in the growth cycle of crops. To add to these long-standing difficulties, the onset of the rains has become even more unpredictable in recent decades, particularly in the drier areas of the region (Hulme 1996).

Yet another problematic element for farmers is rainfall intensity. Typically, rain falls as relatively short, but intense, storms. Unfortunately, some of these storms occur when the soils are at their most vulnerable. In readiness for the onset of the rains, land is prepared by being cleared of vegetation, and ploughed in some cases; intense storms on bare soils can have significant consequences. Soil compaction from the impact of falling raindrops can occur, reducing soil porosity which in turn reduces the aeration of soils. This can lead to waterlogging, resulting in less efficient plant growth. Critically, bare soils are also particularly vulnerable to erosion, either by sheetwash or gully erosion (Jones 1996; Timberlake 1988); either way, valuable topsoil is moved and land productivity may be affected.

A final irony is that while the rainy season is the busiest period of the agricultural calendar for most farmers in the region, it is also the time when the labour force is at its most susceptible to disease, especially malaria (Prothero 2000). The onset of the rains leads to a marked increase in the prevalence of malaria in the region, greatly increasing the likelihood of labour being out of action through illness, thus generating a further brake on productivity.

Because of the relative scarcity of large river valleys with extensive floodplains, the region possesses few large areas of high-quality alluvial soils (this may also be a key reason why there is relatively little major irrigation in the region) (Moss 1992; Osei and Aryeetey-Attoh 1997). However, good quality volcanic soils exist in the plateau areas around the Rift Valley which, if well managed, can (and do) support relatively intensive agriculture (Bernard 1993). This is certainly the case in the Great Lakes region of Uganda, western Tanzania, Rwanda and Burundi, and in Malawi and parts of Zambia. Without doubt, the soils of the region require careful management. In many parts, the soils are deep, reflecting thousands of years of weathering. However, this can lead to the leaching of plant nutrients to depths below the root zone of crops. Furthermore, because of high ambient temperatures, chemical processes are accelerated, such that the oxidisation of organic compounds, for example, is more rapid in Africa than in temperate areas, allowing less time for humus development. Hence, the organic content of soils in the region can be relatively limited. A further problem is the relatively high sand fraction of many soils, which can result in reduced water retention capacity, further reducing fertility levels. For farmers with limited capital for land maintenance or improvements, such soil constraints require very careful management.

Population growth

Some analysts, including influential ones at the World Bank, consider that persistent rates of high population growth in the region continue to be significant in restricting agricultural growth and transformation (Turner *et al.* 1993). Between 1970 and 1997, the population of the region more than doubled from 116 to 245 million, and the populations of several countries continued to grow at 3 per cent per annum in the 1990s (although the impact of AIDS is significantly reducing these rates in some countries such as Zimbabwe, Uganda, Malawi, Zambia and Botswana) (see Chapter 2). These high population growth rates, it is argued, can result in a decline in agricultural production per capita, despite an overall increase in total agricultural output. Even where land availability is not a constraint, the population structure generated by such rapid growth – with up to half the population being children many of whom are at school which limits their contribution to cultivation – makes it difficult for farmers to maintain national per capita production levels. For the agricultural sector to make any meaningful production gains, it is therefore argued, population growth rates need to demonstrate a significant deceleration.

The problem may be exacerbated in some areas by inheritance traditions where land is divided equally between a father's sons (typically). With the fertility and survival rates usual up until the 1990s, it was not uncommon for four or five sons to survive to become inheritors. Under such circumstances, and in countries such as Rwanda, Burundi, Malawi and Zimbabwe, and parts of Kenya, Uganda and Tanzania, where land shortages in some areas are already being felt, it may take only a couple of generations for a consolidated 5-ha farm to become so fragmented that few of the inherited land parcels remain economically viable. A further dimension of this is that larger families place greater demands on limited household resources, with the result that money, which might be usefully invested in farm maintenance or improvements, is diverted into necessary social reproduction, such as school fees, bride price payments and so on.

There still exists in some quarters the Malthusian argument that the growing population places increased pressures on available natural resources, especially on soil fertility and structure (Stocking 1983). Unsustainable agricultural practices, such as overcultivation or animal overstocking, lead to declines in land productivity and reductions in absolute production. To compensate, some people may try to farm even more intensively, but without necessarily having the resources to increase inputs. If so, the result is a downward spiral of declining land productivity and land degradation (Maddox 1996). In some areas, as the population has increased, farmers have migrated into less agriculturally favourable regions, often using the same methods which are quite appropriate to their former farmlands, but which are not easily transferable to new, frequently less productive areas. The result can be increased land degradation, declining fertility levels and soil loss through wind and water erosion. In parts of Kenya (to the east of Mount Kenya, for example) and Tanzania (in some of the coastal regions and in central Tanzania), cultivators have colonised drier, less productive farming areas due to land pressures in their home areas, with a resulting increased degradation (Bernard 1993). Livestock herders have in turn been pushed further out into even drier environments, but still attempt to retain stocking levels that are inappropriate to their new environment, with all the consequences this has for land degradation and also, inevitably, stock losses (Anderson and Broch-Due 1999; Quinlan 1995; Mung'ong'o and Loiske 1995).

It would be misleading, however, to suggest that increasing population growth is the only explanation of these changes and pressures. The acquisition of land in some of the more favourable farming areas by larger farmers with access to capital and, increasingly, multinational companies has pushed out some farmers from these areas and onto less favourable lands. In Kenya, for example, around and to the north of Nairobi and around Lake Naivasha, good quality farmland has been put over to the production of high value horticultural produce for the export market (Barrett *et al.* 1999). In many parts of southern Africa, the demands of commercial agriculture have artificially created land shortages; in parts of Namibia and Botswana, for example, large commercial cattle and game ranches have pressurised small-scale farmers with limited land availability. Moreover, the historical division of farmland on racial lines in South Africa, Zimbabwe and Namibia has left a legacy of extremely uneven land distribution, and land shortages for many small farmers (Bowyer-Bower and Stoneman 2000; Moyo 1995; Cousins 2000; Marcus *et al.* 1996). In these instances, population/land pressures have been brought about less through natural population increase, and more through pressures generated by internal and external economic and political interests. Such land inequalities are highly political, and led at the end of the 1990s to land invasions on a massive scale in Zimbabwe (Worby 2000).

Importantly, there is evidence challenging the view that population increases inevitably lead to land degradation. In the 1930s, Machakos District in Kenya was considered to be an environmental disaster area characterised by famine (Tiffen *et al.* 1994). By the 1990s, however, despite a fivefold increase in population, the environment was in much better condition and agricultural production per capita was higher, as farmers introduced new technologies to take advantage of expanding commercial markets (Box 6.1). Hence, it is important to recognise that the Malthusian argument is not clear-cut and, indeed, is not particularly important in much of the region. Ironically, some areas (for example, parts of central Kenya, Zambia and Botswana) can experience labour shortages at key times, such as land preparation, weeding and harvesting, and this can itself be an important constraint on production.

The political economy environment

A key feature influencing agricultural activity has been government policy choice. During the 1970s, a range of agricultural policies existed among the countries of the region. For example, Tanzania and Mozambique adopted socialist policies favouring the consolidation and/or collectivisaton of peasant agriculture, alongside attempts to develop state-owned commercial farms. It should be noted, however, that in both instances there was a considerable difference between policy and practice (Coulson 1982; Hyden 1980). Kenya, on the other hand, developed policies which focused to an extent on small-scale farmers plugging into a wider agricultural system which favoured free enterprise principles. South Africa, Rhodesia and South West Africa,[1] under white minority regimes, promoted the large-scale commercial farming sector, predominantly white-owned, which attracted substantial subsidies from the state, to the neglect of the small-scale sector.

[1] Rhodesia and South West Africa were the colonial names for today's Zimbabwe and Namibia.

Box 6.1   Challenging assumptions: population and environment in Machakos, Kenya

By the 1930s Machakos district in Kenya had become a highly degraded landscape and was reported as an environmental disaster. The colonial government in Kenya had created a land division in which there were scheduled areas set aside for European farmers and 'native' reserves for black farmers. Machakos was one of the latter, and had become characterised as an area of a crowded and growing population, with increasing livestock numbers, an expansion of shifting cultivation to produce food crops and reductions in the tree stock as they were cut down for fuel and building. The situation was not helped by the droughts of the 1930s when, between 1933 and 1936, six droughts occurred in eight seasons.

Since that time the human population has grown fivefold, livestock numbers have increased from 330,000 units in 1930 to 593,000 in 1989, and the cultivated area of the former reserve has expanded from 15 per cent of the total land area in 1930 to between 50 and 80 per cent in 1978. However, the expected situation of even more degradation has not materialised; indeed quite the opposite, as land is now carefully managed.

The reasons for this go beyond the usual adoption of soil and water conservation methods, although these are important. The importance of marketing cash crops is crucial. Up until 1954, the cultivation of coffee by African farmers was banned, but once the ban was lifted Machakos farmers quickly adopted the crop as it was very profitable and hence started to generate cash surpluses which were then reinvested back in agriculture. In addition, coffee had to be grown on terraces which were then constructed and contributed to reducing soil erosion. The capital generated from coffee margins and remittances from off-farm employment was now available to diversify into other crops, and farmers took the opportunity to develop fruit and vegetable production for the growing urban markets of Nairobi and Mombasa, as well as for the expanding canning industry in Kenya. A critical element of change involved the widespread introduction of oxen as a source of power for ploughing. In addition, oxen produce manure that is then recycled onto the farmland. Livestock are generally stall-fed, and grass is grown on the terrace banks, again contributing to reduced erosion as soils are stabilised. Finally, most farmers now have legal land titles which give considerable security and hence encourage further investment in, and careful management of, the land.

Source: Tiffen *et al.* (1994).

However, during the 1980s, and especially during the 1990s, national agricultural policies have become increasingly similar throughout the region, although some differences inevitably remain. In virtually every country the agricultural sector is now characterised by declining amounts of state interference and production based more on market principles (Campbell and Loxley 1989; Cornia and Helleiner 1994; Gibbon 1995).

Central to this has been the involvement of the International Monetary Fund (IMF) and the World Bank, largely since the mid-1980s, in the development of macro-economic policy in almost every country in the region through structural adjustment programmes (SAPs). Operating at the national macro-economic level, SAPs have had a pervasive influence on agricultural sectors. Typically, a key element of SAPs is their emphasis on the production of primary commodities in which African countries are seen to have a comparative advantage in world trade terms (see Chapter 3).

A further element has been the strategy of abolishing fixed exchange rates and allowing currencies to find more of a 'market' level, a strategy which has the potential to reward exporters (but also of making imports relatively more expensive in local currency terms). Hence, a number of producers of export crops, such as coffee, tea or sugar, have sometimes seen incomes rise, as devalued exchange rates have favoured export farmers. In Tanzania in the late 1980s, for example, government controls on marketing export crops were being relaxed and currency exchange rates were being allowed to 'float' towards their 'market' levels. Arabica coffee prices on international markets at that time fluctuated between about US$2 and 2.30 per kilogram. However, between January and June 1987, the Tanzanian shilling lost value against the US dollar, dropping from about TSh26 (Tanzanian shillings) to about TSh40. For the coffee producer, 1 kg of arabica coffee, although still fetching about US$2 per kilo internationally, now returned TSh80 locally instead of the previous figure of TSh52.

Of particular significance in these policy changes has been the demise of state-controlled marketing boards (see Chapter 3). In Tanzania, for example, every major agricultural commodity used to have its own parastatal marketing board which was responsible for the provision of seeds, and relevant inputs of fertilisers and pesticides to producers; for the purchase and collection of those commodities after harvesting at guaranteed prices; and for the sale of these products on international and domestic markets (Bryceson 1993; Harvey 1988). These boards set the producer prices paid to farmers, and established monopoly purchasing rights, such that it was illegal for producers to sell to anyone other than officials of the authorised marketing board. Tanzania probably had the most widespread range of such boards, but they were common, in various guises, throughout the region. It was argued that such monopoly marketing boards eliminated the exploitative private trader; that they permitted the control of food prices for urban consumers; and that they ensured that foreign exchange earnings for exported agricultural commodities accrued through the state to the producers, and not to capitalist trading companies.

In reality, the performance of these boards was variable. In Tanzania, for example, they were mostly inefficient and, in some cases, corrupt, whereas in Zimbabwe there was a greater degree of efficiency. Logistical difficulties sometimes resulted in crops not being collected after harvesting, or collected so late that they had rotted and lost value. Consequently, many producers bypassed the boards by selling commodities on the parallel market, especially where they were located near international borders with countries with somewhat more liberal marketing strategies (and/or higher prices). Smuggling of coffee from Tanzania into Kenya and Uganda was rife in the 1980s and early 1990s, for example. If the marketing boards created conditions that were intolerable for cash croppers, some producers disengaged from the commercial production of some crops and focused their resources on producing food crops that could be consumed directly within the household. In countries where agricultural exports underpinned the economy, this was evidently disastrous. An important consequence of price control was to distort interregional terms of trade within countries, to the extent that, in some instances, both producer prices paid to farmers and food prices paid by consumers in the cities were being artificially depressed. Consequently, rural producers were in effect subsidising the living costs of urban consumers. For some governments, however, political threats were perceived to come from the cities, not the rural areas, and hence the political priority was to maintain authority there. Reliable and cheap food supplies constituted a strong weapon in this regard, even if it meant disadvantaging farmers in the countryside.

The withdrawal of the state from monopoly marketing has had varying results. In some countries, such as Zambia, steep rises in food prices resulted in significant urban unrest during the late 1980s and mid-1990s; and urban consumers in Zimbabwe have similarly experienced steep price rises for basic foodstuffs throughout the 1990s and this also resulted in riots. Elsewhere, in Tanzania for example, the expected urban unrest from predicted steep rises in urban food prices did not materialise. A key reason lies with the inefficiency of some of the marketing boards themselves, such that with the introduction of competition in marketing, in some instances de facto urban food prices actually came down. For although the Tanzanian state set and controlled official prices for foods, in reality food could rarely be bought by consumers at these cheap prices; consumers were therefore used to buying food at higher black market prices. Hence, with the deregulated market, although food prices were now higher than the official price (and hence some observers saw this as confirmation of higher food prices under deregulation), they were nonetheless lower than the actual prices that consumers had very often had to pay up until then. Maize meal, a staple, had an official price of TSh16 per kilo, but before deregulation consumers generally bought it for about TSh30 per kilo. Immediately after deregulation, which allowed private traders to operate legally, the market price dropped to TSh24–26, representing an increase of TSh8–10 over the official, but rarely available, price. For the vast majority of consumers this therefore represented a real price decrease of TSh4–6. At the same time, producers were being paid more; sometimes as much as TSh4 per kilo more than they had received previously from the National Milling Corporation or from parallel market traders. Unsurprisingly, this stimulated production in some areas, such as in Iringa and Arusha regions, thus further dampening down consumer prices as supply increased. In recent years, in the Tanzanian context at least, there has been a rebalancing of the rural and the urban.

The reorientation towards market-based practices and private traders has also often meant that the pan-territorial pricing policies of the state marketing boards have been dismantled. Such policies ensured that all producers, no matter where located, were paid the same price for their crops by the boards. This was an important promoter of regional equality, since farmers remote from major markets and/or good transport infrastructure still had an incentive to produce for the market and could therefore earn some cash. It also meant that their production was subsidised by producers located nearer the market because the costs of buying 'remote' crops was higher, but not reflected in lower prices. Under market conditions this was a price distortion and, according to neo-liberal economics, therefore 'inefficient'. The advent of private trading of crops inevitably therefore led to many 'poorly' located farmers receiving much lower prices or, worse, not being able to sell at all because there was little or no incentive for the traders to come to their areas. In Malawi, for example, this has been a major problem, with many of the old depots previously operated by ADMARC (the parastatal marketing board which, though corrupt and exploitative, was remarkably efficient in operating pan-territorial marketing arrangements) now closed down. Market pricing has also meant that fertiliser (and many other inputs, such as fuel) prices throughout the region have soared – as subsidies have been removed and devaluation has led to import prices increasing. Particularly where land is being relatively intensively farmed (for example in Malawi or the communal areas of Zimbabwe) such inputs are critical for sustainable (let alone increased) agricultural productivity, but even where better crop prices are being realised, these are sometimes outweighed by input price increases. As a result farmers may stop producing certain crops or their yields may reduce as they skimp on purchased inputs, or eschew them entirely (e.g. Scoones 1997). The impact of market-oriented agricultural policies has therefore been very variable across the region.

An important policy issue to affect agricultural change and transformation is related to the question of land. Many development planners believe that in some areas (for example, Rwanda, Burundi, central Kenya, northern Tanzania, parts of KwaZulu-Natal and the Eastern Cape in South Africa) many farms are too small, often 1–2 ha or less, and/or too fragmented to sustain meaningful agricultural transformation (Kimani and Pickard 1998; Lofchie 1987). The argument is that these plots are too small to generate economies of scale, that they militate against introducing machinery to increase productivity, and that they are too small to produce commercial surpluses. For many African governments, expanding cash crop production is a key priority as, under current global economic conditions, it is the only realistic way of increasing foreign exchange earnings (Dorsey 1999). This policy was strongly advocated under structural adjustment, but may conflict with the immediate interests of small-scale farmers. The reality for governments is that if a wholesale land reform programme is introduced – including land consolidation, the establishment of individual landownership rights (as opposed to the present common usufructary rights, see Chapters 1 and 5) and the creation of bigger farm units – the problem of a dispossessed class of landless labourers will become a major political issue. In times of continuing population growth, of many Africans' traditional attachments to land, and of an urban sector unable to absorb the displaced rural population, many governments are reluctant to follow what would be a radical line of action.

However, the question is more than just one of political realities, no matter how critical these may be. Common land rights with no legal individual title may limit farmers' abilities to access credit from banks, but they do provide an economic and social security for families farming in such areas (Mackenzie 1993). Furthermore, small farms make economic sense under circumstances where labour shortages occur at peak times such as land preparation, weeding and particularly harvesting, as happens in many parts of the region. The small size of many farms may therefore have more to do with managing the economic realities associated with labour availability, and less with land pressures or other such explanations. The persistence of small, fragmented farms, therefore, is likely to be an enduring characteristic of agriculture in much of the region for many years to come.

The question of land has another dimension in Zimbabwe, Namibia and South Africa. In these countries there are significant tensions between large commercial farms on the one hand, and small, family farms on the other. In Zimbabwe, fewer than 5,000 commercial, mainly white, farmers controlled large amounts of some of the most fertile land in the country. Operating at a commercial scale, these farms contributed significantly to foreign exchange earnings through the export of commodity crops, particularly tobacco. Since independence in 1980, the Zimbabwe government greatly increased its support of the smallholder sector, resulting in significantly increased smallholder commercial production, particularly of food crops and cotton (Rukuni and Eicher 1994). Some progress was also made in redressing some of the landholding inequities in Zimbabwe up to 1997, with significant improvements in agricultural production and incomes in resettled areas (Bowyer-Bower and Stoneman 2000; Kinsey 1999). Although fears had been expressed that redistributing land from the large-scale commercial to the smallholder sector would undermine national commercial production, this proved not to be the case under the 'legal'[2] programme up until 1997. In terms of land redistributed, this programme was the largest ever undertaken in sub-Saharan Africa – a point which is too rarely recognised. The redistribution process, which gave smallholders free (albeit leasehold) access to commercial land bought usually at market rates by the government, was slow, however, and political pressures for a more radical approach built up in the late 1990s. This led to mass invasions of commercial farms by disparate groups, particularly from 2000, which were often violent, sometimes led by war veterans, and generally supported, and then fostered, by the government which was facing significant internal political opposition and saw this as one way of (re)gaining popular support. In the short term this did create crises in production of some crops so that in 2002 significant maize imports became necessary (although, as discussed earlier, drought was also a significant factor). It remains to be seen what the long-term outcome of the process will be for Zimbabwe in

---

[2.] The subsequent occupations of farms and the government's determination not to compensate commercial farm owners for their land, declaring that this was Britain's responsibility as the former colonial power, are argued by the Zimbabwean government to have become 'legal' via retrospective legislation passed in Zimbabwe's parliament. However, the commercial farmers, the Zimbabwean opposition and the international community generally do not accept that this is a reasonable or legitimate process.

terms of agricultural productivity, but it was an indication that the historical inequalities in land distribution in southern Africa were potential sources of political *and* economic instability.

In South Africa, the land issue has been handled differently. In 1996 the government approved a land restitution programme designed to allow those forcibly removed from land under apartheid to register claims for the restitution of their former land. Two years were permitted for claims to be lodged, and about 63,000 claims were made (although most of these were urban claims) (Du Toit *et al.* 1998). Progress in processing claims was very slow and in July 2000 the Chief Land Claims Commissioner reported to the Parliament that only 6,500 claims (or 10 per cent of the total) had so far been settled. Interestingly, some of the strongest pressure for the programme comes from some members of the commercial farming community who stand to lose land. There may be two main reasons for this. As compensation is to be paid for restituted land, many see a loss of some land as a source of capital to be used to increase levels of farm mechanisation, now that the extremely cheap and exploitable labour, on which much of the farming system was previously based, is no longer as widely available in the new South Africa. It may also be the case that some farmers are shedding and evicting labour to deny those farmworkers any claim to the land on which they have worked, in some cases over several generations.

Whereas most of the discussion so far has been focused mainly on domestic and national scales, there is little doubt that Africa's engagement in the world economy, and the impacts of globalisation, have significant consequences for farmers in the region. The impact of structural adjustment has already been discussed, but there are other concerns. In particular, Africa's unfavourable and largely uninfluential position in the world trading system remains problematic (Allen and Thompson 1997). The terms of trade for the region's commodity crops continue to generate difficulties, and the problem is especially difficult for those countries which are dependent on only one or two crops for the bulk of their foreign exchange earnings. The price index for coffee on the international markets, for example, demonstrates not only the fluctuating nature of the real commodity price for coffee since 1980, but also the persistent nature of the downward price pressure it experiences (Table 6.7). For Kenya and Uganda, for example, such a trend has serious consequences.

A new development in the region's links with the global economy has been the expansion, during the 1990s in particular, of the production of high-quality, premium-price fruit, vegetables and cut flowers for the export market (Barrett *et al.* 1999) (Box 6.2). Producers in Kenya, Zimbabwe

Table 6.7 Export price index for coffee, 1985–97

	1985	1990	1995	1996	1997
Export price index	81	46	83	65	88

Note: Base 1980 = 100.
Source: Europa Publications (1999).

Box 6.2 Horticultural exports in Kenya

The production of selected premium vegetables, fruits and, increasingly, cut flowers for the export market from a small number of African countries has grown rapidly in the 1990s. For Kenya, such horticultural exports now constitute the country's third largest export commodity after tea and coffee. While vegetable and fruit exports have stabilised at about 26,000 t per annum, exports of cut flowers continue to grow at about 20 per cent per year. The export market is very concentrated, with some 85 per cent of horticultural exports going to Britain, Netherlands, France and Germany in that order of importance. Crucial to this has been the expansion of air cargo capacity, as 93 per cent of all these exports from Kenya are exported by air from Nairobi airport, and are available in shops and supermarkets in western Europe within 48 hours of being picked in Kenya. The remaining 7 per cent of exports travel under refrigerated conditions by sea.

In Kenya, there are two identifiable supply chains. The first is that of wholesalers who specialise in bulk, 'loose' (not pre-packaged) sales of a small group of what are termed 'Asian' vegetables (mainly okra, chillies, aubergines and squashes). They tend to deal with medium-scale, and some small-scale producers, in Kenya and sell to small- or medium-scale traders in Europe. Many operate through family and kinship networks and rely on 'handshake' agreements rather than formal contracts. Because they represent only a small part of the overall trade in horticultural exports, this group of wholesalers can experience problems of accessing air cargo space on planes going north.

The second supply chain is dominated by large European supermarkets who control the bulk of the horticultural export trade from Kenya. There is a high degree of vertical integration in production, in that most of the produce is packed and priced in Kenya, to take advantage of cheaper labour rates, and is delivered to Nairobi airport as a finished product. Contracts between European supermarkets and highly capitalised producers in Kenya are central to the production process. Three big producers (Sulmac, Oserian and Homegrown) now account for 45 per cent of all Kenyan horticultural exports. Thus '[t]he requirements of international regulation, and the need for very large consignments of produce in EU markets, have combined to concentrate export horticulture in the hands of the larger and highly capitalized producers' (Barrett *et al.* 1999: 171).

Even European farmers can struggle to meet the exacting requirements of supermarkets and, in Kenya, this lucrative niche agricultural market is largely bypassing small- and medium-scale farmers. Despite its successes, therefore, it is unlikely that horticultural exports are going to have any meaningful impact on agricultural transformation among Kenya's small-scale farmers.

Source: Barrett *et al.* (1999).

and South Africa have been at the forefront of these developments but, increasingly, producers in Tanzania, Uganda, Zambia and Malawi are also expanding output. Typically, these commodities are grown near to international airports and are flown overnight to the markets of western Europe in particular. There have been significant developments within the vicinity of Nairobi, around Kilimanjaro airport in northern Tanzania between the towns of Arusha and Moshi, and around Upington in the Northern Cape in South Africa, using irrigation water from the Orange River.

Micro-environmental operating conditions for agriculture

Capital formation and availability

An enduring problem for the majority of farmers in the region is access to capital, not only for land improvement schemes, but also to finance annual purchases of inputs such as seed, fertiliser and pesticides, as well as to hire labour at peak times (Wanmali and Islam 1997). Many farmers are only too aware of the potential advantages to be gained from the deployment of capital on their farms, and are frustrated at not being able to access the amounts needed to have any real impact on agricultural production. Capital formation from within the sector itself is limited. There is generally a low savings capacity among small-scale farmers for a number of reasons. Typically, only part of the farm is committed to cash crops and the productivity of these is frequently relatively low. Consequently, the margins on these crops are small, as are the resultant cash surpluses. In addition, given the nature of price fluctuations for primary commodities, farmers cannot easily forecast future incomes in subsequent seasons, thus making long-term, or even medium-term, capital investment commitments difficult. With the SAP-inspired abolition of marketing boards (and guaranteed producer prices), this uncertainty has, if anything, increased. There is also a range of non-agricultural demands on household income, for food not produced on the farm, clothes, household items, school fees and so on, all of which reduce the amount of domestically produced capital available for further agricultural development.

Debt further drains rural savings capacity. The extent of rural debt in the region is very uneven at all geographic scales. Even within the same village community, for example, there may be some households which are economically paralysed by indebtedness, and others where debt is the least of their problems. A key element, at least in parts of eastern Africa, is the extent to which households have borrowed from lenders within the community (for example, traders, shopowners or large landowners), and from whom they may have done so over several generations. A common, but by no means the only, borrowing arrangement is a 50/50 crop share. In southern Africa, this may entail a small farmer cultivating land owned by someone else and paying 50 per cent of the crop proceeds as land rent. In parts of East and central Africa, on the other hand, the farmer may borrow money in advance of the crop being planted, and then repay one-half of the proceeds from the crop when sold several months later. The lender 'earns' income on the difference between the amount lent in the first place and the amount paid after selling the crop by the farmer; for the farmer this difference represents the interest payment on the loan. Theoretically, if the proceeds are less than the original loan, the lender shares the loss with the farmer. In reality, this is rarely, if ever, the case. In these circumstances, this difference is carried over into the following year's loan for the farmer, hence increasing the overall size of the debt. Several years of poor harvests can therefore result in a large amount of debt being accumulated. Clearly, the servicing of this debt reduces the farmer's savings capacity to reinvest in his/her own farm, with all the implications for future production levels.

Banks and agricultural credit organisations represent alternatives to such lending systems. However, the organisation of the formal banking system is such that it does not encourage lending to the small-scale agricultural sector (any more than it does to the small-scale urban informal sector). There is the problem, from the banks' perspective, that the priority for most small-scale farmers in the region is to secure a food supply for the household, and cash crop production is frequently only a second priority. There is a very real concern, therefore, that loans may not be properly serviced and repaid. Some argue that this is compounded by the customary tenure system which exists in much of the region. Land titles are typically vested in the community, tribe or clan, although individuals have security of tenure as long as the land is needed and used (Lawry 1993). Indeed, increasingly, farmers 'sell' land to others, although legally they are selling only the *use* of the land and are being compensated in the sale price for improvements made to that land. This process is currently very active in the Coast Region of Tanzania, for example, where a thriving rural land 'market' has emerged over the last 10 years (Briggs and Mwamfupe 1999) (Box 6.3). However, as far as banks are concerned, farmers do not possess legal land titles and therefore have no collateral against which to secure a loan. The creation of agricultural banks, usually by the state, has been a recognition of this problem, and even small-scale commercial farmers have benefited as loans have been made against a regular income from cash crop production. The Agricultural Finance Corporation (AFC) in Zimbabwe, for example, makes loans to smallholders and the repayments are withheld from the cheques they receive for crops sold to government marketing agencies. Such loans are likely to be less forthcoming, however, where liberalisation of the agriculture sector has ended the government's monopoly of trading in certain crops as this ends the government's assured access to the farmers' incomes from crop sales. For example, AFC loans in Zimbabwe were never available for smallholder horticulture as vegetable sales have always been 'uncontrolled'. However, even with these sources of finance, there has still been a preference to lend to smallholders who are deemed to be essentially or largely commercial – in Zimbabwe such farmers often have 'master farmer' status. This is not only because of potential repayment problems, but also because most governments actively encourage the production of export crops to generate

---

Box 6.3 Peri-urban agricultural land markets: Dar es Salaam

Because land titles in Tanzania are legally vested in the state, theoretically there should be no private land market in operation. However, by the 1990s, it was clear that just such a private land market was in existence in the peri-urban zone of the city of Dar es Salaam and, indeed, around a number of other cities in eastern and southern Africa. Two broad reasons explain the emergence of this peri-urban land market. First, the structural adjustment policies which Tanzania has adopted since the mid-1980s have created an economic climate in which domestic economic confidence has slowly grown. Land in the peri-urban zone of Dar es Salaam has therefore come to represent an attractive investment opportunity.

Second, deficiencies in public land management in Tanzania, and especially Dar es Salaam, have encouraged the evolution of an informal land market. Legally, as all land is owned by the state, individuals are meant to apply to the Ministry of Housing, Lands and Urban Development for 33- or 99-year leases on surveyed land in the peri-urban zone (and elsewhere in the urban area). However, such is the inefficiency of the system that only 6 per cent of all applications between 1978 and 1992 were processed. Buoyed by the emergence of a free market, underpinned by structural adjustment, private deals were, however, being made between farmers and purchasers. *De jure*, such deals were made on the premise that the farmer was being compensated financially for the value of permanent crops, such as trees, on the land and any land improvements he/she might have made. This satisfied the letter of the law. *De facto*, the purchaser was buying the land.

During the decade of the 1990s, it became clear that the peri-urban zone shifted from being a zone of survival for many urban people, based on essentially subsistence agricultural production, to being a zone of investment, based on capital investment not just in agriculture but in other sectors as well. A consequence of this has been that low-income groups, including many farmers, have become increasingly squeezed economically and excluded from peri-urban land, as its monetary and investment value has increased.

Source: Briggs and Mwamfupe (1999).

---

foreign exchange, and agricultural banks are seen as a key enabler in this regard.

For many years, a small but important source of investment capital for farms came from remittances from family migrants to the cities (Potts 2000). In particular, remittances from workers in South Africa and Zimbabwe were a major source of rural investment in Malawi, Swaziland, Lesotho and southern Mozambique, as well as in the rural areas of South Africa and Zimbabwe themselves. It was also similar further north. However, during the 1980s and 1990s, the squeeze on urban incomes, as SAPs took effect, became so intense for many urban people that this source of income to the rural areas could no longer be afforded, and so virtually stopped in many instances. Interestingly, a new source of income has emerged in the 1990s with the rise of non-governmental organisations (NGOs) (Mercer 1999). Such organisations, many operating at the small scale within local communities, have enabled rural people to increase incomes through self-help schemes, for example. Consequently, some communities are starting to address the agricultural investment problem in a locally manageable way. This empowerment of local people at the micro-scale, although still limited in the bigger order of things, is one of the more exciting developments in rural Africa in recent years.

Labour issues

Despite continuing population growth, some areas still experience labour shortages at critical times. Reciprocal labour is one way in which communities deal with these problems, but this system is breaking down in many areas as economic pressures increase. Part of the problem is that some of the potential labour force may be absent as migrant labourers in the urban areas – in parts of southern Africa a migrant labour system was so essential to the economies of white settler regimes that the (preferably circular) migration of young, able-bodied men from rural areas was encouraged and institutionalised from the end of the nineteenth century. This has left a long-term legacy of migrancy which has, over many generations, become deeply incorporated into people's rural livelihoods. Particularly where land shortages are acute, migrant remittances may be essential to survival, and in parts of rural Lesotho and South Africa, for example, incomes derived from agricultural production may contribute very little to rural livelihoods (Bernstein 1996; Potts 2000).

Although out-migration can have a negative consequence in terms of shortage of agricultural labour, it can also have potential benefits in that this is one less mouth to be fed in the rural community, as well as being, as indicated above, a potential source of remittances. Another source of change in the availability of agricultural labour has been primary education. As this has expanded, children's contributions to agricultural production have been reduced. Older children may have been involved previously in some of the more onerous agricultural practices, such as planting, weeding and harvesting, and even younger children fulfilled an important role in scaring pests, such as birds and monkeys, from the growing crops to reduce losses from animal damage.

Crucially, it is now at last recognised that women undertake a major part of the agricultural work in the region (Bryceson 1995; Thomas-Slayter and Rocheleau 1995). In many areas, there is a marked division of labour. Women often focus on producing food crops for the household, particularly staples such as maize, cassava and vegetables (but also selling surpluses as a further source of cash income for the household), while men tend to focus on cash crops. In some areas, men are responsible for land preparation. However, even these distinctions are starting to become blurred, particularly as women are increasingly becoming involved in commercial crop production. Nonetheless, for rural women in Africa, agricultural production is only one of a range of productive activities. They also have responsibility for childcare, the collection of water and wood, cooking, cleaning and other domestic tasks. They may also work as wage-labourers on surrounding farms, if the opportunities exist. In Kikuyuland in Kenya, for example, many rural women, as well as fulfilling all the above tasks, also work as weeders and pickers on nearby tea estates. This broad range of female commitments may well be a significant constraint on farm productivity.

A new factor which now affects labour availability is the spread of HIV/AIDS (Barnett and Blaikie 1992; Daniel 2000; see Chapters 2 and 4). Infection rates in eastern and southern Africa are now the highest in the world and, unlike in Europe and North America, infections are more or less evenly divided between men and women. In 2000, it was estimated that over 21 million Africans were HIV+. Although HIV/AIDS was at first seen as predominantly an urban problem in the region, it is now having a significant impact on the rural areas (Box 6.4). In western Uganda, western Tanzania (especially Kagera region), Rwanda and Burundi, there are now villages where there are few and, in some villages, no people in the 20–40-year age-group. Some are absent as urban migrants, but others have died from AIDS or AIDS-related diseases. Many of these are women, infected by returning husbands on annual leave from the cities. Since women are critical to agricultural production, in these circumstances, the onus for producing food often transferred to surviving grandparents, who not only have had to become fully engaged again in agriculture, but also have to care for orphaned grandchildren. This comes at a time in their lives when they themselves were expecting to be supported by their own, now deceased, adult children. Consequently, agricultural production in some areas is now being adversely affected, as the grandparents do not have the same energy to commit to production as the 'missing generation'. Of great concern, there is now evidence of a secondary impact of AIDS with malnutrition increasing among children in these communities.

Future prospects

It is clear that agricultural production in much of eastern and southern Africa still dominates the economic lives of most people. Nonetheless, the signs are that the present declines in production per capita are unlikely to be arrested in the near future. Even though AIDS deaths are reducing population growth rates, as outlined above, the impact of AIDS is a further constraint on agricultural production so it is probable that the net impact of this tragic epidemic on per capita production will be negative. Similarly, the other constraints discussed above are unlikely to be ameliorated on the scale needed to bring about significant agricultural transformation. Indeed, if anything, in the desire to increase foreign exchange earnings, it is likely that government strategies will become increasingly loaded towards commercial agriculture, at least in the short term, with the result that small-scale subsistence agriculture may become even more marginalised. Hence,

---

Box 6.4 Impact of AIDS on agricultural production: a case study from Uganda

The impact of AIDS on agricultural production is well exemplified by a case study of a 45-year-old woman in Uganda whose husband had recently died from AIDS. The family had about 0.5 ha of land to farm on. With its good quality soils and reliable and plentiful rainfall, the farm was of sufficient size to support the family. A range of crops, including green bananas (a food staple in the area), beans, sweet and Irish potatoes, cassava, tomatoes and chilli peppers, was grown to provide dietary variation for the family.

With the death of her husband, the woman's farming problems multiplied (as did a whole host of non-agricultural problems). Crop production on the farm declined as she had many other demands on her time and so was unable to commit as much time as she would have liked to farming. She stopped growing some of the more labour-intensive crops because of this, and hence dietary variation was reduced. In the past, she and her husband may have had some small amounts of cash to buy in some occasional labour to help out at particular labour bottleneck times. But since her husband's death, income had declined, and this was no longer possible.

The problem was exacerbated, however, as she herself had developed AIDS, further reducing her capacity to work. Her time commitment to farming was thus bound to fall further. But she had no choice – there is no social security safety net in Uganda to fall back on. She was particularly concerned about what was going to happen to her children after her own death. Her deceased husband's brother could not look after them; he was already stretched, looking after a third brother's orphaned children, and simply did not have the capacity or resources to take on any more. Her own parents, although still alive, were simply too old to look after their grandchildren, and she had lost touch with her own sisters.

To cap it all, she was now involved in a land tenure dispute with her deceased husband's family. Land law in this part of Uganda (as, indeed, in many other rural areas throughout eastern and southern Africa) is somewhat confused and ambiguous. On top of everything else, the woman was having to deal with gender bias in relation to land rights in the area, and was faced with legal action by her in-laws, which, if successful, would result in her being dispossessed of her last resource, the farm.

Source: Barnett and Blaikie (1992).

---

the tensions between the commercial and subsistence sectors are unlikely to be reduced for the foreseeable future.

On the other hand, there have been some positive features in recent times, and especially so with regard to the small-scale sector. Without doubt, over the last decade, there has been a greater move towards the empowerment of rural people. Self-help ventures, support and advice by small-scale grassroots NGOs, an increasing representation of African farmers less as passive victims and more as agents of their own destiny, and a valorisation of local (or indigenous) knowledges have all contributed to a more positive view of African agriculture (Leach and Mearns 1996; Osunade 1994). Agriculture in the region is still facing a range of serious issues which have to be addressed before meaningful agricultural transformation can be achieved, but there may at least appear to be the start of a change in approach.

References

Adams, W., Watson, E.E. and **Mutiso, S.K.** (1997) Water, rules and gender: water rights in an indigenous irrigation system, Marakwet, Kenya, *Development and Change* **28**(4): 707–30.

Alderson, M. (1999) Household resilience, food security and recurrent exogenous shocks: a study from the semi-arid communal areas of Zimbabwe. Unpublished PhD thesis, Durham University.

Alderson, M. (2001) Household resilience, food security and recurrent exogenous shocks: a study from the semi-arid communal areas of Zimbabwe. Unpublished PhD thesis, Durham University.

Allen, J. and **Thompson, G.** (1997) Think global, then think again – economic globalisation in context, *Area* **29**: 213–27.

Anderson, D.M. and **Broch-Due, V.** (1999) *The Poor Are Not Us: Poverty and pastoralism in East Africa*. James Currey, Oxford.

Barnett, T. and **Blaikie, P.** (1992) *AIDS in Africa: Its present and future impact*. Belhaven, London.

Barrett, H.R., Ilbery, B.W., Browne, A.W. and **Binns, T.** (1999) Globalisation and the changing food networks of food supply: the importation of fresh horticultural produce from Kenya into the UK, *Transactions of the Institute of British Geographers* **24**: 159–74.

Bell, M., Faulkner, R., Hotchkiss, P., Lambert, R., Roberts, N. and **Windram, A.** (1987) The use of dambos in rural development, with reference to Zimbabwe. Unpublished final report of ODA (Overseas Development Administration) Project R3869.

Bernard, F.E. (1993) Increasing variability in agricultural production: Meru district, Kenya, in the twentieth century, in Turner, B.L., Hyden, G. and Kates, R.W. (eds) *Population Growth and Agricultural Change in Africa*. University Press of Florida, Gainsville: 80–113.

Bernstein, H. (1996) South Africa's agrarian question: extreme and exceptional? *Journal of Peasant Studies*, Special double issue on the agrarian question in South Africa **23**(2/3).

Binns, T. (1994) *Tropical Africa*. Routledge, London.

Bowyer-Bower, T.A.S. and **Stoneman, C.** (eds) (2000) *Land Reform in Zimbabwe: Constraints and prospects.* Ashgate, Aldershot.

Briggs, J. and **Mwamfupe, D.** (1999) The changing nature of the peri-urban zone in Africa: evidence from Dar es Salaam, Tanzania. *Scottish Geographical Journal* 115: 269–82.

Bryceson, D.F. (1993) *Liberalizing Tanzania's Food Trade: The public and private faces of urban marketing policy 1939–88.* James Currey, London; Mkuki na Nyota, Tanzania; Heinemann, United States and Canada.

Bryceson, D.F. (1995) *Women Wielding the Hoe: Lessons from rural Africa for feminist theory and development practice.* Berg, Oxford.

Campbell, B.K. and **Loxley, J.** (1989) *Structural Adjustment in Africa.* Macmillan, Basingstoke.

Cornia, G. and **Helleiner, G.K.** (1994) *From Adjustment to Development in Africa: Conflict, controversy, convergence, consensus?* Macmillan, London.

Coulson, A. (1982) *Tanzania: A political economy.* Clarendon Press, Oxford.

Cousins, B. (ed.) (2000) *At the Crossroads: Land and agrarian reform in South Africa into the 21st century.* National Land Committee/PLAAS, University of the Western Cape, Johannesburg/Cape Town.

Daniel, M.L. (2000) The demographic impact of HIV/AIDS in sub-Saharan Africa, *Geography* 85: 46–55.

Dawson, A. and **O'Hare, G.** (2000) Ocean–atmosphere circulation and global climate: the el Nino Southern Oscillation, *Geography* 85: 193–208.

Dorsey, B. (1999) Agricultural intensification, diversification and commercial production among smallholder coffee growers in Central Kenya, *Economic Geography* 75: 178–95.

Du Toit, A., Makhari, P., Garner, H. and **Roberts, A.** (1998) Restitution review: findings and options (draft report), Programme for Land and Agrarian Studies, Cape Town.

Europa Publications (1999) *Africa South of the Sahara*, 28th edn. Europa Publications, London.

FAO (1999) *FAOSTAT Agricultural Data* [apps.fao.org].

Gibbon, P. (1995) *Liberalised Development in Tanzania.* Nordiska Afrikainstitutet, Uppsala.

Griffiths, J.F. (1972) *Climates of Africa.* Elsevier, Amsterdam.

Harvey, C. (1988) *Agricultural Pricing Policy in Africa: Four country case studies.* Macmillan, London.

Hills, R.C. (1978) The organisation of rainfall in East Africa, *Journal of Tropical Geography* 47: 40–50.

Hulme, M. (ed.) (1996) *Climate Change and Southern Africa.* Climate Research Unit, UEA, and WWF International, Norwich and Gland.

Hyden, G. (1980) *Beyond Ujamaa in Tanzania: Underdevelopment and an uncaptured peasantry.* Heinemann, London.

Jackson, I.J. (1989) *Climate, Water and Agriculture in the Tropics.* Longman, Harlow.

Jones, S. (1996) Discourses on land degradation in the Uluguru Mountains, Tanzania: evolution and influences, *Journal of Rural Studies* **12**: 187–99.

Kimani, K. and **Pickard, J.** (1998) Recent trends and implications of group ranch sub-divisions and fragmentation in Kajiado District, Kenya, *Geographical Journal* **164**: 202–13.

Kinsey, B.H. (1999) Land reform, growth and equity: emerging evidence from Zimbabwe's resettlement programme, *Journal of Southern African Studies* **25**(2): 173–96.

Lawry, S.W. (1993) Transactions in cropland held under customary tenure in Lesotho, in Bassett, T.J. and Crummey, D.E. (eds) *Land in African Agrarian Systems.* University of Wisconsin Press, Madison: 57–74.

Leach, M. and **Mearns, R.** (1996) *The Lie of the Land: Challenging received wisdoms on the African environment.* James Currey, Oxford.

Lofchie, M. (1987) The decline of African agriculture: an internalist perspective, in Glantz, M. (ed.) *Drought and Hunger in Africa: Denying famine a future.* Cambridge University Press, Cambridge: 85–109.

Mackenzie, F. (1993) 'A piece of land never shrinks': reconceptualising land tenure in a smallholding district, Kenya, in Bassett, T.J. and Crummey, D.E. (eds) *Land in African Agrarian Systems.* University of Wisconsin Press, Madison: 194–221.

McMaster, D. (1992) Agricultural development, in Gleave, M.B. (ed.) *Tropical African Development: Geographical perspectives.* Longman, Harlow: 192–222.

Maddox, G. (1996) Environment and population growth in Ugogo, central Tanzania, in Maddox, G., Giblin, J. and Kimambo, I. (eds) *Custodians of the Land: Ecology and culture in the history of Tanzania.* James Currey, London: 43–69.

Marcus, T., Eales, K. and **Wildschut, A.** (1996) *Down to Earth: Land demand in the new South Africa.* University of Natal, Indicator Press and LAPC.

Mercer, C. (1999) Reconceptualising state–society relations in Tanzania: are NGOs making a difference? *Area* **31**: 247–58.

Moss, R.P. (1992) Environmental constraints on development in tropical Africa, in Gleave, M.B. (ed.) *Tropical African Development: Geographical perspectives.* Longman, Harlow: 50–92.

Moyo, S. (1995) *The Land Question in Zimbabwe.* SAPES Books, Harare.

Mung'ong'o, C.G. and **Loiske, V.M.** (1995) Structural adjustment programmes and peasant responses in Tanzania, in Simon, D., van Spengen, W., Dixon, C. and Närman, A. (eds) *Structurally Adjusted Africa: Poverty, debt and basic needs.* Pluto Press, London and Boulder.

Nieuwolt, S. (1979) The East African monsoons and their effects on agriculture, *Geojournal* **3**: 193–200.

Obia, G. (1997) Agricultural development in sub-Saharan Africa, in Aryeetey-Attoh, S. (ed.) *Geography of Sub-Saharan Africa.* Prentice Hall, Saddle River, NJ: 286–324.

Okoth-Ogendo, H.W.O. (1993) Agrarian reform in sub-Saharan Africa: an assessment of state responses to the African agrarian crisis and their implications for agricultural development, in Bassett, T.J. and Crummey, D.E. (eds) *Land in African Agrarian Systems.* University of Wisconsin Press, Madison: 247–73.

Osei, W. and Aryeetey-Attoh, S. (1997) The physical environment, in Aryeetey-Attoh, S. (ed.) *Geography of Sub-Saharan Africa.* Prentice Hall, Upper Saddle River, NJ: 1–34.

Osunade, M. (1994) Community environmental knowledge and land resource surveys in Swaziland, *Singapore Journal of Tropical Geography* **15**: 157–70.

Potts, D. (2000) Worker-peasants and farmer-housewives in Africa: the debate about 'committed' farmers, access to land and agricultural production, *Journal of Southern African Studies*, special issue on Southern African Environments **26**(4): 807–32.

Potts, D. and Marks, S. (2001) Fertility in Southern Africa: the silent revolution, *Journal of Southern African Studies* **27**(2): 189–205.

Prothero, R.M. (2000) Health hazards and wetness in tropical Africa, *Geography* **85**: 335–44.

Quinlan, T. (1995) Grassland degradation and livestock rearing in Lesotho, *Journal of Southern African Studies* **21**: 491–507.

Raikes, P. (1988) *Modernising Hunger: Famine, food surplus and farm policy in the EEC and Africa.* James Currey, London.

Rukuni, M. and Eicher, C.K. (eds) (1994) *Zimbabwe's Agricultural Revolution.* University of Zimbabwe Publications, Harare.

Scoones, I. (1997) Landscapes, fields and soils: understanding the history of soil fertility management in southern Zimbabwe, *Journal of Southern African Studies* **23**(4): 615–34.

Stocking, M. (1983) Farming and environmental degradation in Zambia: the human dimension, *Applied Geography* **3**: 63–78.

Swaziland (2001) Common Country Assessment – Swaziland, 1997 [www.ecs.co.sz/cca/cca_2.htm].

Thomas-Slayter, B. and Rocheleau, D. (1995) *Gender, Environment and Development in Kenya: A grassroots perspective.* Lynne Rienner, London.

Tiffen, M., Mortimore, M. and Gichuki, F. (1994) *More People, Less Erosion: Environmental recovery in Kenya*. John Wiley, Chichester.

Timberlake, L. (1988) *Africa in Crisis: The causes, cures of environmental bankruptcy*. Earthscan, London.

Turner, B.L., Hyden, G. and Kates, R.W. (1993) *Population Growth and Agricultural Change in Africa*. University Press of Florida, Gainsville, Fla.

Wanmali, S. and Islam, Y. (1997) Rural infrastructure and agricultural development in southern Africa, a centre–periphery perspective, *Geographical Journal* **163**: 259–69.

Woodhouse, P., Bernstein, H., Hulme, D. with Pippa Chenevix-Trench (eds) (2000) *African Enclosures? The social dynamics of land and water*. James Currey, Oxford.

Worby, E. (ed.) (2000) Political economy of land in Zimbabwe, special issue of *Journal of Agrarian Change*.

CHAPTER 7

Water resources and development challenges in eastern and southern Africa

Richard Taylor

Maji ya kifufu ni habari ya chungu.

This Swahili proverb, roughly translated as 'water in a coconut shell is like an ocean to an ant', speaks to the importance of context in how we perceive resources. Valuation and use of water clearly depend upon whether one is taking shelter in the humid rainforests of the Congo Basin or wandering the arid landscapes of the Kalahari Desert. The relationship between the availability of water and human demand for this vital resource guides discussion in this chapter. It is a complex relationship in eastern and southern Africa where, like many areas of the world at the beginning of the twenty-first century, increasing demand for water must be reconciled with a resource that not only varies dramatically in time and space but is also highly susceptible to contamination.

The chapter focuses on specific ways in which the distribution of water resources in southern and eastern Africa at different scales affects development prospects and provides challenges to policy-makers concerned with economic and human development. To provide the essential physical context within which water resources are a dynamic element, the chapter first provides a brief review of the fundamental controls exerted by landscape and climate on the spatial and temporal (seasonal) distribution of this resource, followed by discussion of relevant hydrological patterns. The chapter then turns to the vexed question of how to measure water availability in relation to human demand, since this is of central significance to human development prospects. Crude, quantitative estimations of a nation's renewable water resources have their uses as a very rough guide to regional differences but, as will be indicated, they also have important limitations. The use pattern of water in the countries of the region varies widely, and is one indicator of sectoral economic development patterns.

Where supply and demand are estimated to be imbalanced, a nation is deemed to be water-stressed. A widely used measure of this relationship is the water stress index. The pros and cons of national measures as indicators of development-related issues in the different countries of eastern and southern Africa are discussed. For example, underlying current evaluations of the quantity and availability of water resources are fundamental assumptions that available water resources are of acceptable quality and that transboundary waters are shared equitably. Neither of these assumptions hold true however, and the implications are explored in detail. Methods of alleviating water scarcity (e.g. interbasin water transfers, trading virtual water) are considered in the final section. The chapter concludes with a discussion of key challenges facing eastern and southern Africa in trying to meet the demand for water today and in the near future.

Physical setting

A fundamental feature of the landscape in eastern and southern Africa is that it exists as a raised platform. This platform lies predominantly around an elevation of 1,000 m above sea level (masl), though considerably higher altitudes are found in East Africa. The elevated position of the land surface, relative to sea level, is evident from the great escarpment that runs along the coastline of Angola, Namibia and South Africa as shown in Figure 7.1 which also shows the main rivers in the region. This has an important impact on these rivers which tend to be characterised by rapids and waterfalls, thereby reducing their usefulness as modes of transport. Broadly speaking, it also poses a fundamental restriction on the development of cost-efficient irrigation since low-lying floodplains and deltas (which have, for example, lent themselves to smallholder irrigation of rice and other crops on a massive scale in South and South-East Asia) are thereby restricted. On the other hand it enhances hydroelectric potential and one river stands out in this respect – the River Congo. This huge river falls nearly 270 m over a distance of only 240 km before it reaches its mouth (Grove 1989). The sheer volume of water involved means that the energy potential in these falls and cataracts has regional implications, as will be clear as the chapter progresses.

Another central feature of the region shown clearly in Figure 7.1 is the East African Rift System (EARS) which extends from the Afar Depression in Ethiopia through Kenya and Tanzania to southern Malawi where the River Shire discharges into the River Zambezi in Mozambique. A western branch of the EARS exists along the western boundary of Tanzania and continues northward through Burundi and Rwanda to the boundary that divides Uganda and the Democratic Republic of Congo. Developed through tectonic activity, the EARS has had a dramatic impact on the landscape and, hence, hydrology of eastern and southern Africa. The mountains associated with the system influence local rainfall and regional patterns of atmospheric circulation and define, to a large extent, patterns of surface drainage (Beadle 1974; Taylor and Howard 1998). Large lakes throughout eastern and southern Africa result from the collection of river flow in

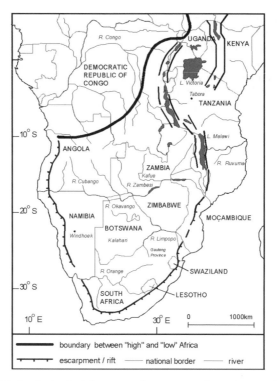

Figure 7.1 Main hydrological features of eastern and southern Africa and the
boundary between 'high' and 'low' Africa
Source: Adapted from Grove (1978).

grabens and other trough-like depressions of the land surface (Ojany 1971;
Taylor and Howard 1999a). Examples include lakes Turkana, Magadi and
Malawi. Along the western branch of the EARS, rifting created lakes Albert,
Edward, Kivu, Tanganyika and Rukwa.

Warping of the land surface in the absence of rifting also determines
surface flows in this region. Lake Victoria was created in this way, as
waters pooling in Lake Victoria eventually eroded an outlet at Jinja and
initiated flow of the Victoria Nile. A similar pattern of downwarping in
the Okavango region of northern Botswana (Chobe fault system) is responsible for disconnecting the river flow of the upper Zambezi from the
Limpopo channel and redirecting it towards the main Zambezi channel
at Victoria Falls, probably during the early Pleistocene (1–2 million years
ago) (Thomas and Shaw 1988).

The climate of eastern and southern Africa is controlled to a large
extent by global patterns of atmospheric circulation and in particular the
movement north and south of the Equator of the inter-tropical convergence
zone (ITCZ). As discussed in Chapter 6, this has enormous implications
for agriculture in eastern and southern Africa as the movement of the
ITCZ determines the annual seasonality of rainfall across tropical Africa.

This is illustrated in Figure 7.2 which indicates how rainfall seasonality varies with latitude in this region. The annual cycle tends to deliver one, distinct, rainy season to latitudes at the southern and northern limits of the ITCZ oscillation, for example in Lusaka and Bulawayo in southern Africa as shown in Table 7.1. In contrast, environments at lower latitudes such as Gulu and Nairobi experience two rainy seasons (i.e. bimodal rainfall distribution). Coastal areas at the south-western tip of the continent (i.e western Cape, South Africa), are beyond the ITCZ oscillation and feature a Mediterranean climate.

Environments at lower latitudes (i.e. between 10° N and 10° S) tend also to receive more rainfall as they are influenced by the atmospheric circulation associated with the ITCZ for a greater proportion of the year (see Table 7.1 where it is evident that annual rainfall tends to decrease with increasing distance from the Equator). Inevitably, atmospheric flows that are responsible for rainfall experienced at specific times and locations in eastern and southern Africa are more complex than the broad patterns presented above. For example, natural variations in the ocean–atmosphere system of the south Pacific, known as the El Niño Southern Oscillation (ENSO), affect weather around the world and there is active research into

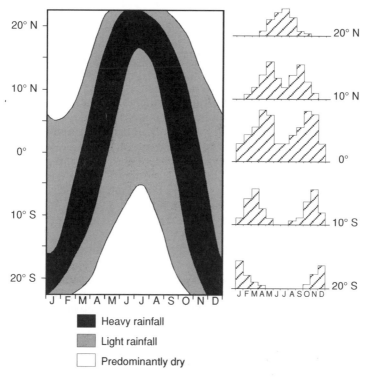

Figure 7.2 The relationship between movement of the ITCZ and seasonal rainfall over tropical eastern and southern Africa
Source: Adapted from McGregor and Nieuwolt (1998).

Table 7.1 Climate data for selected rainfall stations in eastern and southern Africa

Station	Latitude	Longitude	Altitude (masl)	Mean temperature max./min. (°C)	Rainfall (mm)
Gulu, Uganda	02°45′ N	32°20′ E	1,109	29/17	1,470
Yangambi, DRC	00°49′ N	24°29′ E	487	30/20	1,828
Nairobi, Kenya	01°15′ S	36°45′ E	1,798	24/12	1,066
Tabora, Tanzania	05°02′ S	32°49′ E	1,265	29/17	892
Lusaka, Zambia	15°25′ S	28°19′ E	1,274	26/15	837
Bulawayo, Zimbabwe	20°09′ S	28°37′ E	1,344	26/13	589
Tsabong, Botswana	26°03′ S	22°27′ E	962	29/10	271
Kisozi, Burundi	03°33′ S	29°41′ E	2,155	22/11	1,447
Bujumbura, Burundi	03°23′ S	22°21′ E	805	29/19	848
Garissa, Kenya	00°29′ S	39°38′ E	128	34/23	298
Port Elizabeth, South Africa	33°59′ S	25°36′ E	58	22/12	632

Source: Griffiths (1972).

the influence of this phenomenon on rainfall in eastern and southern Africa (Camberlin 1995; Nicholson and Kim 1997; Camberlin *et al.* 2001).

Local features also exert a pronounced effect on rainfall patterns. For example, as shown in Table 7.1, there is increased (orographic) rainfall in mountainous areas such as Kisozi, Burundi, relative to lower-lying areas like Bujumbura, Burundi. Reduced rainfall in eastern Kenya (e.g. Garissa) and Tanzania as well as the Horn of Africa (Somalia, Ethiopia) is believed to stem, in part, from the stripping of moisture from air currents that originate from the Indian Ocean but subsequently pass over the highlands of Madagascar (Kendrew 1961). In contrast to the climates of tropical Africa that are dominated by the movement of the ITCZ, coastal areas at the south-western tip of the continent (i.e. western Cape, South Africa) feature a Mediterranean climate. The strong moderating influence of the ocean reduces seasonal and daily variations in temperature relative to the rest of eastern and southern Africa. Summers are cool and dry, and winters are mild but wet. Rainfall in this region exceeds areas to the north, as shown for Port Elizabeth and Tsabong in Table 7.1, and is strongly dependent upon relief as upland areas (e.g. Cape Fold Mountain Belt) receive significantly greater rainfall.

Upon reaching the land surface, most of the incoming rainfall (70–90 per cent) in eastern and southern Africa is recycled to the atmosphere by evapotranspiration (ET) (Heederik *et al.* 1984; Houston 1990; Howard and Karundu 1992; Taylor and Howard 1996, 1999a). Accurate estimation of ET is, however, extremely difficult. The remainder stays at the land surface and contributes to surface flows via runoff and the discharge of groundwater to stream channels. River flows are strongly influenced therefore by the latitudinal variations in rainfall already described. The rainfall patterns depicted in Figure 7.2 are mirrored in records of maximum and minimum river flows summarised for the Congo and Zambezi river basins in Table 7.2. At low latitudes, the year-round influence of the ITCZ on rainfall gives rise to a fairly consistent and high total discharge for the River Congo at Kinshasa where mean monthly low flow is 55 per cent of the mean monthly peak flow. This contrasts with lower river discharges and more extreme seasonal variations in flow recorded at higher latitudes. For the River Zambezi, mean monthly river flow at Matundo-Cais (Mozambique) is significantly lower, less than a third of the flow in the River Congo if catchment areas for the respective gauges are taken into account (i.e. areally averaged flows). Mean monthly low flow of the River Zambezi is less than half (42 per cent) of the peak flow.

On surfaces of very low relief (e.g. plateaux) in eastern and southern Africa, latitudinal trends in rainfall give rise to seasonally flooded wetlands. These are areas where the flora and fauna have adapted to regular or permanent inundations of the land surface by water. Due, in part, to low surface gradients, seasonal inputs of rainfall dramatically increase flooded surface areas. For example, a doubling and quadrupling of floodplain extent is observed in the Kafue Flats of Zambia and the Okavango Delta of Botswana respectively (see Table 7.3). The increased ET and leakage to the subsurface (i.e. indirect groundwater recharge) that result from the inundation of the floodplain, serve to regulate runoff (Mumeka 1992 cited in Thompson 1996).

Groundwater resources across much of eastern and southern Africa are also rejuvenated by heavy rainfalls associated with the movement of the

Table 7.2 Maximum and minimum, mean monthly discharges for the River Congo and River Zambezi

Basin	Gauged area (km²)	Station (lat., long.)	Discharge[a] max. (m³/s)	Discharge[a] min. (m³/s)
Congo	3,475,000	Kinshasa (04°30′ S, 15°30′ E)	56,081	31,087
Zambezi	940,000	Matundo-Cais (16°09′ S, 33°35′ E)	4,634	1,954

[a] Mean monthly discharge.
Source: UNESCO (1995).

Table 7.3 Selected, seasonally flooded wetlands of eastern and southern Africa including areas of permanent and seasonal inundation

Wetland	Location	Permanent area (km²)	Seasonal area (km²)	Ratio[a]
Bangweulu	Zambia	2,500	8,500	1 : 3.4
Kafue Flats	Zambia	13,000	28,000	1 : 2.2
Okavango Delta	Botswana	3,000	12,000	1 : 4
Upemba	DRC	4,500	8,500	1 : 1.9

[a] Permanent : seasonal.
Sources: Thompson (1996) and references therein.

ITCZ (Adanu 1991; Taylor and Howard 1996, 1999a). The availability of groundwater is very much influenced by the underlying geology which affects the incidence of aquifers – geologic materials that transmit and store groundwater in sufficient quantities to be of use by humans. Productive aquifers include the Karoo sandstones, located primarily in South Africa, as well as the Kalahari and Congo groups of sandstones and limestones that extend roughly along a central longitude (25° E) from southern Africa to the Congo Basin (Kehinde and Loehnert 1989; Wright 1992). Included in the Kalahari group is the vital dolomite aquifer that supplies the city of Lusaka. The dominant geology underlying much of eastern and southern Africa is, however, much less transmissive and consists of weathered and fissured crystalline rocks such as granite, gneiss and quartzite (Chilton and Foster 1995; Taylor and Howard 2000).

The large storage capacity of many groundwater environments, particularly sedimentary aquifers (e.g. sandstone, limestone), gives rise to groundwater flow times that can extend to many thousands of years. Such regional-scale aquifers can consequently store groundwater recharged under climates that may differ from the present day. Indeed, many desert environments overlie massive reserves of groundwater resources referred to as 'fossil groundwater'. For example, groundwater in the Kalahari sands of central Botswana was recharged primarily during the late Pleistocene (11,000–30,000 years ago) when climates were wetter and cooler (de Vries *et al.* 2000). During this time, increased rainfall created palaeolake Makgadikgadi which extended across all of northern Botswana including the Okavango Delta (Thomas and Shaw 1991). In contrast, modern recharge in central Kalahari is minimal (<1 mm per year) but increases to 5 mm per year in eastern Botswana.

Estimation of renewable water resources

The sustainability of any development of water resources depends fundamentally upon whether the resources consumed through such development are renewed. Development and management of water resources are, but for a few exceptions described later, conducted nationally. It follows, therefore, that assessments of the sustainability of water development activities require national estimations of renewable water resources. Shiklomanov (2000) defines renewable water resources as those replenished by the annual turnover of water on the earth. Estimation of renewable water resources on a national basis is, however, highly problematic. A key difficulty is the discrepancy between hydrological and political boundaries.

The most widely cited estimates of renewable water resources for individual nations derive from records of river flow (Gleick 2000; Shiklomanov 2000) and are explicitly equated with annual river 'runoff' in these assessments. The countries of eastern and southern Africa are ranked according to estimates of their renewable water resources (km³ per year) in Table 7.4. This shows the DRC to be exceedingly water-rich, with approximately 2 per cent of the world's renewable freshwater supply (est. 44,800 km³ per year). Smaller countries, such as Burundi and Swaziland, naturally

Table 7.4 Water resources, use and availability for eastern and southern African countries, ranked by annual renewable water resources

Country	Annual resources[a] (km³/year)	Annual water use[b]				Year 2000 population (est. ×10⁶)	Per capita availability[c] (m³/year)
		Total (km³/year)	Domestic (%)	Industrial (%)	Agricultural (%)		
DRC	1,019.0	0.36	61	16	23	51.75	19,690
Mozambique	216.0	0.61	9	2	89	19.56	11,040
Angola	184.0	0.48	14	10	76	13.60	13,529
Zambia	116.0	1.71	16	7	77	9.13	12,700
Tanzania	89.0	1.17	9	2	89	33.69	2,640
Uganda	66.0	0.20	32	8	60	22.46	2,940
South Africa	50.0	13.31	17	11	72	46.26	1,080
Namibia[d]	45.5	0.25	29	3	68	1.73	26,300
Kenya	30.2	2.05	20	4	76	30.34	995
Zimbabwe	20.0	1.22	14	7	79	12.42	1,610
Malawi	18.7	0.94	10	3	86	10.98	1,700
Botswana[d]	14.7	0.11	32	20	48	1.62	9,074
Rwanda	6.3	0.77	5	2	94	7.67	821
Lesotho	5.2	0.05	22	22	56	2.29	2,270
Swaziland	4.5	0.66	2	2	96	0.98	4,600
Burundi	3.6	0.10	36	0	64	6.97	516

[a] Estimates made between 1984 and 1994 except (1970).
[b] Estimates made between 1984 and 1994 except Swaziland (1980) and Uganda (1970).
[c] Availability is based on year 2000 population estimates.
[d] See text for explanation of high per capita figures for Botswana and Namibia, which are misleading.
Source: Gleick (2000).

possess significantly lower volumes of renewable (annual) water. However, although such estimates are commonly reiterated in the literature (Gardner-Outlaw and Engelman 1997; World Resources Institute 2000; World Bank 2001), they must be regarded with great caution as fundamental limitations exist in their estimation. Most importantly, annual river flows, upon which they are based, do not include ET. As a result, soil water – water used by plants, including rainfed and irrigated crops and grazing – is not included in the estimate of a nation's water resources. Furthermore the data in Table 7.4 for agricultural water *use* only refer to water used by *irrigated* agriculture. Given the vital significance of *rainfed* agriculture to livelihoods and national economies in eastern and southern Africa, this is a fundamental problem for analysis of most of the region since the majority of agricultural production is using water other than that shown by these estimates. In other words, in most of the countries in the region, any trends which appear to indicate pressure on 'renewable' water resources (i.e. runoff) do not suggest an imminent crisis for most of the people working in agriculture (assuming that any policies to rectify the situation do not affect soil water). This contrasts with the situation in, for example, North Africa or the Middle East, where large-scale 'modern' irrigation using rivers and groundwater is extremely significant, underpins agricultural production, and is often quite clearly unsustainable (Allan 2001a). This is not to say that irrigated agriculture is not of importance in some eastern and southern African countries (as will be shown below), but the scale of water use in this sector, *in relation to* national renewable resources, does not begin to compare with the situation in North Africa and the Middle East where the rate of annual withdrawal (usually primarily driven by irrigation) is often well in excess of annual internal renewable water resources. The highest rate of withdrawal in southern and eastern Africa is in South Africa, estimated at around 30 per cent, and for most of the other countries it is, as yet, negligible (World Resources Institute 2002). One further gap in current estimations of national water resources is that groundwater which does not contribute to riverflow (i.e. groundwater that discharges to a lake or wetland or directly to the sea), is excluded from estimates that are based on annual river 'runoff'.

Equitable sharing of transboundary riverflow is also assumed in national water resource calculations since half of the inflow from the adjacent, upstream riparian nation is included in each country's tally (Shiklomanov 2000). Such an assumption is discussed further below but is unsupported by experience. Accurate estimations of riverflow are, furthermore, constrained by the limited distribution and time series of hydrological measurements in eastern and southern Africa that commonly compel indirect assessment (e.g. correlation models) of water resources. The reliability of estimates derived from indirect assessments is usually untested yet such data are commonly included in compendia of water resources (e.g. World Resources Institute 2000; World Bank 2001). Finally, it is worth noting that estimation of renewable water resources does not include fossil groundwater resources which, though not renewable or sustainable, may be locally important (e.g. Kalahari sands).

For national measures of water resources to be more representative and useful, improvements are required in both hydrological monitoring and methods of estimating national renewable water resources. The adherence to political boundaries might usefully be abandoned in favour of altern-ative models such as catchment or basin-wide water management. In addition, the description of a nation's renewable water resources by a single value is inherently unable to represent temporal and spatial variations in water resources. Such variations, largely determined by tectonics and movement of the ITCZ in eastern and southern Africa, are considerable. Strong interannual and interdecadal variabilities in national water resources also exist and arise from global phenomena (e.g. ENSO, orbital insolation).

Use of water resources

Water is unique among natural resources in that its consumption, both directly and indirectly, in the production of food, is fundamental to life. Water is also used to generate electricity (hydroelectric power) and in the manufacture of many of the products that people use. Human use of water can consequently be classified into three broad classes: domestic, agricultural and industrial.

Domestic use

Domestic uses of water include household needs (drinking, cooking and washing) as well as municipal and commercial activities. In urban areas of eastern and southern Africa, domestic access to water is commonly gained from piped supplies in which water is drawn from a centralised reservoir and delivered to consumers via a standpipe or tap. Sources of water to reservoirs include lakes, river dams, springs (i.e. gravity schemes) and wells. In rural areas, water is commonly collected from individual, point sources that are supplied by groundwater (e.g. well, spring), rainfall (e.g. rooftop catchment) and surface water (e.g. lake, river, wetland). Urban population growth in the region in combination with lack of financial capacity has overwhelmed many urban, piped supplies and led, in humid areas, to the development of point sources of water in urban and peri-urban environments. Seasonal changes in the availability of water also force consumers to draw water from a mix of sources through the year. Household collection, primarily conducted by women, often involves great ingenuity in meeting water needs. As the Swahili proverb states, *mwanawaki ni maji* – women are water.

Estimation of the amount of water used domestically is an exercise about which there is commonly little agreement due, in part, to the absence of reliable data in unmetered areas. Under the United Nations International Drinking Water Supply and Sanitation Decade (IDWSSD) from 1981 to 1990 (renewed from 1991 to 2000), a minimum, domestic re-quirement of 20 l per person per day was suggested. Estimates of national, domestic consumption in Table 7.4 assume rates of 50–100 l per person per day in most parts of eastern and southern Africa (Gleick 2000; Shiklomanov

2000). A key limitation to such 'blanket' estimates of domestic water use is that consumption is unrelated to access. For instance, in refugee camps and water-scarce regions, per capita (person) consumption may be as little as 5 l per day whereas in areas of Europe and North America with in-house taps consumption is estimated between 500 and 800 l per person per day.

Agricultural use

Similar to global patterns, most water in eastern and southern Africa is used for agriculture – in all the countries except the DRC and Botswana more than 50 per cent of annual water use is estimated to be used in this sector (see Table 7.4). Agricultural use here refers to both irrigation of cropland and watering of livestock. Estimates, however, are imprecise and based upon both available records of water used for irrigation and comparative analyses involving countries featuring similar climates, economies and land cover (Shiklomanov 2000). Recent efforts to improve food security in areas that experience low or seasonal rainfall (e.g. Kenya, Zimbabwe, South Africa), typically involve increasing the area of arable land under irrigation. Table 7.5 shows, in the left-hand columns, the countries of the region ranked by total agricultural water use. South Africa clearly stands out, as here irrigated agriculture is a very important element of the commercial farming sector. This sector is dominated by white

Table 7.5 Countries ranked by total agricultural water use and industrial water use

Country	Annual water use (km^3/yr)			
	Agricultural	Total		Industrial
South Africa	9.583	13.31	South Africa	1.464
Kenya	1.558	2.05	Zambia	0.120
Zambia	1.317	1.71	Zimbabwe	0.085
Tanzania	1.041	1.17	Kenya	0.082
Zimbabwe	0.964	1.22	DRC	0.058
Malawi	0.808	0.94	Angola	0.048
Rwanda	0.724	0.77	Malawi	0.028
Swaziland	0.636	0.66	Tanzania	0.023
Mozambique	0.543	0.61	Botswana	0.022
Angola	0.365	0.48	Uganda	0.016
Namibia	0.170	0.25	Rwanda	0.015
Uganda	0.120	0.20	Swaziland	0.013
DRC	0.083	0.36	Mozambique	0.012
Burundi	0.064	0.10	Lesotho	0.011
Botswana	0.053	0.11	Namibia	0.008
Lesotho	0.028	0.05	Burundi	0.000

Notes: As for Table 7.4.
Source: calculated from Table 7.4.

farmers, due to the legacy of apartheid, and the provision of this water was heavily subsidised under the apartheid regime, giving this sector a major productive advantage. Very similar patterns can be found in Zimbabwe, which is also near the top of the table. Its total use of water by irrigated agriculture is much lower than South Africa's, but use in both countries is roughly proportional to the number of commercial farmers in each, estimated at about 4,000 in Zimbabwe (before the large-scale evictions from commercial farms) and 50,000 in South Africa. However, in most of the countries irrigated agriculture uses less than 1 km^3 of water per year and, as in most of Africa where it is estimated that less than 5 per cent of cropland is irrigated (Khroda 1996), large-scale irrigation is limited (although its share of national water use can still be rather significant). It is useful to recognise however that 'traditional' smallholder forms of irrigation such as use of dambos (wetlands) in Zimbabwe, Zambia and Malawi, are widespread and important for local food security, but rarely included in statistics on irrigated agriculture at either the national or global level (Bell *et al.* 1987).

Countries seeking to improve their self-sufficiency in food production are expected to increase their agricultural use of water (i.e. through large-scale irrigation). Such use is intensified by the high evapotranspirative flux over many parts of eastern and southern Africa that currently requires, per unit output, application of quantities of water (e.g. 15,000–20,000 m^3 per hectare) that are up to twice that applied in temperate latitudes. Despite improvements in the efficiency of irrigation, large-scale agricultural use of water necessarily depends upon exploitation of groundwater and/or surface water reservoirs. Due to strong seasonality in riverflow, particularly at high latitudes in southern Africa (Table 7.2), large dams are often constructed to store peak flows for use during low-flow periods. Examples include the River Zambezi which is dammed both at Kariba in Zambia/Zimbabwe and Cahora Bassa in Mozambique. Large dams require, however, the resettlement of people living in areas flooded upstream and induce significant evaporative water losses from the standing water. Furthermore, lower and more regulated flows disrupt floodplain agriculture as well as fishing and animal husbandry downstream.

Traditional approaches to agriculture in eastern and southern Africa utilise seasonal river flows and may combine flood agriculture with fishing and dryland grazing for livestock (Adams 1992; Scudder *et al.* 1993). In sloping areas of moderate to high relief, small-scale 'valley dams' may be constructed using local materials to trap seasonal rainfall for watering cattle and irrigating adjacent lands. In floodplain areas of low relief (e.g. Okavango Delta in Botswana, Kafue Flats in Zambia), farming follows an annual sequence of events which begins with planting on the floodplain prior to flooding. When rains come, livestock are directed to graze in upland areas while cultivation on the floodplain is conducted during and immediately after flooding. As flood waters recede, stored soil moisture and water retained in pools (e.g. bunds) are used for another growing season. Livestock are then moved to lowland areas for grazing. Improvement and expansion of such farming methods, sympathetic to local ecology, are likely to prove a far more sustainable use of water for agriculture than large-scale dams and 'high-tech' irrigation. Such developments would also greatly decrease the 'modernising' expectation that a scale shift is needed or inevitable in the use of water resources for agriculture in many countries in the region – bearing in mind that soil water is not included in such analysis.

Industrial use

Water used for industrial purposes in eastern and southern Africa is influenced by mining. This is reflected in the proportion of water used for industry (>10 per cent) by countries such as Botswana, the DRC, Angola and South Africa (shown in Table 7.4). Table 7.5, which ranks the countries of the region by total industrial water use, also indicates this factor, with Zambia shown to be the second largest user. The relative importance of the non-mining industrial sector (including processing of minerals) in South Africa, Zimbabwe, Zambia and Kenya is also indicated in this table. Regionally, manufacturing sectors are not well developed. Indeed, some are now stagnating or in decline due to trade liberalisation (see Chapters 3 and 10) so that, in many countries, industrial water use represents less than 5 per cent of total consumption whereas, globally, industry accounts for 23 per cent of total water use (Falkenmark and Widstrand 1992). Yet again it should be noted that these national estimates are crude and based upon both available records of industrial withdrawals of water and comparative analyses involving countries featuring similar industries and economies (Shiklomanov 2000).

Hydroelectric power (HEP) generation induces significant evaporative losses that are also reflected in estimates of industrial water use for both Uganda (Owen Falls Dam) and Zambia (Kariba Dam) in Table 7.4. With the exception of South Africa, current demand for electricity is low, as only a small fraction (<10 per cent) of the population of most nations in eastern and southern Africa has household access to electricity (SAD-ELEC 1996 cited in Horvei 1998). Large increases in HEP generation in southern Africa are expected due to both rising demand, that is associated urbanisation and economic development, and the high potential for HEP generation along many watercourses including, most importantly, the River Congo in the DRC. To date, only a small fraction (2–3 per cent) of the potential for HEP generation has been developed in the DRC (SADC 1996). Development of the Inga Falls on the River Congo alone could generate 45,000 MW of electricity (Horvei 1998), sufficient to satisfy the entire electricity demand of the southern African regional grouping (SADC) in 2000 (Pitman and Hudson 1997).

The capacity of the River Congo to generate HEP is of great regional significance and is widely regarded as a primary reason why SADC was keen for the DRC to join the grouping once a seemingly sympathetic change of regime (i.e. from Mobutu to Laurent Kabila) occurred in 1997. The DRC is also part of the Southern Africa Power Pool (SAPP), a parastatal utility corporation of the Southern African Development Community that was launched in 1995 to facilitate development of a regional transmission network and infrastructure in power generation (Horvei 1998). SAPP seeks, among other objectives, to address imbalances in the demand and supply of electricity and to promote co-operation and cross-border trading in electricity among the region's utilities. The corporation will undoubtedly be tested by current proposals to develop further HEP capacity on the River Zambezi at Batoko Gorge downstream of Victoria Falls on the border between Zimbabwe and Zambia, and at Tete in Mozambique (Mepanda Uncua Dam). Both projects are controversial, not least because of their significant ecological (and human) costs of development (World Commission on Dams, 2000). Poor regulation of flow along the Zambezi by existing dams (Kariba, Cahora Bassa) already affects both the efficiency of HEP generation and downstream communities when floodgates on both dams are opened simultaneously.

Supply and demand – the water stress index

The 'water stress index', devised by Falkenmark *et al.* (1989), attempts to relate the supply of renewable water resources to demand. Although the simplicity of its derivation (reviewed below) has since been criticised by the authors themselves, the index remains a widely cited measure of water availability (Coetzee 1994; Khroda 1996; Ong'wen 1996; Gardner-Outlaw and Engelman 1997; Gleick 2000). Under the water stress index, per capita water demand is estimated and related to water resources that are available nationally. Domestic demand for drinking, washing and cooking is set to 100 l per capita per day. Note that this is well above the actual average use in eastern and southern Africa today. Demand for irrigated agriculture and industry is estimated, on a per capita basis, to be 20 times domestic demand (again, this is well above the existing multiple for most of the region as indicated by Table 7.4). Total per capita demand is calculated, therefore, at 840 m^3 per year. According to the water stress index, a nation that has sufficient resources for more than twice the estimated demand of each citizen (i.e. >1,700 m^3 per capita per year) is designated as relatively sufficient in water resources. Such countries are expected to experience no, or only intermittent or localised, water shortages. Nations with less than this amount are said to be under conditions of water stress where chronic and widespread water supply problems are expected. Nations with renewable water resources of <1,000 m^3 per capita per year are said to be under conditions of water scarcity in which chronic water shortages are expected to cause serious environmental degradation and hinder economic development.

Use of the water stress index is intended to highlight current water problems and to anticipate future water shortages. Using population estimates for the year 2000, per capita water availability for each country in eastern and southern Africa has been listed in the last column of Table 7.4. Most countries are relatively sufficient with respect to water resources. According to this measure, however, South Africa and Zimbabwe are under conditions of water stress, and Burundi, Rwanda and Kenya are facing water scarcity. Although the simplifications, omissions and assumptions involved mean that the representativity of estimated water demand is dubious, the index is easily comprehensible and allows for simple, relative comparisons between nations. According to the index, water shortages should be expected in densely populated countries such as Rwanda and Burundi and nations with low quantities of renewable water resources relative to their populations such as Kenya, South Africa and Zimbabwe. Several important limitations to the water stress index merit, however, further discussion.

The index is, by definition, linked to estimates of population. Due to the difficulties of estimating population growth accurately, predictions of water stress, like the weather, become increasingly speculative the further that one tries to predict into the future. This is ably demonstrated by Gardner-Outlaw and Engelman (1997) who report global predictions of the number of people living under conditions of water stress or scarcity in 2050 ranging from 2 billion to 3.5 billion depending upon which prediction of population is applied. It is also, as noted before, based on estimations which ignore soil water and the sectoral composition of the economy. In countries like Rwanda and Burundi, for example, where industrial demand for water is, and is likely to remain for the foreseeable future, relatively insignificant, and most people's livelihoods depend on soil water, use of the index will significantly underestimate the availability of freshwater resources. Critically, the water stress index bears no relation to the *provision* of water resources in a country. A country like Uganda may be relatively sufficient in renewable water resources (2,940 m^3 per capita per year, Table 7.4) while half of its population is without access to potable water (Bellamy 2000; WHO 2000). Criticism of population-based predictions of water stress and scarcity does not, however, deny the global crisis of limited access to potable water (and sanitation).

As already mentioned, current appraisals of a nation's renewable water resources (Gleick 2000; Shiklomanov 2000) and, hence, water stress index, neither consider nor represent spatial and temporal variability in the supply and demand of water. This shortcoming is particularly revealing in the case of Namibia. Despite 'possessing' a significant quantity of renewable water resources (45.5 km^3 per year) and, seemingly, the region's highest per capita water availability of 26,300 m^3 per year, river flow in Namibia is highly seasonal and confined almost entirely to its borders (e.g. Rivers Orange, Cunene and Cubango). Groundwater resources are also limited. Urbanisation, mining developments and a degree of industrial development have, therefore, led to critical water shortages and overpumping of fossil groundwater. In other words, Namibia evidently suffers conditions of water scarcity, rather than the relative sufficiency in water resources suggested by the water stress index. At the end of 1996, for example, Windhoek, the capital city, was experiencing major water problems and the dams supplying the city were expected to run dry by April 1997 if significant rains did not fall. The Department of Water Affairs was forced to pump water from a disused vanadium mine at Berg Aukas near Grootfontein, which was estimated to be able to supply 4 million m^3 of water for one year only. A very similar argument can be made for Botswana, which appears far from water-stressed in Table 7.4, but again a geographical mismatch between urban water demand and water resources has necessitated significant investment in the North/South Carrier, a pipeline from the north of the country (Turton 2000). At another level of analysis, it might simply be observed that both countries are dominated by deserts. The failure of the water stress index to capture the significance of soil water for most people in southern and eastern Africa is thereby well illustrated.

There are two further assumptions that underlie assessments of a nation's renewable water resources and water stress index: the acceptable quality of water resources and equitable sharing of transboundary water resources. These are considered in turn below.

Quality of water resources

Current estimations of renewable water resources (Gleick 2000; Shiklomanov 2000) and assessments of the water stress index (Falkenmark *et al.* 1989) do not consider the quality of water that is available or required. This is a glaring oversight considering both human activity (e.g. sewage, agriculture) and natural processes (e.g. rock weathering) can significantly impair the quality of water resources for future use. Indeed, deterioration of water quality poses a significant constraint to the availability of freshwater resources in eastern and southern Africa. Note that freshwater is typically defined as water with less than 1 g of dissolved solids per litre.

Water quality is a relative measure that is assessed against criteria that vary according to intended use (e.g. domestic, agricultural, industrial). The best-defined water quality guidelines are those for drinking water (WHO 1993). In a general sense, safe or potable water for human consumption is defined as that which is free from micro-organisms or chemical substances in concentrations that are able to cause illness (Lloyd and Helmer 1991). The principal threat to the quality of domestic water supplies in most areas of eastern and southern Africa is sewage. Gastrointestinal illness (diarrhoeal disease) derived from acute consumption of faecally contaminated water is a major cause of infant mortality across eastern and southern Africa. According to the World Health Organisation (2000), diarrhoeal diseases account globally for 3.4 millions deaths each year.

Containment of sewage through adequate sanitation (e.g. pit latrines, septic tanks or sewerage) and protection of water sources from surface-washed wastes are central to efforts to prevent sewage contamination of water sources and incidence of water-related disease. In terms of disease prevention, studies in Botswana and the DRC also highlight the importance of good hygiene to prevent incidental contamination of household water after collection (Kaltenthaler and Drasar 1996; Manun'Ebo *et al.* 1997). Recent, national statistics of access to safe water and sanitation are presented in Table 7.6. Data derive from a recent, global assessment of water supply and sanitation (WHO 2000). The purpose of the study was to produce reliable, representative statistics, as it was recognised that rapid and implausible changes in the coverage of water supply and sanitation for individual countries have been observed from one assessment to the next as highlighted in Box 4.3. The data indicate that, despite population growth,

Table 7.6 Percentage of population with access to safe water and adequate sanitation in 1990 and 2000

Country	Access to safe water		Access to sanitation	
	1990	*2000*	*1990*	*2000*
Angola	n.d.	38	n.d.	44
Botswana	95	n.d.	61	n.d.
Burundi	65	n.d.	89	n.d.
DRC	n.d.	45	n.d.	20
Kenya	40	49	84	86
Lesotho	n.d.	91	n.d.	92
Malawi	49	57	73	77
Mozambique	n.d.	60	n.d.	43
Namibia	72	77	33	41
Rwanda	n.d.	41	n.d.	8
South Africa	n.d.	86	n.d.	86
Swaziland	n.d.	n.d.	n.d.	n.d.
Tanzania	50	54	88	90
Uganda	44	50	84	75
Zambia	52	64	63	78
Zimbabwe	77	85	64	68

Notes: n.d. no data.
Source: WHO (2000).

modest increases in the coverage of water supply and sanitation have occurred in several countries between 1990 and 2000. The effect of instability during this period on the provision of these basic services is, however, evident in other countries such as Rwanda and the DRC.

Groundwater is an important source of safe water because the soil matrix serves both to filter sewage-derived (faecal) pathogens such as bacteria and protozoa and adsorb smaller, viral pathogens. Microbial contamination of groundwater sources still occurs (e.g., Barrett and Howard 2002) but the extent and incidence are significantly reduced relative to exposed, surface-water sources (e.g. rivers, lakes). On account of its generally superior microbial quality and distribution in eastern and southern Africa, groundwater is currently being developed for low-cost (i.e. untreated) town water supplies (e.g. small town water and sanitation projects supported by the World Bank), though the sustainability of these schemes is unresolved. Much larger groundwater-fed urban water supplies exist in both Namibia (Windhoek) and Tanzania (Dodoma).

Chemical contamination of drinking water also results from sewage. Nitrate and other nitrogen-bearing compounds (nitrite) that are leached from sewage, can lead to illnesses (e.g. methaemoglobinaemia, liver cancers) that, like other chemical contaminants in water, arise from chronic consumption at elevated levels. Apart from the impacts on human health, leaching of nitrogen (N) and other nutrients like phosphorus (P) can lead to algal blooms (i.e. eutrophication) and undesirable growth in aquatic plants (e.g. hyacinth, duckweed). These growths consume available dissolved oxygen and render surface waters unfit for many aquatic species (e.g. fish) and many uses including drinking water. Similar to sewage contamination, application of fertilisers on agricultural lands inevitably leads to excess nutrients (N, P) in the soil that are subsequently leached by runoff and groundwater recharge to surface waters.

On lands that are intensively farmed in Kenya, Zimbabwe and South Africa, pesticides have regularly been applied to control insects and weeds. Such chemical controls are also occasionally employed to halt the advance of invasive plant species such as water hyacinth. However, the properties of toxicity and persistence that constitute a good pesticide such as DDT (dichloro-diphenyl trichloroethane), aldrin or atrazine also serve ultimately to degrade aquatic environments and severely constrain future use of contaminated waters. Very few data exist to assess the extent to which freshwater resources have been degraded by pesticide use in eastern and southern Africa. Similarly, industrial contamination is largely unmonitored due, in part, to the difficulty and expense of analysing industrial contaminants such as solvents and petroleum products in water. Although both intensive agriculture and industrial activity are currently limited in many areas of eastern and southern Africa, careful monitoring will be required to ensure that expected increases in agricultural and industrial activity do not constrain future availability of water of acceptable quality.

Undesirable levels of chemical solutes that result naturally in groundwater from the weathering of materials at the earth's surface, also restrict the availability of usable water resources. High levels of fluoride in groundwater-fed water supplies cause fluorosis (mottling of the teeth) and result from the weathering of volcanic rocks like those associated with the East African Rift System. Weathering of crystalline rocks such as granite and gneiss often yields iron-rich groundwater that is unpalatable and stains clothes and pots (Taylor and Howard 1995). The extent of the iron problem is evident from the distribution of reddish-brown (iron-bearing) lateritic soils across eastern and southern Africa that similarly derive from the weathering of crystalline rocks (Taylor and Howard 1999b). Rock weathering and, hence, leaching of metals into aquatic environments can be enhanced through the exposure of (unweathered) rock surfaces by mining (e.g. Bugenyi 1982; Bugenyi and Lutalo-Bosa 1990; Muwanga 1997).

Evident from this discussion is that the presumption that available water resources in eastern and southern Africa are of acceptable or usable quality is invalid as both human activity and natural processes can significantly impair the quality of water resources for future use.

Hydropolitics of transboundary watercourses

Assessment of renewable water resources and, hence, the water stress index presume equitable sharing of renewable water resources among riparian nations that share a transboundary watercourse. This assumption is, however, commonly threatened by two basic but fundamentally opposed positions that, in one form or another, characterise most riparian debates:

territorial sovereignty and riverine integrity. Territorial sovereignty is the claim that a nation has absolute rights to water flowing through its territory. In contrast, riverine integrity is the claim that every riparian is entitled to the natural flow of a river system crossing its borders. According to Wolf (1999), claims are usually determined by hydrography (the origin of a river or aquifer) and chronology (how long the river or aquifer has been used). For example, an upstream country, like Zambia in the case of the Zambezi Basin, may assert its right to use riverflow as it sees fit, whereas a downstream country like Zimbabwe or Mozambique may, through long-developed dependency, demand its right to the unobstructed, natural flow of the River Zambezi. Clearly, conflicts over running water are asymmetrical as the upstream riparian controls the quantity and quality of flow.

In eastern and southern Africa, major river basins are shared, usually by several countries (see Figure 7.1 and Table 7.7). The potential for conflicts over shared water resources exists as several countries are currently under conditions of water stress (South Africa, Zimbabwe, Kenya) and experiencing local water shortages (e.g. Botswana, Namibia). The importance of clear legislation in conflict resolution is demonstrated by the case of Namibia where the Orange River forms its southern boundary with South Africa (Figure 7.1). From 1847 to 1915, the precise location of the boundary bounced from one side of the river to the other until 1915, when it became an open boundary between the Union of South Africa and its mandated dependency, South West Africa (Barnard 2000). Following Namibia's independence in 1990, the boundary was (re)imposed and the new constitution (subsequently used as a model for post-apartheid South Africa) stipulated that Namibia's southern boundary would extend to the middle of the river. This was contested by South Africa which had

Table 7.7 Selected, shared watercourses in eastern and southern Africa

Watershed	Area (km²)	Percentage of watershed within individual countries
Congo	3,699,100	DRC (62.4), CAR[a] (10.9), Angola (7.9), Congo (6.7), Zambia (4.8), Tanzania (4.5), Cameroon (2.3)
Limpopo	415,500	South Africa (44.3), Mozambique (21.0), Botswana (19.6), Zimbabwe (15.1)
Nile	3,038,100	Sudan (63.7), Ethiopia (11.8), Egypt (9.1), Uganda (7.9), Tanzania (4.0), Kenya (1.7), DRC (0.7), Rwanda (0.7), Burundi (0.4), Eritrea (0.1)
Okavango	708,600	Botswana (50.7), Namibia (25.0), Angola (21.2), Zimbabwe (3.2)
Orange	947,700	South Africa (59.7), Namibia (25.4), Botswana (12,8), Lesotho (2.1)
Zambezi	1,388,200	Zambia (41.6), Angola (18.4), Zimbabwe (15.6), Mozambique (11.8), Malawi (8.0), Tanzania (2.0), Botswana (1.4), Namibia (1.2), DRC (0.1)

[a] Central African Republic.
Source: Gleick (2000).

concerns over mineral and grazing rights on islands of the Orange River. However, in 2000, the South African and Namibian governments agreed to respect Namibia's boundaries as outlined in its constitution.

To promote regional co-operation in the development and management of water resources, the 'Protocol on Shared Watercourse Systems in the Southern African Development Community (SADC)' was instigated in 1995 and implemented in 1998 following ratification by most members including: Angola, Botswana, Lesotho, Malawi, Mozambique, South Africa, Swaziland, Tanzania, Zambia and Zimbabwe. The protocol enshrines a set of principles that include:

- right of each member state to utilise shared watercourses;
- maintenance of a balance between development and conservation;
- collaboration between riparian member states on development affecting shared watercourses;
- free exchange of relevant resources information between riparian countries;
- equitable exploitation.

Similar agreements have been set up for individual basins such as the Nile Basin Initiative (UNDUGU) and Zambezi River Action Plan (ZACPLAN). Chenje and Johnson (1996) provide a useful summary of water agreements for shared watercourses throughout eastern and southern Africa. However, most international water agreements lack the specificity to make unequivocal allocations of resources. This is ably demonstrated by the legally imprecise language of the 1997 UN Convention on the Law of Non-Navigational Uses of International Watercourses, whereby countries are to 'use water in an equitable and reasonable manner . . . in order to achieve optimal and sustainable utilisation . . . in a manner such that other riparians do not suffer significant harm'. Furthermore, international agreements such as the 1997 UN Convention embody no supranational element to take recourse (e.g. sanctions) against individual riparian nations that violate international agreements. As a result, nations can choose to respect the convention when it serves their interests and ignore the convention when it does not. The new National Water Act of South Africa (Box 7.1) divorces water rights from landownership and may serve as a model for international, basin-wide management of water resources.

Alleviating water scarcity – interbasin water transfers and 'virtual water'

Water scarcity impedes development both in terms of time wasted for household water collection and constraints to agricultural and industrial growth. Alleviation of water scarcity in southern Africa (e.g. Botswana, Namibia, South Africa) has, to date, focused on engineered solutions in the form of interbasin water transfers (Kuffner 1993) whereby water is brought over large distances from one basin to another for consumption – 'making rivers flow uphill' (Coetzee 1994). A clear example is the Lesotho Highlands Water Project, the largest interbasin transfer of water in sub-Saharan

Box 7.1   South Africa's National Water Act: a radical departure in water policy

In 1998, the South African government introduced new legislation, embodied in the National Water Act, that specifies the development, use and management of water resources in South Africa. The act is both socially and environmentally progressive and reflects a culture of human rights, enshrined in the country's new constitution. These include the right to have access to sufficient food and water and the right to have the environment protected. To uphold these rights and enable allocation of water according to priorities of equity, redistribution and improvement of living conditions (redressing past imbalances of the apartheid era), the Act takes the sensible, though radical, step of decoupling water rights and, hence, water allocations from landownership. Consequently, landowners no longer have an automatic (riparian) claim to the flow of streams, rivers and groundwater running through their property. Water is considered to belong to the nation as a whole and is deemed to be indivisible, so no legal distinction is made between surface water and groundwater. This encourages a more holistic, ecological approach to the management of water resources whereby the national government acts as a steward ensuring that 'development, apportionment, management and use of the resources are carried out using criteria of public interest, sustainability, equity and efficiency of use' (van Tonder 1999). A hierarchy of access is imposed in which priority allocation goes to the *water reserve* comprising a *basic human needs reserve* – the basic right to a minimum amount (and quality) of water; and an *ecological reserve* – the minimum quantity and quality of water for ecosystem health.

Other innovations in the National Water Act include the establishment of water management bodies and areas that take into account catchment boundaries and socio-economic conditions, and promote stakeholder participation. Operating on a river-basin scale, *catchment management agencies* will be created that are not only sensitive to local resources and needs but also enjoy community-level participation through the involvement of *water user associations*. Once established, these agencies are expected to prepare catchment management strategies, consistent with the national water strategy, and allocate water use by granting licences. The fees charged for such licences (with specific limits of time and quantity) are intended to fund water resource management and development activities so that users are, in effect, charged the full financial costs of providing access to water. Controversial aspects of the Act include the issue of compensation to those refused a licence or given a licence of lesser use. The onus is, however, on the user to show that any restrictions in the name of protecting the *water reserve* would create an economic disadvantage even when appropriate conservation measures have been put in place. This recognises the potential economic consequences of protecting the water reserve but also strives to facilitate sound approaches to water use.

Despite the best intentions of the Act, its implementation has raised several practical challenges. First, human rights are expensive to implement. This is recognised explicitly in the new constitution and National Water Act which are replete with provisos like the 'progressive realisation of rights within available resources'. Fundamental commitments like the standard of free access to 50 l per person per day have been revised to 20 l per person per day. In terms of resource allocation, significant difficulties exist in the accurate estimation of available water resources and resources dedicated to ecological reserves. On a basin scale, it is unclear how all the stakeholders (e.g. water user associations) will effectively participate in the operation of catchment management agencies. A radical new approach such as this is bound to take time to implement, however, and given the significance accorded to water as a key environmental resource in the twenty-first century, the progress of South Africa's experiment which tries to address many of the fundamental environmental and economic problems associated with orthodox water legislation will be closely analysed.

Sources: deLange (2001), Hamann and O'Riordan (2000), Hughes (2001), van Tonder (1999), Wright and Xu (2000).

Africa. The first phase of this project is onstream, and the total scheme, scheduled for completion in 2020, should divert 2.2 km^3 of water from the headwaters of the River Orange in Lesotho each year via a network of dams and tunnels to augment the water supply of Gauteng Province (Waites 2000), the industrial heartland of South Africa (see Box 7.2).

Further interbasin water transfers (IWTs) have been proposed and include a range of options to divert flow from the River Zambezi to alleviate real and perceived water shortages in Botswana, Zimbabwe, Namibia and even South Africa where projected increases in demand will exceed the increased capacity gained by the Lesotho Highlands Water Project (Scudder *et al.* 1993). In Namibia, augmentation of water supplies in central and eastern areas of the country is planned from the diversion of riverflow (0.1–0.2 km^3 per year) from the River Cubango at Rundu (upstream of the confluence with the River Cuito that forms the River Okavango) along the Eastern National Water Carrier (Eales *et al.* 1996). Although the SADC Protocol on Shared Watercourse Systems (yet unsigned by Namibia) serves as a basis to mediate political consequences of IWTs in the region, ecological impacts remain poorly known (Snaddon *et al.* 1998). Indeed, it is questionable whether IWTs serve as long-term, sustainable solutions to water scarcity since such solutions do not address water demand or the environmental impacts in the area from which water is diverted. Efforts to address demand have, however, been instigated in Windhoek (Namibia) where water shortages have led to increased efficiency in water use and the reuse of wastewater. Bulawayo in Zimbabwe has a long history of efficient water use and is a model in some respects. A much more expensive solution, desalinisation, has also been initiated in Walvis Bay, Namibia's main port.

An interesting, alternative solution that faces up to, rather than denies, national water scarcity is the concept of trading in 'virtual water' (Allan 2001b). The concept, developed by the Water Issues Group at the School for Oriental and African Studies (University of London), simply recognises that water used for irrigation in the agricultural sector usually greatly exceeds a nation's other uses (domestic and industrial), as is clearly illustrated for

Box 7.2   Lesotho Highlands Water Project (LHWP)

The LHWP diverts riverflow from the headwaters of the River Orange (Senqu) in the Maluti Highlands of Lesotho through a network of dams and tunnels to the Vaal River basin in Gauteng Province, South Africa. It constitutes the largest interbasin water transfer in sub-Saharan Africa. The project, largely financed by the Republic of South Africa but supported by the World Bank and other donors will, when complete, bring 2.2 km^3 per year of water to the industrial heartland to South Africa and supply enough electricity to Lesotho (72 MW) through hydroelectric power generation to make Lesotho self-sufficient in this regard. A treaty in 1986 between the apartheid government of South Africa and the military regime of General Metsing Lekhanya in Lesotho provided the basis for the LHWP. This followed a military coup in January 1986 which not only deposed the government of Leabua Jonathan which had demanded control over the flow of water to South Africa but also resulted in the lifting of a trade embargo by South Africa that had crippled Lesotho's economy. Apartheid South Africa, defined as water-stressed, was in great need of more water for its mining and industrial heartland of the Witwatersrand and its support for this change of regime in Lesotho was, at least in part, related to this water issue.

Phase 1A of the LHWP, construction of the Katse dam (182 m in height) on the River Malibamatso, began in 1989 and was completed in 1995. Water flows from the dam by gravity along a 45 km tunnel (to Muela) and thence to Gauteng. It also drives Lesotho's newly constructed hydroelectric power station, although this is merely a by-product of the scheme as a whole. Progress on Phase 1B of the LHWP has, however, been delayed as water conservation measures in South Africa postpone the need (demand) for water many years into the future. Development, to date, has furthermore been marred by accusations of bribery by construction firms and labour unrest. Despite heavy investments in social and environmental programmes under the LHWP, serious human and environmental concerns associated with this development also exist. It is estimated that following construction of the Katse dam, 24,000 people lost their farms and access to communal grazing. A further 7,400 people are expected to have been affected by the Muela power station. Compensation to people affected by the LHWP, promised in the 1986 treaty, has been delayed or is grossly inadequate. Environmental impact assessments also paid little attention to downstream effects. Diversion of 40 per cent of annual flows in the headwaters of the River Orange (Senqu) in Lesotho significantly reduces flow in the River Orange that sustains irrigation schemes in the Eastern Cape of South Africa. The lack of seasonality in riverflow that is created by dam storage, greatly impairs traditional flood recession and floodplain agriculture downstream.

Sources: Horta (1995), Waites (2000).

Figure 7.3 Location of places referred to in Chapter 7

eastern and southern African countries in Tables 7.4 and 7.5. A water-deficient nation can, therefore, alleviate its perceived condition of water stress by reducing its demand for irrigation water through the importation of food from another country relatively sufficient in renewable water resources. Although building a nation's dependence on the importation of food – trading in virtual water – is inherently incongruous with political goals if these include self-sufficiency in food production, the sensibility of this solution to water scarcity on hydrological and ecological grounds is obvious. Trading in virtual water highlights another key limitation of the water stress index for assessing water scarcity. Because per capita demand is fixed in the index, a country's water stress index will remain unchanged by the importation of food (virtual water) even though the demand for water has, in fact, been reduced. However, many countries in the region still have the option of increasing their agricultural production using soil water.

Concluding discussion – future challenges

This chapter has endeavoured to describe the complex relationship between the availability of water and human demand for this vital resource in eastern and southern Africa. The discrepancy between the distribution of water resources and human demand that inhibits development in many countries, will probably continue in the near future even within countries that are (at least in theory) relatively sufficient in water resources (e.g. Namibia). Indeed, perhaps more important than actual water scarcity in this region is the endemic poverty that restricts capacity to adapt to, and thereby overcome, water shortages and conditions of water scarcity. There is a need, furthermore, for a better understanding of water resources in the region, gained through improved monitoring and research, and improved local capacity to develop and manage water resources. Key challenges include the delivery of potable water to rapidly growing urban populations as well as managing the impacts of global climate change and extreme climatic events (e.g. drought, flooding) on the availability of water resources. More representative assessments of renewable water resources including their relation to demand are also required. Indeed, the scale of the water crisis remains unclear as current macroscale (national) estimations of water resources and their availability (relative to demand) are unreliable.

The transboundary nature of water resources combined with the water scarcity experienced in several countries in eastern and southern Africa require integrated, basin-wide approaches to water management in order to use resources in the most efficient manner and to prevent conflict over shared resources. Recent implementation of basin-wide management plans (e.g. ZACPLAN and UNDUGU) and development of a regional water protocol by the Southern African Development Community are hydrologically and politically sensible. In terms of domestic (household) water supplies, the emergence of community-based water development and management initiatives, that are more accountable to the communities that they serve, is encouraging and will ideally result in improved access to potable water facilities. Indeed, it would be difficult to be less efficient than the top-down, engineering solutions to community water supplies that characterised development from the 1960s to the 1980s. It remains to be seen whether this and other current trends such as the privatisation of water supplies will ultimately serve the interests of the very poor, most in need of improved access to potable water. Nevertheless, both the push towards integrated, basin-wide management of water resources and community-based development and management of household water supplies in eastern and southern Africa are consistent with global recommendations contained in Chapter 18 of United Nation's Agenda 21, 'Protection of the quality and supply of freshwater resources: application of integrated approaches to the development, management and use of water resources'. Time will tell whether we all stay on the same track.

Acknowledgements

The author is grateful to numerous colleagues across eastern and southern Africa for surely teaching me more than they ever learned from me. Research support from the International Development Research Centre (Ottawa), Natural Sciences and Engineering Research Council of Canada and Department for International Development (UK) is also greatly appreciated. Editorial comments from Debby Potts, Martin Todd and Mark Hughes improved the accuracy and clarity of this manuscript.

References

Adams, W.M. (1992) *Wasting the Rain: Rivers, people and planning in Africa.* Earthscan, London: 256.

Adanu, E.A. (1991) Source and recharge of groundwater in the basement terrain of Zaria-Kaduna area, Nigeria: applying stable isotopes, *Journal of African Earth Sciences* **13**: 229–34.

Allan, J.A.A. (2001a) *The Middle East Water Question: Hydropolitics and the global economy.* IB Tauris, London.

Allan, J.A.A. (2001b) Virtual water – economically invisible and politically silent, a way to solve strategic water problems, *International Water and Irrigation* **21**: 39–41.

Barnard, W.S. (2000) 'Cheaper than fences': the functional evolution of the Lower Orange River boundary, *Erde* **131**(2): 205–20.

Barrett, M.H. and **Howard, A.G.** (2002) Urban groundwater and sanitation – developed and developing countries, in Howard, K.W.F. and Israfilov, R.G. (eds) *Current Problems of Hydrogeology in Urban Areas, Urban Agglomerates and Industrial Centres.* Kluwer Dordecht: 39–56.

Beadle, L.C. (1974) *The Inland Waters of Tropical Africa.* Longman, New York.

Bell, M., Faulkner, R., Hotchkiss, P., Lambert, R., Roberts, N. and **Windram, A.** (1987) *The Use of Dambos in Rural Development with Reference to Zimbabwe.* Final Report of ODA Project R3869, Loughborough University, Loughborough.

Bellamy, C. (2000) *The State of the World's Children 2000.* UNICEF, New York.

Bugenyi, F.W.B. (1982) Copper pollution studies in Lakes George and Edward, Uganda: the distribution of Cu, Cd and Fe in water and sediments, *Environmental Pollution* (Series B) **3**(2): 129–38.

Bugenyi, F.W.B. and **Lutalo-Bosa, A.J.** (1990) Likely effects of salinity on acute copper toxicity to the fisheries of the Lake George–Edward Basin, *Hydrobiologia* **208**(1–2): 39–44.

223

Camberlin, P. (1995) June–September rainfall in northeastern Africa and atmospheric signals over the tropics: a zonal perspective, *International Journal of Climatology* **15**: 773–83.

Camberlin, P., Janicot, S. and **Poccard, I.** (2001) Seasonality and atmospheric dynamics of the teleconnections between African rainfall and tropical sea-surface temperature: Atlantic vs. ENSO, *International Journal of Climatology* **21**: 973–1005.

Chenje, M. and **Johnson, P.** (1996) *Water in Southern Africa*. SADC/IUCN/SARDC, Harare: 238.

Chilton, P.J. and **Foster, S.S.D.** (1995) Hydrogeological characteristics and water-supply potential of basement aquifers in tropical Africa, *Hydrogeology Journal* **3**: 3–49.

Coetzee, H. (1994) Southern African water issues, in Venter, M. (ed.), *Prospects for Progress – Critical choices for Southern Africa*. Longman, Cape Town: 142–51.

De Lange, M. (2001) Water law and human rights – roles and responsibilities. *Water Science and Technology* **43**(4): 143–50.

De Vries, J.J., Selaolo, E.T. and **Beekman, H.E.** (2000) Groundwater recharge in the Kalahari, with reference to paleo-hydrologic conditions, *Journal of Hydrology* **238**: 110–23.

Eales, K., Forster, S. and **Du Mhango, L.** (1996) Strain, water demand and supply direction in the most stressed water systems of Lesotho, Namibia, South Africa and Swaziland, in Rached, E., Rathgeber, E. and Brooks, D. (eds) *Water Management in Africa and the Middle East: Challenges and opportunities*. IDRC, Ottawa: 166–202.

Falkenmark, M., Lundqvist, J. and **Widstrand, C.** (1989) Macro-scale scarcity requires micro-scale approaches, *Natural Resources Forum* **13**: 258–67.

Falkenmark, M. and **Widstrand, C.** (1992) *Population and Water Resources: A delicate balance*. Population Bulletin, Washington: 35.

Gardner-Outlaw, T. and **Engelman, R.** (1997) *Sustaining Water, Easing Scarcity*. Population Action International, Washington.

Gleick, P. (2000) *The World's Water 2000–2001: Biennial report on freshwater resources*. Island Press, Washington.

Griffiths, J.F. (1972) Climates of Africa. *World Survey of Climatology*, vol. 10. Elsevier, Amsterdam, New York.

Grove, A.T. (1978) *Africa south of the Sahara*, 3rd edn. Oxford University Press, Oxford.

Grove, A.T. (1989) *The Changing Geography of Africa*, 2nd edn. Oxford University Press, Oxford.

Hamann, R. and **O'Riordan, T.** (2000) Resource management in South Africa, *South African Geographical Journal* **82**(2): 23–34.

Heederik, J.P., Gathuru, N., Majanga, F.I. and **van Dongen, P.G.** (1984) Water resources assessment study in Kiambu District, Kenya. IAHS Publication No. 144: *Challenges in African Hydrology and Water Resources*, Proceedings of the Harare Symposium: 95–104.

Horta, K. (1995) The mountain kingdom's white oil: the Lesotho Highlands Water Project, *The Ecologist* **25**(6): 227–31.

Horvei, T. (1998) Powering the region: South Africa in the Southern African Power Pool, in Simon, D. (ed.) *South Africa in Southern Africa*. James Currey, Oxford: 146–63.

Houston, J.F.T. (1990) Rainfall–runoff–recharge relationships in the basement rocks of Zimbabwe, in Lerner, D., Isaar, A.S. and Simmers, I. (eds) *Groundwater Recharge: A guide to understanding and estimating natural recharge*. IAH vol. 8, Heise, Hannover: 271–83.

Howard, K.W.F. and **Karundu, J.** (1992) Constraints on the exploitation of basement aquifers in East Africa – water balance implications and the role of the regolith, *Journal of Hydrology* **139**: 183–96.

Hughes, D.A. (2001) Providing hydrological information and data analysis tools for the determination of ecological instream flow requirements for South African rivers, *Journal of Hydrology* **241**: 140–51.

Kaltenthaler, E.C. and **Drasar, B.S.** (1996) Understanding of hygiene behaviour and diarrhoea in two villages in Botswana, *Journal of Diarrhoeal Disease Research* **14**: 75–80.

Kehinde, M.O. and **Loehnert, E.P.** (1989) Review of Africa groundwater resources, *Journal of African Earth Sciences* **9**: 179–85.

Kendrew, W.G. (1961) *The Climates of the Continents*. Oxford University Press, Oxford.

Khroda, G. (1996) Strain, social and environmental consequences, and water management in the most stressed water systems in Africa, in Rached, E., Rathgeber, E. and Brooks, D. (eds) *Water Management in Africa and the Middle East: Challenges and opportunities*. IDRC, Ottawa: 120–52.

Kuffner, U. (1993) Water transfer and distribution schemes, *Water International* **18**: 30–4.

Lloyd, B. and **Helmer, R.** (1991) *Surveillance of Drinking Water Quality in Rural Areas*. Longman, London.

McGregor, G.R. and **Nieuwolt, S.** (1998) *Tropical Climatology*, 2nd edn. Wiley, Chichester: 339.

Manun'Ebo, M., Cousesn, S., Haggerty, P., Kalengaie, M., Ashworth, A. and **Kirkwood, B.** (1997) Measuring hygiene practices: a comparison of questionnaires with direct observation in rural Zaïre, *Tropical Medicine and International Health* **2**: 1015–21.

225

Muwanga, A. (1997) Environmental impact of copper mining at Kilembe, Uganda: a geochemical investigation of heavy metal pollution of drainage waters, stream sediment and soils in the Kilembe valley in relation to mine-waste disposal, *Braunschweiger geoswiss Arb.* **21**: 140.

Nicholson, S.E. and **Kim, J.** (1997) The relationship between El Niño Southern Oscillation and African rainfall, *International Journal of Climatology* **17**: 117–35.

Nyamweru, C. (1980) *Rifts and volcanoes – a study of the East African rift system.* Nelson, Nairobi: 128.

Nyamweru, C. (1996) The African Rift System, in Adams, W., Goudie, A. and Orme, A. (eds) *The Physical Geography of Africa.* Oxford University Press, Oxford: 18–33.

Ojany, F. (1971) Drainage evolution in Kenya, in Ominde, S.H. (ed.) *Studies in East African Geography and Development.* Heinemann, London: 137–45.

Ong'wen, O. (1996) NGO experience, intervention, and challenges in water strain, demand and supply management in Africa, in Rached, E., Rathengeber, E. and Brooks, D. (eds) *Water Management in Africa and the Middle East: Challenges and opportunities.* IDRC, Ottawa: 274–89.

Pitman W.V. and **Hudson, J.** (1997) Regional Water resources: prospects for trade and co-operation, in Kritzinger-van Niekerk, L. (ed.) *Towards Strengthening Multisectoral Linkages in SADC,* Development Bank of Southern Africa, Development Paper No. 33, March.

Scudder, T. *et al.* (1993) *The IUCN Review of the Southern Okavango Integrated Water Development Project.* IUCN, Gland.

Shiklomanov, I. (2000) Appraisal and assessment of world water resources, *Water International* **25**: 11–32.

Snaddon, C.D., Wishart, M.J. and **Davies, B.R**. (1998) Some implications of inter-basin water transfers for river ecosystem functioning resources management in southern Africa, *Aquatic Ecosystem Health and Management* **1**: 159–82.

Southern African Development Community (1996) *SADC Energy Co-operation Policy and Strategy.* SADC Energy Sector – TAU, Luanda.

Taylor, R.G. and **Howard, K.W.F.** (1995) Groundwater quality in rural Uganda: hydrochemical considerations for the development of aquifers within the basement complex of Africa, in McCall, J. and Nash, H. (eds) *Groundwater Quality.* Chapman and Hall, London and New York.

Taylor, R.G. and **Howard, K.W.F.** (1996) Groundwater recharge in the Victoria Nile basin of East Africa: support for the soil-moisture balance method using stable isotope and flow modelling studies, *Journal of Hydrology* **180**: 31–53.

Taylor, R.G. and Howard, K.W.F. (1998) Post-Palaeozoic evolution of weathered landsurfaces in Uganda by tectonically controlled cycles of deep weathering and stripping, *Geomorphology* 25: 173–92.

Taylor, R.G. and Howard, K.W.F. (1999a) The influence of tectonic setting on the hydrological characteristics of deeply weathered terrains: evidence from Uganda, *Journal of Hydrology* 218: 44–71.

Taylor, R.G. and Howard, K.W.F. (1999b) Lithological evidence for the evolution of weathered mantles in Uganda by tectonically controlled cycles of deep weathering and stripping, *Catena* 35: 65–94.

Taylor, R.G. and Howard, K.W.F. (2000) A tectono-geomorphic model of the hydrogeology of deeply weathered crystalline rock: evidence from Uganda, *Hydrogeology Journal* 8: 279–94.

Thomas, D.S.G. and Shaw, P.A. (1988) Late Cenozoic drainage evolution in the Zambezi Basin: geomorphological evidence from the Kalahari Rim, *Journal of African Earth Sciences* 7: 611–18.

Thomas, D.S.G. and Shaw, P.A. (1991) *The Kalahari Environment.* Cambridge, Cambridge University Press: 284.

Thompson, J. (1996) Africa's floodplains: a hydrological overview, in Acreman, M.C. and Hollis, G.E. (eds) *Water Management and Wetlands in Sub-Saharan Africa.* IUCN, Gland: 5–20.

Turton, A. (2000) Precipitation, people, pipelines, and power in Southern Africa: towards a 'Virtual Water' – based political ecology discourse, in Stott, P. and Sullivan, S. (eds) *Political Ecology: power, myth and science* (Arnold, London): 132–53.

UNESCO (1995) *Discharge of Selected Rivers of Africa.* UNESCO, Paris.

van Tonder, G. (1999) Groundwater management under the new National Water Act in South Africa, *Hydrogeology Journal* 7: 421–2.

Waites, B. (2000) The Lesotho Highlands Water Project. *Geography* 85(4): 369–74.

WHO (1993) *Guidelines for Drinking-Water Quality.* Vol. 1 *Recommendations*, 2nd edn. World Health Organisation, Geneva.

WHO (2000) *Global Water Supply and Sanitation Assessment 2000 Report* [www.who.int./water-sanitation-health/globalassessment/globaltoc. htm]

Wolf, A.T. (1999) Criteria for equitable allocations: the heart of international water conflict, *Natural Resources Forum* 23: 3–30.

World Bank (2001) *African Development Indicators.* World Bank, Washington.

World Commission on Dams (2000) *Dams and Development: A new framework for decision making.* Earthscan, London.

World Resources Institute (2000) *World Resources 2000–2001: People and ecosystems – the fraying web of life.* World Resources Institute, Washington.

World Resources Institute (2002) Table FW.1 Freshwater resources and withdrawals, [earthtrends.wri.org].

Wright, E.P. (1992) The hydrogeology of crystalline basement aquifers, in Wright, E.P. and Burgess, W.G. (eds) Hydrology of Crystalline Basement Aquafiers in Africa. *Geological Special Publication* **66**: 1–27.

Wright, K.A. and **Xu, Y.** (2000) A water balance approach to the sustainable management of groundwater in South Africa, *Water SA* **26**(2): 167–70.

CHAPTER 8

Desertification in eastern and southern Africa

Tanya Bowyer-Bower

Introduction

The term 'desertification' was first used by the French forester, Aubreville (1949), in response to his observations while working on the south side of the Sahara that the desert was spreading inexorably southwards into the savannas and forests of West Africa (Goudie 2001). In his work Aubreville described the spread of cultivation resulting in forested regions being transformed into savanna, which in turn resulted in desert-like conditions should their fragility be disregarded (Glantz and Orlovsky 1983).

Since this time there has been considerable discussion about what the term really means, and indeed the extent to which the phenomenon exists at all (Stiles 1995). This chapter reviews this controversy, clarifies present understanding that has developed from the controversy and highlights factors considered to be crucial to the implications of desertification for the sustainable development of eastern and southern Africa.

Developments in the understanding of desertification

Since Aubreville's contributions desertification has received widespread attention, particularly in Africa where 73 per cent of the continent is dryland (Hopkins and Jones 1983) and thus potentially prone to desertification. The 1972 Stockholm Conference on the Human Environment was responsible for focusing attention on the need to sustain all environments. With drylands comprising 40 per cent of the world land surface, in 1974 the UN called for global action on desertification, with Resolution 3337(XXIX) recommending a UN Conference on Desertification (UNCOD) to assess the phenomenon, which took place in Nairobi in 1977 (UNDS 1977). This resulted in the UN Plan of Action to Combat Desertification (PACD), consisting of 26 recommendations that provided guidelines for many national plans of action. These included both social and environmental measures and partnerships between local partners and national and international agencies (UNCOD 1978).

UNCOD defined desertification in 1977 as

> the diminution or destruction of the biological potential of the land, . . . [that] can lead ultimately to desert-like conditions. It is an aspect of the widespread deterioration of ecosystems, and has diminished or destroyed the biological potential, i.e., plant and animal production, for multiple use purposes at a time when increased productivity is needed to support growing populations in the quest of development.
>
> (UNDS 1977)

During and since UNCOD many publications have painted a gloomy, if not apocalyptic, picture of the dwindling productivity of African drylands in particular, attributed to various processes collectively understood to be desertification. For example, in Timberlake's *Africa in Crisis* (1985), it is stated that 'Africa has taken too much from its land. It has overdrawn its environmental accounts', and the result for much of the continent has been 'environmental bankruptcy' (Timberlake 1985). According to Darkoh (1993), this environmental crisis stems from intricate processes of land degradation whereby the biological potential of the continent and its ability to support populations are severely diminished. He regards the main causes of such degradation as drought, desiccation and human activities. Drought is defined in climatic terms as protracted rainfall failure which is usually short term (one to two years) and, in ecological terms, as a dry period from which an ecosystem often recovers rapidly after the rains return. Desiccation is described as a process of aridification resulting from a dry period lasting in the order of decades (e.g. climate change, or a climatic fluctuation that leads to positive feedback mechanisms of soil erosion reducing the productive potential of the land). Human activities include overcultivation, overgrazing, deforestation, poor irrigation practices and any other inappropriate land-use and human management of ecosystems. The rapidly increasing human and animal populations leading to increased overexploitation of water, land, forest and pasture resources are also emphasised.

The areas most affected by these processes are argued to be those where average annual rainfall is between 100 and 600 mm (i.e. drylands) and where the ecosystems are largely supporting crop and livestock farming activities (Darkoh 1993). In other words, desertification need not be spread from deserts by hot dry winds, but instead could occur anywhere where land was overexploited. Darkoh estimates that such drylands comprise 65 per cent of Africa as a whole (about one-third of the world's drylands) with 34 per cent of the area of the continent being under threat of desertification, with the subsequent erosion and degradation of productive lands resulting in food insecurity.

In the 1980s a host of working definitions and interpretations of what constituted desertification developed in the literature. Stiles (1995) estimates that there are over 100 definitions/interpretations. A review by Glantz and Orlovsky (1983) distinguished two main contrasting interpretations, with desertification being recognised as either a process of change (i.e. desertification-as-process – incremental changes from a favoured or preferred state with respect to quality of, for example, vegetation, moisture availability, soils, atmospheric phenomena, societal value or ecological stability), or the end result of a process of change (i.e. desertification-as-event, e.g. extended desert margins).

Darkoh's views of desertification, as exemplified above, are evidently an example of the first interpretation. An influential report by Warren and Agnew (1988) provides an example of the second interpretation. This report identified desertification as being the state of conversion of land to a desert-like condition (i.e. one of *eliminated* productivity). The report distinguished this from processes that merely *diminished* productivity (without necessarily producing deserts) which it instead classified as land degradation. Such differing interpretations have been a major cause of the considerable controversy over the extent of desertification (Glantz and Orlovsky 1983), since land areas undergoing change to desert-like conditions are much more limited in extent than those experiencing more temporary land degradation.

A further controversy has arisen from the apparent misuse and misrepresentation of reports of the extent of desertification (Stiles 1995). For example, an unpublished memo from Lamprey (1975), who at the time was undertaking a review of pastoral systems in the Chalbi Desert in Kenya during the drought years of the early 1970s, suggested the average rate of desert advance for the area was 5.5 km per year. Three periods of assessment of rates and extent of desertification were also undertaken and reported by the UN (UNCOD 1977; UNEP 1984; UNEP 1991). These assessments, however, contrasted with remote sensing studies undertaken in the Kordofan Province of Sudan, as reported by Hellden (1984, 1988, 1991) and Olsson (1985, 1993). This research found no evidence for the type and extent of long-lasting desertification suggested by Lamprey and in the further UN reports. Warren and Agnew (1988) and subsequently Thomas and Middleton (1994) have highlighted this lack of evidence and suggested that the extent of the desertification problem had been exaggerated and that the concept of desertification had been misrepresented. It is feared by some, however, that views of this nature, which seemingly arise partly from a lack of clarity of what is meant by desertification, reduce the perceived importance of desertification as a global problem and thereby reduce commitment to tackling its causes, with negative implications for much needed development in badly affected areas such as eastern and southern Africa (Stiles 1995).

These and other criticisms prompted the UN to reconsider the official definition of desertification, and an ad hoc consultative meeting of experts convened in 1990 concluded there was no point in distinguishing between the two interpretations of desertification that were pervading the literature. This, in combination with further debate in the lead-up to the 1992

UN Conference on Environment and Development (UNCED, also known as the 'Rio Summit' or 'Earth Summit'), resulted in an internationally negotiated definition of desertification approved by all participating governments and included in the Agenda 21 output of UNCED (1992) which states that:

> Desertification is land degradation in arid, semi-arid and dry subhumid areas resulting from various factors, including climatic variations and human activities.

UNCED also called for the adoption of an 'international convention to combat desertification in those countries experiencing serious drought and/or desertification, particularly in Africa . . . through effective action at all levels, supported by international co-operation and partnership agreements, in the framework of an integrated approach which is consistent with Agenda 21 (UNCED 1992, Article 2, Part 1)'.

Negotiation to establish the UN Convention to Combat Desertification (UNCCD, eventually signed in 1997), was protracted and considerable. A contributing factor was the controversy that had surrounded the validity of the desertification issue through the 1980s, and the feeling that land degradation in drylands (the new working definition of desertification) was such a very broad topic that it was already being addressed by donor countries at a variety of levels and in a variety of ways. Rather than commit new and additional financial resources to this cause, donor countries instead held out in the negotiations for more efficient coordination, and more appropriate use, of existing funds (Mwangi 1994).

Under the UNCCD (1995), all signatory parties have an obligation to 'adopt an integrated approach addressing the physical, biological, and socio-economic aspects of the process of desertification and drought' and this is to be implemented through action programmes. At the national level, these should address the underlying causes of desertification and drought and identify measures to prevent and reverse it. It is intended that national programmes will be complemented by sub-regional and regional programmes. Box 8.1 presents the framework for the action plan to combat desertification in Africa, produced by the UNCCD (1995).

Desertification today

As demonstrated above, desertification is now the name for land degradation in drylands (UNCED 1992; UNCCD 1995). To comprehend fully the contemporary implications of desertification for development in eastern and southern Africa at the local level it is necessary first to clarify the terms 'drylands' and 'land degradation', as both phenomena themselves are far from clear.

Drylands

Nowadays the term 'dryland' is used to refer collectively to all dry areas, where dryness is interpreted as significant periods of moisture deficiency

Box 8.1   UNCCD (1995) Combating Desertification in Africa

Desertification has its greatest impact in Africa. Two-thirds of the continent is desert or drylands. There are extensive agricultural drylands, almost three-quarters of which are already degraded to some degree. The region is afflicted by frequent and severe droughts. Many African countries are landlocked, have widespread poverty, need external assistance, and depend heavily on natural resources for subsistence. They have difficult socio-economic conditions, insufficient institutional and legal frameworks, incomplete infrastructure, and weak scientific, technical and educational capacities. These difficult circumstances explain why African countries put so much effort into convincing the international community of the need for a 'Convention to Combat Desertification in Those Countries Experiencing Serious Drought and/or Desertification, Particularly in Africa'.

Africa's desertification is strongly linked to poverty, migration and food security. In many African countries, combating desertification and promoting development are virtually one and the same due to the social and economic importance of natural resources and agriculture. When people live in poverty they have little choice but to overexploit the land. When the land eventually becomes uneconomic to farm, these people are often forced into internal and cross-border migrations, which in turn can further strain the environment and cause social and political tensions and conflicts. (The link with migration was important to the international community's recognition of desertification as a truly global problem, like climate change or biodiversity loss.) Food security can ultimately be put at risk when people already living on the edge face severe droughts and other calamities.

The Regional Implementation Annex for Africa outlines a strategy for action. This Annex is the most detailed and thorough of the four regional annexes to the Convention. Its proposals for National Action Programmes benefited from early attention when parties adopted a resolution on urgent measures for Africa which entered into force in June 1994, some two-and-a-half years before the Convention itself.

National Action Programmes strongly emphasize awareness raising . . . Already, some 30 African countries have organised national awareness-raising seminars. The seminars gather together a wide range of stakeholders to discuss the Convention and its philosophy and how to apply it to national circumstances. In some countries, local-level seminars have also been held to bring the message even closer to the actors in the field.

. . . and wide participation and consultation. All stakeholders need to be consulted when Action Programmes are drafted. Many countries have set up coordinating bodies and focal points to ensure that this happens. Some have used national forums to officially launch the resulting Programme. The participation of non-governmental organisations (NGOs) is particularly important and their valuable contribution to the process has been widely recognised.

Sub-regional Action Programmes have also been launched. The existing sub-regional organisations entrusted with coordinating these Programmes are the

Arab Maghreb Union for northern Africa, the Permanent Inter-State Committee for Drought Control in the Sahel (CILSS) for the west, the Intergovernmental Authority on Development (IGAD) for the east, and the Southern African Development Community (SADC) for the south. While community-based organisations are very important actors in the process of elaborating National Action Programmes, specialised intergovernmental organisations feature as main partners in designing Sub-regional Action Programmes. When possible these Programmes seek synergies with other regional objectives. For example, a project for connecting all sub-regional organisations to each other and to their respective member states via electronic systems will contribute to the strengthening of the regional communications network.

A Regional Action Programme is also being developed. A Regional Coordination Unit has been endorsed by the Organisation of African Unity (OAU) and will be hosted by the African Development Bank in Abidjan. The Pan African Conference on the Implementation of the Convention to Combat Desertification, held in Ouagadougou, Burkina Faso, in March 1997 decided that the Regional Action Plan should draw on the output of seven thematic workshops. Starting in 1998, these workshops will look into prospects for establishing networks so as to promote agroforestry and soil conservation; rangelands use and fodder crops; the integrated management of international river, lake and hydrogeological basins; ecological monitoring, natural resources mapping, remote sensing and early warning systems; new and renewable energy sources and technologies; sustainable agricultural farming systems; and enabling environments and capacity building.

African countries are off to a good start, but the real work still lies ahead. To succeed, affected countries must ensure that combating desertification is given top priority. They must actively promote an enabling environment by adopting appropriate legal, political, economic, financial and social measures. For instance, they may need to change their rules on land-use and ownership, further decentralize government administration and strengthen political rights at the local level. Meanwhile, external partners will have to prove themselves fully committed to the principles of the Convention by entering into productive partnerships with affected countries. Greater efforts, including capacity building and financial support, are also needed to enable NGOs and civil society to remain active throughout the implementation stage.

Source: www.unccd.int/publicinfo/factsheets UNCCD Fact Sheet 11.

(Grainger 1990). Thus, rather than being a function of rainfall input alone, this 'dryness' is dependent upon the timing and amount of rainfall inputs relative to solar radiation. Solar radiation causes evapotranspiration (an output) of some of the rainfall and the extent of moisture deficiency is defined as the difference between evapotranspiration and precipitation over a time span.

As precursors to this basic concept a number of indices of degree of dryness (i.e. aridity) have been used (e.g. Koppen 1931; Meigs 1953; UNESCO 1977). Different schemes use different criteria, the precise land area existing in each category thus varying somewhat with the index used. Categories of extents of dryness accepted as comprising 'dryland' are generally hyperarid, arid, semi-arid and subhumid. The FAO/UNESCO (1977) bioclimatic aridity index and FAO (1981) climatic classification of categories of 'dryland' are presented in Table 8.1 as examples. Further discussion on the classification of drylands can be found in Heathcote (1983), Beaumont (1989), Grainger (1990) and Mainguet (1999).

Drylands are the dominant ecosystem in the countries of eastern and southern Africa as shown in Table 8.2. Of the 16 countries covered in this volume, 10 have more than half of their land area classified as drylands. Botswana and Namibia are entirely so classified and the proportion of non-dryland in Zimbabwe and South Africa is almost negligible. Only 7 per cent of Swaziland and 13 per cent of Kenya and less than one-fifth of Tanzania and Mozambique are non-drylands. The significance of this factor for development in these countries cannot be overemphasised. In all but South Africa the majority of the population is rural, and agriculture remains the basis (although not the sole source) of most people's liveli-hoods. This means that unirrigated agriculture is always a risky business in these countries, and people are highly vulnerable to unpredictable rainfall events. This essential environmental context strongly influences decision-making by smallholder farmers who, logically, are highly risk-averse since the outcomes of poor, or unlucky, decisions about crop mix and technol-ogy and livestock management can be, literally, fatal. In South Africa the contribution of agriculture to most people's livelihoods is very much less, even in African rural areas (Bernstein 1996), due to the long-standing apart-heid legacies of land shortage and the underdevelopment of African agri-culture. Here the economy is based largely on mining and industry, but nevertheless shortage of water is a serious issue for sustained urban-based development as exemplified by its need for the Lesotho Highland Water Project (see Chapter 7). There are important variations within these coun-tries – nearly all the drylands in Swaziland, for example, are subhumid drylands where the chances for non-irrigated cropping are rather better than in the semi-arid and arid zones. On the other hand, in Botswana, Namibia, South Africa and Kenya most of the drylands fall into this drier end of the aridity spectrum and the ramifications of their environmental constraints are thus more serious.

In Zambia, Uganda, Angola, Malawi, Rwanda, Burundi and the DRC most of the land is non-dryland. Indeed the DRC is at the opposite end of the environmental spectrum in this respect to, for example, Namibia, and (lack of) water is not a major constraint on human development there. However, Zambia, Uganda and Angola all have more than one-third of their land classified as dryland, as is more than one-quarter of Malawi's land area. Overall then, debates and development issues related to drylands are of extreme significance to the region.

Common to all these areas of dryland are periods of moisture defi-ciency as a normal long-term climatic phenomenon. A further most dis-tinctive feature, particularly with regard to their degradation (as is explained in more detail later), is the strong positive relationship between aridity and rainfall variability (see also Chapters 6 and 7). The interannual variability

Table 8.1 Examples of classifications of degrees of dryness

FAO/UNESCO (1977) Bioclimatic Aridity Index			Land uses relative to local resource potential[c]	FAO (1981) Climatic Classification	
Dryland zone	P^a/PET^b Index	Vegetation description		(a) Rainfall (mm)	(b) No. of humid months[d]
Hyperarid	<0.03	Extreme deserts without vegetation except for some ephemerals and xerophytic bushes in beds and wadis	Inhabited only in oases	100–200	0–1
Arid	0.03–0.20	Barren areas and areas of sparse vegetation, including perennial and annual plants	Pastoral nomadism is possible but not rainfed agriculture	200–400	2–3
Semi-arid	0.20–0.50	Steppe and tropical bush, perennial plants are most frequent here	More sedentary pastoralism is possible, seasonal rainfed agriculture is unreliable	400–600	3–5
Subhumid	0.50–0.75	Tropical savanna, sometimes covered in bush occasionally without trees, sometimes with dry forest	Seasonal rainfed agriculture now more reliable. But all land-uses still need to adapt to seasonal drought	600–1200	4–7

[a] P = annual precipitation.
[b] PET = potential evapotranspiration (quantity of water lost by evaporation directly from the soil and by transpiration of a plant cover).
[c] i.e. not accounting for land-uses transposed onto the local landscape by using resources imported from more productive environments elsewhere.
[d] months in which P is greater than 0.35 PET.

Table 8.2 Drylands in eastern and southern Africa (countries ranked by extent of dryland)

		Land type		Aridity type			
		Total dryland	Non-dryland	Hyperarid	Arid	Semi-arid	Subhumid
AFRICA	%	72.5	27.5	24.0	18.9	15.3	14.3
	million km²	21.3	8.1	7.1	5.5	4.5	
Botswana	%	100	0	0	3.7	92.5	4.2
	km²	575,000	0	0	32,775	526,125	16,100
Namibia	%	100	0	10.9	40.4	44	4.7
	km²	822,021	0	89,600	332,096	361,689	38,636
Zimbabwe	%	97.3	2.7	0	0	34.5	62.8
	km²	379,770	10,532	0	0	134,657	245,113
South Africa	%	97.1	2.9	0.5	32.3	25.3	39
	km²	1,817,302	34,864	6,269	395,494	308,850	476,689
Swaziland	%	92.9	7.1	0	0	0	92.9
	km²	16,165	1,235	0	0	0	16,165
Kenya	%	87	13	0	35.1	37.4	14.5
	km²	506,862	75,738	0	204,493	217,892	84,477
Tanzania	%	82.5	17.5	0	0	23.7	58.8
	km²	777,627	164,951	0	0	223,391	554,236
Mozambique	%	80.8	19.2	0	0	14	66.8
	km²	634,249	150,712	0	0	109,895	524,354
Lesotho	%	65.4	34.6	0	0	0	65.4
	km²	19,842	10,498	0	0	0	19,842
Zambia	%	40.8	59.2	0	0	9.4	31.4
	km²	307,069	445,551	0	0	70,746	236,323
Uganda	%	39.8	60.2	0	0	6.7	33.1
	km²	94,271	142,589	0	0	15,870	78,401
Angola	%	39.2	60.8	0.5	2.8	8.2	27.7
	km²	488,707	757,993	6,234	34,908	102,229	345,336
Malawi	%	27.1	72.9	0	0	0	27.1
	km²	25,496	68,586	0	0	0	25,496
Rwanda	%	21.9	78.1	0	0	0	21.9
	km²	5,766	20,564	0	0	0	5,766
DRC	%	0.2	98.8	0	0	0	0.2
	km²	4,691	2,340,718	0	0	0	4,691

Source: Hopkins and Jones (1983).

of rains is typically greater than 40 per cent in hyperarid regions, 30–40 per cent in arid regions, 20–30 per cent in semi-arid regions, and less than 20 per cent in subhumid regions (UNESCO 1977). This high natural variability manifests itself in a number of ways – the higher the coefficient of variation in rainfall occurrence the more normal it is for any one or a combination of the following:

- the rainy season to commence later or earlier than the 'normal' start time;
- the rainy season to end later or earlier than the 'normal' finish time;
- the periodicity of rains within a rainy season to vary through the season and between seasons;
- the intensity of rains to vary during and between seasons;
- the amount of rain occurring in the rainy season (i.e. the amount of moisture deficiency) to vary between seasons.

Given that the climatic definition of drought is 'a period of below average rainfall', it follows that droughts are a normal occurrence in drylands, the occurrence of which depends upon what has been identified as 'the norm' for rainfall occurrence in the area for the period of time in question. It should be noted that 'drought' is a widely used term, and the 'climatic drought' referred to here is to be distinguished from, for example, agricultural drought (a lack of available moisture at the time at which crops need it), and water supply drought (deficiency in the supply of water relative to the demand for it), etc. For a more detailed discussion of definitions of drought refer, for example, to Agnew and Anderson (1992). The main point to be made here is that the major significance of the normality of periods of climatic drought in drylands for desertification is the requirement that all human activities in drylands that are to any extent dependent upon, or otherwise affected by, rainfall input must also be well adapted to this natural variability if they are to be sustainable in the longer term.

Land degradation

The term 'degradation' generally means a loss in quality, and thus the term 'land degradation' suggests a loss in the quality of the land. A loss has to be relative to something. Some consider there are absolutes in nature, and thus land degradation means a loss from an 'original better state'. Others view nature as a finite dynamic resource subject to human activities, the quality of which is judged by its ability to support the human activities being undertaken. Thus, the loss of quality implied by the term 'land degradation' is a loss in the quality of the land relative to the human activities being undertaken. As activities change with evolving and changing knowledge, technology, aspirations, priorities, preferences, needs, etc., what is interpreted as land degradation will equally evolve and change, and be somewhat both site and time and activity specific. Typical perceptions of the likely patterns of such change (each of which can constitute a desertification process) for the main dryland land-use activities are illustrated schematically in Figure 8.1.

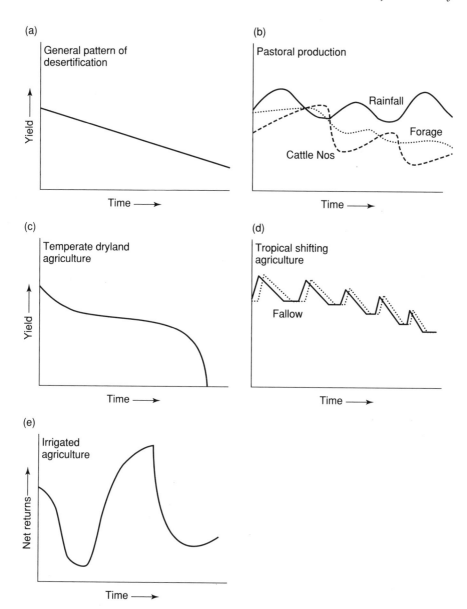

Figure 8.1 Perceptions of patterns of change in drylands

(a) general pattern; (b) approximation to situation under pastoral production;
(c) approximation to pattern in drylands where persistent cropping depletes
the soil nutrients, reduces organic matter content and causes erosion; (d) pattern
where drylands shifting cultivation has degenerated; (e) a generalisation of what
happens in irrigation projects: a pattern of rising yields, breakdown possible,
and so on
Source: From Barrow (1991).

Figure 8.2 Scenarios of land degradation cause and effect

─────► cause effect; ┄┄┄┄► possible feedback (may act to exacerbate degradation or may tend to diminish cause–effect relationships or may have no significant effect). *Causes can be local, external (possibly global) or a combination of both. Environmental = biogeophysical or biochemical; human = social, cultural, political
Source: From Barrow (1991).

This decline may be brought about by changes in the physical land system itself caused, for example, by climate change, and thus potentially unrelated to and possibly beyond the control of human activities (Charney 1975; Williams and Balling 1995). Equally it may be brought about by change in human activities due, for example, to social, economic, cultural or political changes producing a feedback on the physical land system

(Otterman 1974; Blaikie 1985; Blaikie and Brookfield 1987); or a combination of both physical and human causes which may be manifest at any variety of spatial and temporal scales relative to the degradation that is resulting. A schematic representation of this complexity of possible causal combinations that could be responsible for any one instance of land degradation is illustrated in Figure 8.2. Various theses often put forward to explain why land degradation is occurring, particularly in developing countries such as in eastern and southern Africa, are summarised in Table 8.3.

The decline in quality that land degradation constitutes may take the form of, for example, pollution, contamination, soil fertility decline, soil erosion, soil crusting, biomass reduction, and aridification, affecting the ability of the land to provide the goods and services the human activities are attempting to gain from it. If land is viewed as a system of interrelated parts, with inputs and outputs linked by throughputs involving positive and negative feedbacks, it is clear that, particularly given the advances in scientific and technological knowledge that have occurred worldwide, no land degradation is ultimately irreversible when the wherewithal for the necessary interventions to reverse the process are available. Land degradation of concern, however, can occur when at a particular time and place the wherewithal is not available to those involved, or the timescale and/or extent of the processes required to bring about recovery are infeasible relative to the needs and/or ability of those involved. For a fuller discussion of land degradation refer, for example, to Blaikie and Brookfield (1987) and Barrow (1991).

Table 8.3 Categories of explanations of land degradation

1.	Natural disaster theses	Blame degradation on biogeophysical causes or 'acts of god'
2.	Population change (neo-Malthusian) theses	Argue degradation occurs when population growth exceeds critical environmental parameters (and possibly in some cases when population decreases)
3.	Underdevelopment theses	Argue resources are exploited to benefit the world economy or developed countries, leaving little profit to manage or restore the degraded environment
4.	International theses	Suggest that taxation and other forces interfere with market and trigger overexploitation
5.	Legacy of colonialism theses	Argue that trade links, communications, rural–urban linkages, etc., 'hangovers' from the past, promote poorly managed resource exploitation
6.	Inappropriate technology and advice theses	Argue that the wrong strategies and techniques are promoted resulting in land degradation
7.	Ignorance theses	Blame lack of knowledge for degradation, e.g. use of CFCs and their effect on the ozone layer
8.	Attitude theses	Blame people's/institutions' attitudes for degradation

Source: Barrow (1991).

The findings of an assessment of the extent of land areas experiencing desertification in countries of eastern and southern Africa undertaken in the late 1980s (Ahmad and Kassas 1987) is presented in Table 8.4.

Contemporary implications of desertification for development in eastern and southern Africa

Even when the concept of desertification is defined and explained as above, many contemporary researchers remain uncomfortable with the notion that desertification in eastern and southern Africa necessarily involves change from a favoured state to a less favoured state. In this argument researchers point to the high natural variability of tropical drylands (particularly with regard to rainfall input both spatially and temporally), an increasing awareness of manifestations of a misapplication of Western assumptions regarding ecological dynamics to semi-arid and arid environments, and the often inherent adaptability of indigenous land-use practices to land degradation which, when unrecognised, underestimates the sustainability of local livelihoods (Behnke and Scoones 1993; Sullivan 1996; Mazzucato and Niemeijer 2001). These arguments are largely based on the following observations:

- Since rainfall in such environments is extremely patchy in both temporal and spatial extent, land productivity naturally fluctuates considerably. Observations of change in environmental quality are therefore highly relative to spatial and temporal scale and may reflect nothing more than the natural dynamics of these environments (Walker 1985; Scoones 1995).
- Some studies suggest that many of the factors considered to directly promote desertification, such as rapid increase in human populations or overstocking of rangelands (e.g. Hardin 1968; Ehrlich and Ehrlich 1990), are in some instances counteracted by empirical evidence (e.g. Tiffen *et al.* 1994; Ward *et al.* 1998).
- Other studies focus on an underestimation of the ability of local farmers to continue to sustain livelihoods from land showing significant visual signs of degradation (e.g. suggesting the need for experts to discriminate more carefully between a 'naturally bad state, a temporary bad state and a degraded state of land' (Mazzucato and Niemeijer 2001)). This illustrates the importance of redefining and re-evaluating valid indicators of degradation when assessing extents, rates and severities of desertification.

Boxes 8.2 and 8.3 provide case study material on desertification from Zimbabwe and Namibia. The Zimbabwe study draws on an aerial survey of the spatial extent of soil erosion, which indicated that this was a major problem in parts of the country, with patterns correlated to population density which, in turn, at the time of the survey were still mainly determined by the extremely unequal division of land between African smallholders and the (formerly entirely white-owned) commercial farming areas. The unequal racial division of land which is also a legacy of white settler

Table 8.4 Extent of desertification, by land-use type, for selected countries in eastern and southern Africa

Country	Rangeland			Rainfed cropland			Irrigated land		
	Total land area	Area affected by desertification		Total land area	Area affected by desertification		Total land area	Area affected by desertification	
	'000 ha	'000 ha	%	'000 ha	'000 ha	%	'000 ha	'000 ha	%
Kenya	22,000	21,000	95.5	300	270	90.0	20	1.1	5.5
Uganda	375	350	93.3	0	0	00.0	4	0.2	5.0
Botswana	50,000	10,000	20.0	30	20	66.6	2	0.2	10.0
Namibia	66,000	16,500	25.0	10	5	50.0	8	0.1	1.3
South Africa	45,000	38,000	84.4	1,000	650	65.0	860	46.0	5.3
Tanzania	28,000	14,000	50.0	2,400	1,900	79.2	40	4.0	10.0
Zimbabwe	7,500	4,000	53.0	300	150	50.0	0	0.0	0.0

Source: Ahmad and Kassas (1987).

Box 8.2   Evidence for desertification: case study from Zimbabwe

Interpretation of visual signs of land surface erosion was undertaken in the early 1980s in Zimbabwe using 1 : 25,000 scale aerial photographs. The severity of erosion, defined in terms of the proportion of the sampled site which showed evidence of erosion in the photographs, was identified for 1,000 1 ha plots per 45 km^2 area, and the whole of the country was covered. Whereas gullying and streambank erosion were easily identified, only advanced states of sheetwash and rilling could be identified from the methodology used, and the recorded extents of eroded land from this analysis were thus considered to be conservative. A total of 4.7 per cent of Zimbabwe's land area was identified as having visible evidence of erosion via this means. The results of erosion were then correlated with a number of human and physical factors (population density, land tenure, cropland area, rock outcrops, natural region and erosion hazard). Of the factors used in the analysis population density (using data obtained from the 1982 population census) had the largest influence, showing a strong positive correlation with amount of erosion present. Second to this, and indeed interdependent with it, was land tenure. Three categories of land tenure were identified – communal agricultural land, commercial agricultural land and non-agricultural land. The considerably higher population densities of communal compared with commercial land, a legacy of unequal racial land division under white settler rule, was identified clearly, and the greater amounts of erosion correspondingly associated with the communal lands was clear from the results: it was reported that the commercial land, with less than one-quarter of the rural population (an average density of 7.6 people per km^2) showed no pattern of increasing erosion with increasing population density (with even the most densely populated areas – over 30 people per km^2 – experiencing a mean erosion extent of only 2.2–2.5 per cent, with such areas generally occurring in Zimbabwe's non-dryland areas). In contrast the communal areas, which are mainly on the country's poorest quality land (classified as Natural Regions IV and V, the main determinant being low rainfall), had an average population density of 25.5 people per km^2. In the communal areas there was a clear positive correlation between erosion and population density and a mean erosion extent of 12.8–15.3 per cent was measured. A measure of erosion hazard, based on an evaluation of erosivity of rainfall, erodibility of soils, plant cover and average slope (devised by Stocking 1975) was also correlated with presence of erosion. Interestingly, very little correlation was determined. While it is often assumed that the greater the risk of erosion, the more extensive the land degradation, and the lower the erosion risk the more limited the erosion, this study illustrated a poor correspondence between degrees of erosion risk and the extent of land degradation (i.e. there was a discrepancy between potential erosion and actual erosion). Instead the erosion pattern in Zimbabwe resulting from primarily a function of human factors, most notably population density and the inherited racial division of land between African communal areas and commercial farming areas. From this the authors of the report concluded that human factors were the major causes of land degradation, with physical factors being of lesser significance, and that increasing population density and/or communal land tenure would result in land degradation becoming more widespread. They therefore argued that these human factors should be the main focus of policies to control desertification.

Source: Whitlow (1988), Whitlow and Campbell (1989).

Box 8.3   Is desertification occurring in Namibia? Counter-evidence from two different land management regimes

Field research was undertaken in the 1990s in a heavily grazed communal area (the Otjimbingwe) in Namibia to assess the evidence for desertification (signs of significant long-term decline in the productive potential of the land). Vegetation and soil parameters were measured that could indicate declines in the quality of the land for grazing. The results for the heavily grazed communal grazing area were compared with those for less heavily grazed and more formally managed large-scale commercial farms that surround this communal area.

The area is a typical African dryland region traditionally considered prone to desertification. It has an average annual rainfall of 165 mm with a 70 per cent coefficient of variation. It is in a desert savanna transition zone, 117,000 ha in size, and borders the Namib Desert. Stocking rates in the communal area were 18.99 ha per LSU (large stock unit = a 450 kg cow), with an unfenced interior and no regulation of grazing. This was considerably higher than the 27.6 ha per LSU recommended for this area, and very considerably higher than the 40–378 ha per LSU of the surrounding commercial farms which have fenced interiors and stock rotation camps.

A report from a European trader (Andersson) from 1856 described 'Hereros flocking to the area in great numbers with their cattle to the banks of the Swakop, the result of which was that every blade of grass was consumed for miles around on both sides of the river.' During colonial times more pressure was put on the surrounding areas by commercial agriculture and even higher stocking rates were experienced (at one stage 6.8 ha per LSU being recorded), with access to traditional dry season grazing lands further north being blocked by commercial settlement as well. In this communal area crop farming was also traditionally practised along the Swakop River in the rainy season (for example in 1949, 70 ha were planted to wheat with a mean harvest of 24 t per year). However, this ceased in 1977 when the Swakop River was dammed upstream. By 1981, 2,500 people were recorded as living in this communal area, and some 5,000–6,000 lived there by the 1990s (the increase during this period being partly attributed to in-migration since independence from South Africa). It is estimated that in 1952, 244 families had 406 ha per family for their stock. In 1996, 472 families had 248 ha per family.

It is therefore clear that the Otjimbingwe communal area has long had, and still has, high livestock densities in relation to local commercial practices, obstructed seasonal grazing migration, and reduced livelihood options (the conditions for cultivation having been removed). The study therefore expected that the vegetation and soil parameters in this communal area would indicate that the quality of the land for grazing would be lower than that in the surrounding commercial farming land, confirming the 'tragedy of the commons' scenario of Hardin (1968). The study assumed that the communal areas had poor management strategies and that without individual landownership it would experience worse degradation and loss of productivity (i.e. desertification) than privately owned areas where it is assumed that vested interests for longer-term care of the environment will shape better land management practices.

The vegetation parameters measured were grass height and density, percentage cover, species richness and species diversity. The soil parameters measured were soil organic carbon content, total nitrogen and total phosphorus. In addition seeds of radish were planted and their growth used as a measure of comparative soil quality. These parameters were measured in an extremely wet, a wet and a dry year. No significant difference was found in grass height, species richness or percentage of plant cover during the wet season. During the dry season grass height was significantly higher in the commercial areas than the communal area. Soil organic carbon content was significantly higher in the communal area compared with commercial areas, thought to be either a legacy of past seasonal flooding of the Swakop River (prior to the dam built in 1977), or a *positive* benefit of the vastly higher stocking units (through more dung being produced). There was no significant difference in total nitrogen or phosphorous content of the soils or seed germination between the communal and commercial sites.

This research thus determined that despite the clear differences in human and livestock population densities, the soil parameters measured indicated no lower future productive potential (i.e. no indication of desertification) in the communal area compared with the surrounding commercial areas. This result could be due to the fact that all areas have been equally overgrazed, or a threshold effect of heavy grazing (long-term effects not being linearly related to stocking rate) could be masking the effect of heavier grazing on the communal areas. One effect of grazing on the availability of grass in the dry season was ascertained – higher stocking rates in the communal area resulting in significantly less grass being available. This suggests the lower stocking strategies of the commercial farmers may benefit them in that they have more grass per animal for longer into the dry season than the communal farmers.

From this the study concluded that communal farming is not more destructive to the natural environment than commercial farming. In spite of the area possessing the characteristics often assumed to result in the 'tragedy of the commons' scenario, it was not occurring here. Instead, the resilience of this desert margin area in Namibia in the face of increasing human and livestock pressure is highlighted, suggesting that further credence should be given to the critics of the notion that rampant desertification is occurring (e.g. Behnke and Abel 1996).

The researchers point out that this study does not conclude that degradation and/or desertification have *not* occurred, but rather that the high inherent variability in rainfall tends to mask the relatively small impacts by pastoralism in such regions. This supports the notion already put forward by, for example, Ellis and Swift (1988) and Sullivan (1996), that rainfall in arid regions is the major driving variable and has the ability to 'recharge' a system that suffers heavy grazing pressure. It does not exclude, however, the possibility that slow long-term degradation has occurred in both communal and commercial areas, which can only be ascertained more clearly by longer-term field research.

It is also noted that no measure was taken of the nutritional content of grass, or of livestock health. Other variables such as these should thus, perhaps, be the focus of further research to clarify the nature of the interventions required for positive development in the region to be sustained.

Source: Ward *et al.* (1998).

rule in Namibia also provides the context of the second study, which assessed the evidence for environmental degradation in two adjoining commercial farming and African smallholder areas and found no important differences between the two. These studies indicate how contradictory the evidence can be and the reasons why the debate over land degradation and desertification still continues. Overall, however, mainstream opinion remains that desertification is a real and widespread phenomenon in eastern and southern Africa, and is intimately connected with human activities and demands on land. It is also generally agreed that the phenomenon is in need of considerable further research to help improve the sustainability of global environments and the livelihoods of those living in drylands in particular (UNCCD 1995). In order to assess the realities of desertification, it is also necessary to examine further the causal hypotheses which are given to explain the desertification process.

Causes of desertification

Deterioration in land quality is considered to arise from complex interactions between human activities, vegetation and soil dynamics, and climatic regimes. For example, removal of vegetation by cultivation and pastoral activities may lead to soil exposure to wind and rain, increasing soil erosion and constraining the re-establishment of vegetation cover (Hudson 1995; Morgan 1995; Mainguet 1999). The lower productivity of such land could imply increased pressure on sparse vegetation, leading to a positive feedback effect and a spiralling decrease in land productivity and potential (e.g. Solbrig 1991). If erosion is severe, vegetation may not be able to re-establish even when the land is abandoned and human activities discontinued. Furthermore, postulations of further positive feedbacks implicate climate changes as arising from such interactions (e.g. Charney *et al.* 1977).

The albedo effect hypothesises that the removal of vegetation changes the proportion of solar radiation absorbed and reflected from the land surface. Since vegetation absorbs more radiation and bare soil reflects more, temperatures at the land surface would be expected to be higher for land under vegetation cover than for bare land. Lower temperatures at the land surface resulting from desertification would mean there would be less sensible and latent heat transfer to the atmosphere, which in turn would produce less rain-giving convective cloud (Charney *et al.* 1977). The albedo effect has been tested in various sites but results are contradictory. Nevertheless, if such a process were able to affect climatic regimes, its consequences for human activities would be considerable (Warren 1993).

Contemporary critiques of the realities of desertification have not challenged the obvious plausibility of such causal hypotheses of desertification. Rather, they challenge the notion that such processes are the *inevitable* result of human activities, citing various factors which may mitigate against such deleterious positive feedback processes. Central to these arguments is the proposition that certain observations of social and environmental change are not necessarily indicators of the kinds of processes implied by the desertification hypotheses. Instead, observations of change may indicate any of a wide range of environmental and social phenomena, from the natural dynamics of climate and ecological regimes, to the oppression and marginalisation of particular social groups. The former is the realm of environmental science, the latter the realm of political ecology (Bryant and Bailey 1997; Stott and Sullivan 2000).

Desertification indicators

Since both environmental and social parameters are extremely hetero-geneous across different regions, the degree to which changes indicate a desertification process is intimately tied to local conditions. Of particular importance is the degree to which changes indicate local problems in terms of people's livelihoods or global problems of widespread environmental degradation. Because of the widespread pressure to develop simple and preferably inexpensive indicators of desertification, changes in food pro-duction, farm gate prices or family income may be taken as indicating the occurrence or threat of imminent desertification. Since negative changes in these indicators would be expected to reflect increased land pressure, no direct measurement of land change is required because it is assumed that even if such observations are not the result of desertification, they will inevitably contribute to its cause (see, for example, Hambly and Angura 1996).

The consequence of such debates has been a re-emphasis by NGOs, gov-ernments and international agencies on developing meaningful indicators of land degradation and desertification. Ideally, such indicators should be explicit about the relationship between observations and predictions, the spatial and temporal scales over which these apply, and the particular communities at risk from perceived environmental changes. Calling any change which is not desirable for human beings 'desertification' may have value in terms of raising development investment, but is of negative value as the scientific basis of desertification research and management (Hambly and Angura 1996). Thus, causal links between observations and pre-dictions must incorporate indicators which are observed to be strongly related to postulated processes. Given the complexity of environmental and social dynamics this is an almost overwhelming task.

Conclusion

Whether or not one can conclude that specific environmental changes indicate a desertification process (still perceived as a more permanent long-term change) or a local fluctuation (less permanent shorter-term change) in climate or productivity, the consequences of environmental changes for present livelihoods, and thus for the positive development of eastern and southern Africa, may be devastating. Establishing the nature of the dynamics is of considerable importance for how such changes should be managed. For example, resource shortages arising from climatic fluctua-tions, war or social inequities will imply a different mix of management strategies than those arising from long-term or permanent diminution of biological resources. It is also noted that, while many agencies are keen to develop desertification indicators, they are not all fully committed to delineating the possible causes and consequences of various observations, relying rather on assumptions of plausible but unsupported links between observations and postulated processes. Focusing research to develop understanding of these aspects of the phenomena are obviously an important next step.

Meanwhile it remains acknowledged that desertification is a complex phenomenon requiring the expertise of researchers in such disciplines as climatology, soil science, meteorology, hydrology, range science, agronomy and veterinary medicine, as well as geography, political science, economics and anthropology. It has been defined in many different ways by researchers in these and other disciplines, as well as from many national and bureaucratic (institutional) perspectives, each emphasising different aspects of the phenomenon. Desertification is complex in terms of the processes involved and measurement indicators. It remains debatable whether desertification is a trend to a more negative state or rather a characteristic of certain environments in which it has been an ongoing and sustainable process for millennia.

Despite the controversy that has dominated this subject for the last two decades, it clearly remains an issue crucial to the sustainability of the dominant land-uses in eastern and southern African countries. It is hoped that the UNCCD will provide the clarity and framework required for more effective action to secure, if not improve, the productivity of these dryland areas. More effective research links between those interested in long-term environmental change in drylands and those more concerned with contemporary land degradation, and the impact of human activity, are now being effected. For example in August, 2001 a multidisciplinary conference was held in Upington, South Africa on 'Dryland Change 2001' and this was followed in January 2002 by the Developing Areas Research Group of the Institute of British Geographers/Royal Geographical Society holding a session on 'Landscapes of change: socio-environmental interactions in developing areas' at the annual IBG conference. Papers from these two events were to be published in the RGS/IBG's *Geographical Journal*, starting with the September 2002 issue with further papers to appear in the March and September 2003 issues (Thomas *et al.* 2002). The first issue contained three papers on dryland issues in southern Africa (Dougill *et al.* 2002; van Rooyen 2002; Reed and Dougill 2002), and the editorial (Thomas *et al.* 2002: 193) emphasised how two key themes had emerged from the 'Dryland Change 2001' proceedings:

First was the complexity of past dryland environmental responses to global climatic changes, an understanding of which is highly significant for predicting how drylands will respond to future changes, including the effects of global warming. Second was the need to replace many 'conventional wisdoms' of human agencies of environmental change with an appropriately researched empirical understanding of human–dryland relationships.

Within the region what remains uncontroversial is that dryland areas are a significant component of the resource base and that processes leading to change within them need to be more fully understood to ensure their productivity is sustained so that they can contribute to, rather than detract from, the region's economic and human development.

Useful websites

http://www.iied.org Website of the International Institute for Environment and Development (IIED), London, including details of their Drylands Programme, including a drylands publication series (Drylands Issue Papers) and a quarterly newsletter: *HARAMATA: Bulletin of the Drylands: People, policies, programmes.*

http://www.odi.org.uk Website of the Overseas Development Institute, an independent body focusing on development. Refer, for example, to the Land Degradation in Sub-Saharan Africa project of the Rural Policy and Environment Group.

http://www.unccd.int Website of the United Nations Convention to Combat Desertification.

http://www.undp.org Website of the United Nations Development Programme, e.g. giving details of the UN Drylands Development Centre.

http://www.unep.org Website of the United Nations Environment Programme. Refer for example to DEDC/PAC – the UNEP Drylands Ecosystem and Desertification Control Programme Activity Centre.

References

Agnew, C. and **Anderson, E.** (1992) *Water Resources in the Arid Realm.* Routledge, London and New York.

Ahmad, Y.J. and **Kassas, M.** (1987) *Desertification: Financial support for the biosphere.* UNEP, Hodder and Stoughton, London.

Andersson, C.J. (1956) *Lake Ngami; Or explorations and discoveries, during four years' wanderings in the wilds of south western Africa,* 2nd edn. Hurst and Blackett, London.

Aubreville, A. (1949) *Climats, forêts et désertification de l'Afrique tropicale.* Société d'Éditions Geographiques, Maritimes at Coloniales, Paris.

Barrow, C.J. (1991) *Land Degradation – Development and breakdown of terrestrial environments.* Cambridge University Press, Cambridge.

Beaumont, P. (1989) *Drylands: Environmental management and development.* Routledge, London and New York.

Behnke, R. and **Abel, N.** (1996) Revisited: the overstocking controversy in semiarid Africa, *World Animal Review* **87**: 4–27.

Behnke, R.H. and **Scoones, I.** (1993) Rethinking range ecology: implications for rangeland management in Africa, in Behnke, R.H., Scoones, I. and Kerven, C. (eds) *Range Ecology at Disequilibrium: New methods of natural variability and pastoral adaptation in African savannas.* ODI, London: 1–30.

Bernstein, H. (1996) South Africa's agrarian question: extreme and exceptional? *Journal of Peasant Studies,* special double issue on The Agrarian Question in South Africa **23**(2/3).

Blaikie, P. (1985) *The Political Economy of Soil Erosion in Developing Countries.* Longman Development Series, London and New York.

Blaikie, P. and **Brookfield, H.** (1987) *Land Degradation and Society.* Methuen, London.

Bryant, R.L. and **Bailey, S.** (1997) *Third World Political Ecology.* Routledge, London and New York.

Charney, J.G. (1975) Dynamics of deserts and droughts in the Sahel, *Quarterly Journal of the Royal Meteorlogical Society* **101**: 193–202.

Charney, J., Quirk, W.J., Crow, S.H. and **Kornfield, J.** (1977) A comparative study of the effects of albedo change on drought in the semi-arid regions, *Journal of Atmospheric Sciences* **34**: 1366–85.

Darkoh, M.B.K. (1993) Desertification: the scourge of Africa, *Tiempo,* Issue 8 [www.cru.uea.ac.uk/cru/tiempo/issue08/desert.htm].

Dougill, A., Twyman, C., Thomas, D and **Sporton, D.** (2002) Soil degradation assessment in mixed farming systems of southern Africa: use of nutrient balance studies for participatory degradation monitoring, *Geographical Journal* **168**(3): 195–210.

Ellis, J.E. and **Swift, D.M.** (1988) Stability of African pastoral ecosystems: alternate paradigms and implications for development, *Journal of Range Management* **41**: 450–9.

Ehrlich, P. and **Ehrlich, A.** (1990) *The Population Explosion.* Simon and Schuster, New York.

FAO (1981) *An Ecosystem Classification of Inter-tropical Africa.* Plant production and protection paper No. 31. FAO, Rome.

FAO–UNESCO (1971–79) *Carte des sols du monde,* vols 1–10. FAO, Paris.

Glantz, M.H. and **Orlovsky, N.** (1983) Desertification: a review of the concept, *Desertification Control Bulletin* **9**: 15–22.

Goudie, A.S. (2001) *The Arid Expansion; Special report – spreading deserts.* Guardian Unlimited [www.guardianunlimited.co.uk/desertification].

Grainger, A. (1990) *The Threatening Desert: Controlling desertification.* IIED, Earthscan, London.

Hambly, H. and **Angura, T.O.** (1996) *Grassroots Indicators for Desertification: Experience and perspectives from Eastern and Southern Africa.* IDRC, Canada.

Hardin, G. (1968) The tragedy of the commons, *Science* **162**: 1234–48.

Heathcote, R.L. (1983) *The Arid Lands: Their use and abuse*. Longman, London.

Hellden, U. (1984) *Drought Impact Monitoring: A remote sensing study of desertification in Kordofan, Sudan*, Rapporter och Notiser, No. 61, Lund Universtets Naturgeografiska Institution, Lund.

Hellden, U. (1988) Desertification monitoring: is the desert encroaching? *Desertification Control Bulletin* **17**: 8–12.

Hellden, U. (1991) Desertification – time for an assessment? *Ambio* **20**(8): 372–83.

Hopkins, S.T. and **Jones, D.E.** (1983) *Research Guide to the Arid Lands of the World*. Oryx Press, Phoenix.

Hudson, N. (1995) *Soil Conservation*. Batsford, London.

Koppen, W. (1931) *Die Klimate der Erde*. Walter der Gruyter Co., Berlin.

Lamprey, H. (1975) Report on the desert encroachment reconnaissance in northern Sudan, 21 October to 10 November 1975. UNESCO/UNEP mimeo, Nairobi.

Mainguet, M. (1999) *Aridity: Droughts and human development*. Springer-Verlag, New York.

Mazzucato, V. and **Niemeijer, D.** (2001) *Overestimating Land Degradation, Underestimating Farmers in the Sahel*. Drylands Issue Paper. IIED, London.

Meigs, P. (1953) *World Distribution of Arid and Semi-arid Homoclimates*. UNESCO Arid Zone programme 1: 203–210.

Morgan, R.P.C. (1995) *Soil Erosion and Conservation*. Longman, Harlow and New York.

Mwangi, W. (1994) *The desertification convention: 'not just another document'*, IDRC, Vol. 22: 2, Canada.

Olsson, L. (1985) *An Integrated Study of Desertification*. University of Lund, Lund.

Olsson, L. (1993) Desertification in Africa – a critique and an alternative approach, *GeoJournal* **31**(1): 23–31.

Otterman, L. (1974) Baring high-albedo soils by overgrazing: a hypothesized desertification mechanism, *Science* **186**(4163): 531–53.

Reed, M. and **Dougill, A.** (2002) Participatory selection process for indicators of rangeland condition in the Kalahari, *Geographical Journal* **168**(3): 224–34.

Scoones, I. (ed.) (1995) *Living with Uncertainty: New directions in pastoral development in Africa*. IT Publications, London.

Solbrig, O. (ed.) (1991) Savannah modelling for global change, *Biology International* (Special Issue) **24**: 1–47.

Stiles, D. (1995) Desertification is not a myth, *Desertification Control Bulletin* **26**: 29–36.

Stocking, M.A. (1975) Soil erosion potential: the overview, in *Engineering Handbook*. Department of Conservation and Extension, Salisbury: 1–7.

Stott, P. and **Sullivan, S.** (eds) (2000) *Political Ecology: Science, myth and power*. Arnold Press, London.

Sullivan, S. (1996) Towards a non-equilibrium ecology: perspectives from an arid land, *Journal of Biogeography* **23**: 1–5.

Thomas, D., Twyman, C. and **Harris, F.** (2002) Sustainable development in drylands: geographical contributions to a better understanding of people–environment relationships, *Geographical Journal* **168**(3): 193–4.

Thomas, D.S.G. and **Middleton, N.** (1994) *Desertification: Exploding the myth*. Wiley, Chichester.

Tiffen, M., Mortimore, M. and **Gichuki, F.** (1994) *More People, Less Erosion: Environmental recovery in Kenya*. Wiley, Chichester.

Timberlake, L. (1985) *Africa in Crisis: The causes, the cures of environmental bankruptcy*. Earthscan Publication (IIED), London.

UNCCD (1995) *UN Convention to Combat Desertification in Countries Experiencing Serious Drought and/or Desertification, Particularly in Africa*. UN Secretariat of the Convention to Combat Desertification, United Nations.

UNCED (1992) *Agenda 21: Programme of Action for Sustainable Development*. Rio de Janiero, United Nations.

UNCOD (1978) *United Nations Conference on Desertification, Round-up, Plan of Action, and Resolutions*. United Nations, New York.

UNDS (1977) *Desertification: Its causes and consequences*. United Nations Desertification Secretariat, Pergamon Press, Oxford.

UNEP (1984) *General Assessment of Progress in the Implementation of the Plan of Action to Combat Desertification 1978–1984*. UNEP/GC.12/9, Nairobi.

UNEP (1991) *Status of Desertification and Implementation of the UN Plan of Action to Combat Desertification*. UNEP/GCSS.III3, Nairobi.

UNESCO (1977) *Map of the World Distribution of Arid Regions*. MAB Technical Note 7, Paris.

van Rooyen, M. (2002) Management of old field vegetation in the Namaqua National Park, South Africa: conflicting demands of conservation and tourism, *Geographical Journal* **168**(3): 211–23.

Walker, B. (1985) Structure and function of savannas: an overview. In Mott, J. and Tothill, J. (eds) *Management of the World Savannas*. Australian Academy of Sciences, Canberra: 83–91.

Ward, D., Ngairorue, B.T., Kathen, J., Samuels, R. and **Ofran, Y.** (1998) Land degradation is not a necessary outcome of communal pastoralism in arid Namibia, *Journal of Arid Environments* **40**: 357–71.

Warren, A. (1993) Desertification as a global environmental issue, *GeoJournal* **31**: 11–14.

Warren, A. and **Agnew, C.** (1988) *An Assessment of Desertification and Land Degradation in Arid and Semi-arid Areas.* Dryland Issues Paper No. 2, IIED, London.

Whitlow, R. (1988) Potential versus actual erosion in Zimbabwe, *Applied Geography* **8**: 87–100.

Whitlow, R. and **Campbell, B.** (1989) Factors influencing erosion in Zimbabwe: a statistical analysis, *Journal of Environmental Management* **29**: 17–29.

Williams, M.A. and **Balling, R.C.** (1995) *Interactions of Desertification and Climate.* Arnold, London.

CHAPTER 9

Geographies of governance and regional politics

Marcus Power

Introduction: geographical analysis and African politics

> Maps, frankly, have been a disaster for Africa.
>
> (Rostron 2000: 56)

> [T]he most fundamental of all changes in the next century and a half will be the boundaries of what constitutes Africa itself. Where does Africa end?
>
> (Mazrui 1993: 150)

In the vast literature on African politics it has become almost a cliché to suggest that boundaries are inherently artificial in the sense that they separate African peoples not divided naturally by some 'fundamental law' of human organisation. Spatial boundaries are expensive to construct and often impossible to police over long distances, while they are not always particularly conspicuous on the ground (Nugent and Asiwaju 1996). Despite the ambiguities of boundaries and their lack of congruence with events 'on the ground', they serve principally to demarcate political space, to create either citizens or subjects. The fluidity of African boundaries and the regularity with which they are disputed and contested illustrates something of the complexity of representing 'patterns' of governance and politics in eastern and southern Africa. Most boundaries in the region do not correspond to the cultural boundaries of particular ethnic, regional and national communities, and so the political power that is inscribed by borders frequently criss-crosses complex and dynamic cultural spaces.

The specific historical processes underlying the construction of African boundaries vary quite substantially as do the central forms of authority that take responsibility for policing them. Indeed the very uncertainty and contestation that engulf many African boundaries today can be taken as important indications of the wider difficulties associated with understanding and interpreting African politics (Simon 1996a). South Africa's border with Mozambique, for example, became a vast electrified fence during the apartheid era in an attempt to stem illegal migration and, at the end of the twentieth century, Namibia allowed the MPLA government of Angola, led by President Eduardo dos Santos, to use Namibian territory to launch attacks on UNITA rebels across the border in Angola. African boundaries have often served then both as 'zones of interaction' as well as clearly demarcated political partitions. The boundaries that divide the countries of eastern and southern Africa are therefore characterised by several ambiguous features and purposes, dividing as well as uniting nations from one another and simultaneously serving as barriers and junctions and as organs of defence and attack (Strassaldo 1989).

Frontier areas or 'borderlands' in Africa have been managed by different political forces and institutions so as to maximise different functions at different times. It is necessary therefore to broaden the concept of 'boundary', to encompass not only formal frontiers between states, but the whole idea of territoriality and the way in which this is understood within different cultures or used to make competing demands for social and political control (Clapham 1996). Boundaries mean different things to different people in different contexts and many situations of conflict have consequently arisen. Different conflicts have to be contextualised in different ways and in this sense a critical geographical perspective on the study of international relations and conflicts is useful. This is concerned with the complex interconnections of power, territory and conflict between and within nations and states. Adopting a critical geopolitical perspective, it is necessary to begin by distinguishing between nations and states. Nations are defined on the basis of culture, religion and ethnicity, for example, whereas states represent political units whose authority is bounded by other political spaces. The 'critical geopolitics' literature explores the rights and abilities of each state to control the territory encompassed by its boundaries or gain international recognition for its territorial authority. Relationships between power and territory can be observed at all geographical scales:

> The study of geopolitics involves considerations of territory, power and conflict between nations and states. In international relations, control of territory usually increases power, while increased power can expand control of territory. More powerful states exercise direct or indirect territorial control over weaker ones. In many cases, power implies direct, formal political sovereignty over designated territories.
>
> (Braden and Shelley 2000: 9)

These close interrelationships between power and territory have also been described as the problematic of 'geo-power' (Ó Tuathail 1996). Each state uses power in some way to articulate its claims to particular territories, to express its domination territorially. The concept of 'geo-power' then reminds us that the power of states is profoundly geographical, their very authority dependent on a geographical definition of where that power begins and ends. Going beyond a simplistic consideration of the geography of international relations, it is thus necessary to examine how international politics have been *geographically produced*. Rather than assuming a fully formed state system and state-delimited territories, the concern of a 'critical geopolitics' perspective is with the power struggle between different societies over the right to speak sovereignly about geography, space and territory (Ó Tuathail 1996: 11). As Luke (1993) has argued, states have sought to establish their power by *in-state-ing* themselves in space, imprinting a mark of their territorial presence. The form of instatement of power and authority through political space varies considerably, reflecting the variety of historical struggles that have gone into the creation and maintenance of states as coherent territories and identities seeking international legitimacy (Ó Tuathail 1996: 12). The late nineteenth century and the 'scramble for Africa' is a useful case in point. This very concern for geopolitical knowledges first emerged among the competing colonial powers which sought to dominate African space towards the end of the nineteenth century. During the 'age of empire' several European powers sought to extend their territories into Africa, to surround what was considered unclaimed space and enclose it within a larger empire. The 'scramble for Africa' can be understood as an example of one important historical form of geo-power:

> a new horizon of geo-power was surfacing as the last pockets of unclaimed and unstated space were surrounded and enclosed within the colonizing projects of expansionist empires and territorializing states.
>
> (Ó Tuathail 1996: 15)

This approach has in some senses drawn from the post-structuralist work of Michel Foucault, specifically focusing on the governmental apparatuses that produce the territorialisation of space. Foucault's work raises questions about the power of states and governments, about how states 'produce' and 'manage' their subjects. With reference to African contexts, this work suggests the need to focus on the ways in which states and governments construct and attempt to popularise their authority. Through the apparatus of government, states can thus produce and manage territorialised space through an ensemble of technologies of power. It is this 'governmentalization of geography' (Ó Tuathail 1996: 7) that must be understood if we are to make to sense of geopolitics in eastern and southern Africa. It is also important to consider how a 'geopolitical imagination' is constituted in the making of national, regional and international 'development' (Slater 1993). How do states and governments imagine geopolitical space within and beyond their borders and how do they attempt to manage particular territorial spaces?

Another more recent example of geo-power comes from the 'total strategy' of political destabilisation adopted by the South African state during the apartheid era. Apart from direct military destabilisation in neighbouring countries, this also included economic and diplomatic interventions, the manipulation of aid provision and interference in the transport systems of other southern African states (Hanlon 1984, 1991). This definition of where political power begins and ends was blurred by the apartheid state which directly and indirectly intervened in political affairs across southern Africa. In order to spread this destabilisation South Africa supported and trained armed rebel movements, interventions which have had continuing implications in terms of the creation of enduring political forces such as Unita in Angola or Renamo in Mozambique (Hanlon 1984, 1986; Minter 1994). Namibia was considered simply an extension of South Africa and consequently suffered illegal occupation by its powerful neighbour, delaying independence until 1990. Military incursions in the name of destabilising neighbouring states and territories were also made by South Africa into Zimbabwe and Zambia and a détente was forged with President Banda's Malawi, deepening the scale of destabilisation in Mozambique (which shares a border with both countries). Smaller countries like Lesotho and Botswana were particularly unable to resist the continual involvement of South Africa in their internal political affairs, with South Africa supporting a military coup in Lesotho in 1986 for example (see Box 9.1).

The lines on maps of Africa which were drawn and sometimes redrawn during the nineteenth and twentieth centuries also represent important historical attempts to demarcate competing spheres of territorial authority: Belgian, French, German or Portuguese for example. A key objective was to classify populations in order that they may be governed by colonial and imperial states. Many contemporary discussions of the continent's political affairs include a 'political map of Africa', yet all maps of Africa are in some way political (and very quickly become outdated), offering only a 'snapshot' representation of dynamic territories and shifting power relations. According to Rostron (2000: 56), 'maps are power' and the settling of the political map of Africa during the colonial period has had enduring consequences so that everyone, 'black and white, [is] still trapped by these outdated co-ordinates'. Thus, colonial boundaries which originally fulfilled practical functions, such as the regulation of colonial trade and the implementation of colonial policy, have in most cases survived the demise of colonialism and have bequeathed important historical legacies which continue to shape African politics in the 'post-colonial' period (Griffiths 1993). One interesting exception is Walvis Bay in Namibia which, at independence in 1990, remained under white minority South African rule (Du Pisani 1994). The South Africans argued that, since Walvis Bay had been annexed to the Cape Colony in the nineteenth century, long before independent South Africa had been given the mandate over South West Africa (Namibia) after the First World War by the League of Nations, independence for Namibia did not affect their rights over this port. The claim was not only dubious, but logistically absurd since Walvis Bay was Namibia's main port. In 1992, however, a Joint Administrative Authority was established and in 1994, shortly before the first democratic elections in South Africa, the port enclave was reintegrated into Namibia (Simon 1996a).

Debates about post-colonialism have considerable relevance in any attempt to understand issues of politics and governance in Africa today. The 'post-colonial' is relevant here because it raises questions about the historical formation of states, nationalism and nationalities in ways which

Box 9.1   Lesotho and Swaziland: in the shadow of South Africa

The southern African kingdoms of Lesotho and Swaziland have many common political features and also some very significant dissimilarities, making for an interesting comparison of their post-colonial histories. Both experienced colonial domination by the same power – Britain – and were (like Botswana) considered British High Commission Territories (HCTs) until they attained political independence in 1966 (Lesotho) and 1968 (Swaziland). Neither country's independence required an armed liberation struggle but both were to find the scope of this freedom considerably limited by the powerful presence of apartheid South Africa.

During the era of South African destabilisation in the 1980s, Swaziland was considered a conservative, ideological ally of the apartheid regime, which used a mixture of 'stick and carrot' tactics to keep Swaziland in line. These included the signing of a security pact in 1982 and an abortive land deal in which South Africa proposed to transfer the territories of Ingwavuma and Kangwane to Swaziland, thereby significantly increasing Swaziland's territory, and simultaneously conveniently removing a significant number of black South African citizens from South Africa's complex political equations under apartheid. The state in Lesotho, however, became increasingly critical of apartheid and offered support to South African liberation movements as well as supporting the imposition of international sanctions on South Africa. In response South Africa imposed its own political and economic sanctions against Lesotho, intervening directly in a military coup of 1986 and commissioning several SADF (South African Defence Force) attacks on the country. According to Matlosa (1998: 322), 'Lesotho and Swaziland have historically remained client or hostage states for South Africa, very much akin to South Africa's own Bantustans.'

The emergence of a post-apartheid state, one which has adopted the role of 'peacemaker', has consequently recast the relationships with these former 'hostage' states in a number of important ways. Nonetheless both remain landlocked countries with few natural resources. Lesotho is quite literally engulfed by South Africa and remains a significant labour reserve for the South African mines. The economies of both states remain inextricably tied to South Africa and are in a sense 'artificially divided [from South Africa] by porous borders'. Lesotho has experienced dictatorship of both civilian and military varieties, and when the military refused to recognise the election results of 1998 South Africa once again intervened in the country's internal political affairs. These events underlined the fragility of the democratic transition in Lesotho, where the initial transition from the military dictatorship was stage-managed by the military itself. Under King Mswati III, Swaziland, on the other hand, remains an 'aristocratic autocracy', and has put in place only quite limited and cosmetic political reforms despite the growing strength of pro-democracy forces in civil society.

Sources: Salmond (1997), Sejanamane (1996), Matlosa (1998).

are highly relevant to Africa and the study of African political space. The way in which Africa (and African politics in particular) is represented is also a key theme here. Many contemporary examples of writing about African politics are shaped by the myth of Africa as the 'dark continent' (Jarosz 1992) which 'homogenises and flattens peoples and places' in constructing 'Africa' as a singular, unitary, political space. Crush (1995) and Watts (1993) have written of the need to 'decolonise' 'development geography' and its scripting of African spaces and places, pointing to the partiality and situatedness of all histories and geographies of African affairs. Thus each attempt to write about the history or geography of African politics offers only 'partial truths' and reflects a subjective viewpoint which can be contested. Other authors concerned with 'development' and geography have also raised questions about the specific need to consider issues of cultural identity which emerge from post-colonial literatures, to question how nations, national identities and national boundaries (cultural and political) are constructed in particular contexts. What was the contribution, for example, of Julius Kambarage Nyerere, 'one of the greatest African nationalist and post-colonial leaders' (Allen and Burgess 1999: 315), to the changing styles of political leadership in Africa?

This chapter adopts a 'holistic' view of politics in the region, not simply concentrating on the 'state-in-society' approach of comparative politics which focuses on the social and political interactions between states and societies (Smith 1996). Ethnicity, class and gender are three important social relations which must also be considered. According to Kamrava (1999: 30), who outlines a multidisciplinary paradigm of 'Third World' politics, there are four key elements that must be considered:

> They include culture in general and political culture in specific; political economy, especially in relation to the economic causes and effects of the state–society interaction; international influences, both overt and subtle, diplomatic and political and otherwise; and the grey area of uncertainty and unpredictability that is the inevitable outcome of historical accidents, individual initiatives and unintended consequences.

The first part of the chapter considers these various aspects of African politics, looking at particular examples from across the region of eastern and southern Africa which illustrate their contemporary relevance. The second section focuses on the particular theme of democratisation, exploring its connections to experiences of conflict in Angola and the DRC or to wider issues of empowerment and development in Kenya and South Africa.

Key themes and issues in national, regional and international (geo)politics

> In each of these polities, the various elements of politics – state, society, political culture and predicaments – have a different relationship with one another, in turn reinforcing and sustaining that particular pattern of political rule.
>
> (Kamrava 1999: 4)

Africanists have put enormous energy into describing and cataloguing the state in sub-Saharan Africa, variously describing these states as 'failed' or 'collapsed', as 'patrimonial' or 'clientelist', 'kleptocratic' or 'predatory'/ 'extractive' and even 'vampire-like' (Frimpong-Ansah 1991; Englebert 1997; Goldsmith 2000). According to Rodrianja (1996: 25), the failures of African post-colonial nation states have been so widespread that it 'no longer requires any demonstration', despite major regional variations and the emergence of a 'New South Africa'.

The notion of a 'vampire-like' state drastically illustrates the idea of the state as fundamentally extractive, draining away the very lifeblood of the nation and its people. Some states are seen to profit from the labour and surplus of their own people than by being the clients of other more powerful states, by constructing a male-dominated social order concentrated in the hands of a few or simply because the desire to steal is seen as instinctive in some state officials and 'kleptocratic institutions'.

None of the categories listed above can adequately account for the complexities of statehood in Africa, however, and although they may have applied in particular settings at particular times, they are far from universal. To attach a particular label to a state is to ignore the variations of ideas and opinions that can exist within the state apparatus: there are often conflicts within the state and opposition to corrupt practices, for example (Hyden and Bratton 1992). What is clear, however, is that where states fail or even collapse altogether, the implications do not stop at the limit of its political boundaries but can spill over into neighbouring territories and regions with often quite devastating consequences.

Regional examples include the collapse of President Mobutu's Zaire in 1996/97. The result of an unprecedented correspondence of domestic, regional and international interests, the crisis in Zaire (subsequently renamed the Democratic Republic of the Congo (DRC)) had substantial and continuing consequences for the shape and form of politics in the region (see Box 9.2). Angola has also been hugely affected by external events. Both the ending of the Cold War and South Africa's transition to democracy in 1994 had major implications for its internal politics. Up until the end of the 1980s a variety of Western countries had supported Unita in order to weaken what was perceived as a communist presence in Africa (the ruling MPLA) with strong links to Castro's Cuba and the USSR. South Africa also provided considerable support to Unita, seeking not only to destabilise but also to capitalise on this instability in terms of gaining access to key resources like diamonds and oil.

Western support for Zaire hinged on Cold War perceptions of its strategic significance in terms of its size and location within Africa and its important mineral wealth. Mobutu played on these perceptions throughout his reign, stressing the regional consequences should he be deposed. Once the Cold War was over, however, his arguments lost much of their weight and by 1996 his regime had lost most national and international support. An opposition alliance under the leadership of Laurent Désiré Kabila was established to force Mobutu to step down (see Box 9.2).

Frantz Fanon's description of this country as the 'trigger of Africa' that could determine the political contours of much of the continent has been

Box 9.2   The fall of Mobutu and the emergence of the DRC

The former Belgian Congo gained independence in 1960 and President Mobutu Sese Seko became the first President of the Republic of Congo in 1965, later renaming the country Zaire. It is a huge country, the second largest in Africa, with very significant mineral resources, including many of strategic global significance. The Zairean state was therefore sustained during the Cold War with Western support as a bulwark against communism and as a source of raw materials. The US poured over US$400 mn. in weapons and training into Zaire during Mobutu's 32-year reign in addition to the US$250–300 mn. it invested in the anti-government forces of Unita in neighbouring socialist-ruled Angola. The genocide in Rwanda in 1994 began when an estimated 800,000 Tutsi and Tutsi sympathisers were murdered and a Tutsi-led force with strong Ugandan backing then took power, sending an estimated 1 million Hutu refugees fleeing into Zaire in 1994. Shortly after, reports emerged of friction between eastern Zaire's Tutsi community, newly arrived Hutu militants and the local Zairean authorities. Mobutu also faced growing internal opposition from Laurent Kabila's Alliance of Democratic Forces for the Liberation of the Congo (AFDL), supported by Rwanda and Uganda. At first Kabila's forces faced limited opposition from demoralised and unpaid Zairean soldiers, but as they closed in on Kinshasa they faced tougher resistance from Unita troops from Angola. These troops were sent by their leader, Jonas Savimbi, to defend Unita's crucial bases in Zaire, which had been used since 1974–75 (when Angola gained independence from Portugal) to launch attacks against Angola's MPLA government in Luanda and to provide supply bases. As civil war began in Zaire in 1996 a range of perspectives emerged on how to interpret the events. Some observers pointed to an irredentist attempt by Uganda and Rwanda to extend their hegemony in the region. President Mandela, on the other hand, celebrated a spontaneous revolutionary uprising by the oppressed peoples of Zaire to sweep away Mobutu's dictatorship. The anti-government forces were victorious and a new nation, the DRC, was forged from the ashes of the former Zaire. Growing popular disaffection with Mobutu from within Zaire was an important part of his fall from power, as was the implosion of the old regime as Mobutu's health failed and support from Western powers dissipated once the Cold War had ended.

Laurent Kabila came to power in mid-1997. The DRC's vast natural resource base has remained a source of conflict and contention. It has been a major factor in the new war which broke out in 1998 (see Box 6.4) against forces backed by Uganda and Rwanda, now opposed to Kabila, a conflict involving seven different African countries with a devastating cost in Congolese lives. The intervention of troops from Zimbabwe, Namibia and Angola in August 2000 to support Kabila was crucial to the defence of his regime as rebel forces made dramatic advances across the country and closed on Kinshasa. In July 1999 in Lusaka a peace agreement was agreed and a plan set up for a joint military commission to identify and neutralise rebels accused of attacking

Rwanda, Burundi and Uganda from the DRC. Yet the agreement was never implemented. In 2001 Laurent Kabila was assassinated and replaced with his son, Joseph. The Lusaka agreement was revisited but the future of the country seemed very uncertain as it was by then essentially divided between two rebel factions (supported, *inter alia*, by Rwanda and Uganda) and the government. In 2002, however, most of the foreign troops supporting the various internal forces were pulled out but unfortunately the DRC's government remained too weak to exert control over the areas where the internal rebels held sway, and where fighting continued. This difficulty is greatly exacerbated by the DRC's vast size, for these eastern, rebel areas are virtually a 'continent away' (Regional Roundup 2002) from Kinshasa. The threat that this continued instability posed for regional neighbours, especially Rwanda, left the future of the DRC still very uncertain and poses key questions about the ability of any government to exert its sovereignty within its inherited borders and, ultimately, to hold the state intact.

Sources: Amnesty International UK (1998), Hartung and Moix (2000), Hanghton (1998) McNulty (1999), Regional Roundup (2002).

---

further vindicated, however, and Zaire's strategic significance has yet again been proven, even in the absence of Cold War machinations. Since Kabila came to power, conflict in the DRC has had such major regional and continental implications for the political map of Africa that some observers have spoken of the conflict(s) as 'Africa's great war' (Shearer 1999). Others talk about the possibility of a new Afrocentric sphere of influence. Seven different African states have become embroiled in war in the DRC (Rwanda, Uganda, Namibia, Zimbabwe, Angola, Burundi). Like many of the recent conflicts in the region there are no easy answers or explanations for this tragic situation, particularly given that in the first few years of the formation of a new state it has been difficult to make reliable assessments of the situation 'on the ground' in the DRC. One important point is that the chief protagonists, Rwanda, Uganda and Angola, have been fighting, at least in part, to preserve *their own* regimes against insurgents from mounting attacks from bases in Congolese territory. The Rwandan state intervened after accusing Kabila of supporting the Hutu extremist rebel group, the Interahamwe, and for discriminating against the Tutsis that had helped him to power (see Box 9.2). Uganda claimed that it had entered the war to fight Ugandan rebels who had bases in Congolese territory, while the Zimbabweans joined in primarily because of personal relationships and economic promises between leaders (Human Rights Watch 2000). Angola, with one of the biggest armies on the continent, was concerned to shut down support for Unita rebels with bases in the DRC, and its armed forces have actively engaged Unita rebels on Congolese territory. Namibia and Zambia also had their own specific reasons for being part of the war (although these were often far from clear). In the Namibian case this was possibly related to the age-old friendship between

Eduardo Dos Santos and Sam Nujoma, forged during the Namibian liberation struggle. Corruption among military and political elites (e.g. in Zambia) also made the provision of logistical support to Unita more likely in that there were significant profits to be made.

In January 2002 the seven heads of state of the countries involved in the conflict came together in New York to discuss the Lusaka agreement which had been signed in July 1999 but not implemented due to the lack of political will and motivation (Hartung and Moix 2000). A peacekeeping force of 5,500 UN troops was assigned to the region by the UN Security Council in February 2000 although this was never fully assembled. The assassination of Kabila by one of his bodyguards in January 2001 led many to conclude that this might hasten the end of the war, although some commentators argued that the killing had been sponsored by Angola (unhappy with Kabila's role in the peace process) or by senior generals within his own military. His son, Joseph Kabila, took over the presidency and (unlike his father) accepted Ketumile Masire (a former President of Botswana) as mediator, thereby partly reviving the peace process. By the end of 2002 most of the external military forces in the conflict had pulled out, but difficult problems still remained (see Box 9.2). This drew an eighth country from the region into the DRC debacle, as South Africa's President, Thabo Mbeki, became involved in an attempt to broker a peace between Kabila's government and the two largest domestic rebel groups, which were still backed respectively by Rwanda and Uganda.[1] At the time of writing, the outcome of this initiative, which would require further international support, was uncertain.

The conflicts in Angola and the DRC have also had wider regional implications in the sense that both have contributed to forced population migration and internal displacement on a major scale, cutting across 'national territorial boundaries' with remarkable fluidity (see below). Some 1.3 million people fleeing conflicts in Rwanda, Burundi and the DRC sought refuge in western Tanzania, for example, between 1993 and 1998 where 'in some areas refugees outnumbered locals five to one' (UNHCR 2000).

Other recent conflicts in the region also raise questions about the territorial authority of the state. In Mozambique, there was a one-party state in the initial post-independence era, governed by the Marxist-Leninist Frelimo party. The authority of this party and its presidents, Samora Machel and then Joaquim Chissano, was significantly eroded during the course of a 16-year civil war against Renamo, a rebel movement which was armed and trained by apartheid South Africa (Minter 1994). The settlement that ended this war provided for a resumption of some powers by traditional authorities as part of a programme of democratic decentralisation, though the extent, nature and relevance of traditional authorities in

---

[1] Mbeki proposed an interim government including representatives of all three parties and a general election within two years. South Africa also offered to dispatch 1,500 troops to help oversee a ceasefire along with the 3,000 UN observers already in the country (Regional Roundup 2002).

local governance are still being debated (Alexander 1997; West and Kloeck-Jenson 1999).

Frelimo had initially tried to deny the importance of local traditions, which were seen as divisive and a threat to the modernisation of the countryside promised by Marxism-Leninism. Traditional authorities, for example, were seen as a colonial invention. There was some justification for this view as hundreds of years of Portuguese influence in the country had transformed these institutions so that they often performed many tasks for the colonial state (e.g. tax collection, labour recruitment) which were at odds with their supposedly 'traditional' and 'local' roles. Lineage leaders who would not conform to the colonial imperatives were often ruthlessly deposed and replaced by more 'cooperative' men. However, the state's efforts in this respect were a significant cause of popular disenchantment with the Frelimo government as many Mozambicans viewed this strategy as an attempt to centralise political control and deny the legacies of the past. Renamo seized upon this disaffection with the state's marginalisation of traditional authorities as a source of political mobilisation, an opportunity to illustrate that Frelimo had abandoned its African heritage and its cultural traditions in favour of a foreign 'communist' ideology. The recent debates about traditional authority in post-war Mozambique reflect an attempt by Frelimo to illustrate a new concern to respect cultural diversity and to explore the possibility of decentralising power away from the central government in Maputo.

In Angola, the civil war between the MPLA government and Unita rebels, which began in 1975 when the country became independent from Portugal, has also considerably undermined the capacity of the state to express, territorially, its authority and power in recent decades. There has been no simple geography to the spheres of influence that both sides have developed during this complex and dynamic war, though access to mineral resources and transit routes have been particularly important. Unita had particularly strong associations with the central highlands of Angola and the city of Huambo, and its military HQ of Jamba in the south, and the MPLA with Luanda and its hinterland, and the coastal regions. The fact that the country's oil is produced offshore and in the northern exclave of Cabinda has been very convenient for the MPLA as it rendered these resources peculiarly difficult for Unita to annex, unlike the inland diamond mines.

At the turn of the century Angola's civil war had raged for 25 years, with only two short periods of relative peace: in 1991 when a ceasefire was brokered by the UN as the end of the Cold War (as in the DRC) lessened the international pressures which had fostered the conflict. The country's first multiparty elections were held in September 1992 but Unita refused to accept the MPLA's win and the war resumed. A further period of relative peace was established at the end of 1994 when regional actors helped to bring about the signing of the Lusaka Protocol. By 1998, however, all-out war was being waged again, and many analysts and the MPLA were taken aback by the extent to which Unita had managed to rearm and the effectiveness of its campaigns which led to it establishing territorial control (or at least preventing effective government control) over much of the country for a while (Simon 1998a,b). Eventually however, in 2000, Unita lost its strongholds in Andulo and Bailundo in the central highlands to the MPLA and Jonas Savimbi could also no longer rely on the Malanje corridor which had provided relatively easy access in and out of the DRC for moving weapons and diamonds. Nevertheless, Unita's ability to wage a long-term, low-level, and still debilitating, guerrilla conflict appeared to remain. It was only Savimbi's death in a government attack in February 2002 which finally brought peace to this beleaguered nation. The speed with which a ceasefire was implemented, and an inevitable political jockeying for position began, highlighted the extraordinary influence that this one man's thirst for power had had on Angola's conflict. For the general population of Angola, the civil war has been disastrous (e.g. Simon 2001): official estimates in 2000 suggested that some 65 per cent of the population were living below the poverty line, despite the country's oil and diamond wealth (Ministério do Planeamento cited in *Africa Hoje*, May 2000) and UNICEF claimed the country had the dubious distinction of being the worst place in the world to be a child.

The complexity of statehood in the region was also powerfully illustrated at the beginning of Zimbabwe's twentieth year of independence. The authority, popular support and power of President Robert Mugabe's ZANU-PF party were radically shaken by a referendum in February 2000 when 54 per cent of Zimbabweans voted against a new draft constitution produced by a government-sponsored institution. This result partly reflected the electorate's frustration with the ruling party, with the political stagnation they had produced, with ineffectual structural adjustment plans and delays in delivering much needed land and other socially democratic reforms (Dorman 2000). Zimbabwean involvement with the war in the DRC also seemed for many to become a symbol of the regime's distance from the concerns of the average voter. In response to the shock of the referendum result, Mugabe used the politically delicate land reform issue and anti-white sentiment in a desperate appeal to national populism and this led to considerable instability and upheaval across the country. He insisted that the transfer of the majority of white-owned land must occur, without full financial compensation, thereby overturning two decades of legal, but slow, land reform, and persisted with this policy despite threats of international sanctions (which he called 'regional or block gangsterism'). Britain was accused of using its influence to perpetuate 'old colonial injustices' (BBC 2000).

Despite the opposition victory in the referendum, the ruling party managed to cling on to power in the parliamentary elections held in June 2000, with some 65 per cent of the electorate casting a vote. In the run-up to the vote, Mugabe created a climate of terror through intimidation where opposition members were tortured and even murdered, where workers at white-owned farms were forced to attend night-time rallies and where voters were cautioned that a vote for the MDC would be severely punished. However, Mugabe's regime was stunned by the size of the vote for the opposition MDC, a party formed only a year prior to the elections. After 20 years of huge parliamentary majorities, ZANU-PF's victory was reduced to a few seats.

Ethnicity and political identities

There are interesting and important similarities between the political processes apparent in Zimbabwe and those of Kenya, particularly around the
issue of political intimidation. Southall (1999) described the Kenya African
National Union (KANU) of President Daniel Arap Moi as a 'kleptocracy'
which was engulfed, as a class elite, in a series of financial scandals
around the time of the general election of December 1997 (Ajulu 1998). Yet
KANU was re-elected, despite widespread dissatisfaction with the post-
colonial state in Kenya, partly because of divisions among the opposition
and partly, as in Zimbabwe, because of violent intimidation of the opposition. A key dimension to the 1997 elections in Kenya was that voting
patterns coincided to a certain extent with patterns of ethnicity. The Kalenjin
tended to vote for KANU and Moi, while the Kikuyu tended to vote for
the Democratic Party and the Luo for the National Development Party of
Africa. For some observers the significance of ethnicity as a crucial force
within Kenyan politics is self-evident: Southall (1999: 98) writes of the
politicking of the 'Kikuyu lobby' and a 'Kalenjin political elite'. On the
other hand, Bayart (1993: 55) has warned against reading too much into
the importance of ethnicity in African politics, arguing that: '[i]n Africa
ethnicity is almost never absent from politics, yet at the same time it does
not provide its most basic fabric'. For Bayart, tribalism is less an actual
political force than a 'shadowy theatre' through which competition for the
acquisition of wealth, power and accumulation is expressed and played
out. He also argues that ethnicity is inseparable from the very structure of
the state, simultaneously a product of history and the process of accumulation, but never fixed either in time or space. In referring to the 'politics of
the belly' in Africa, Bayart is trying to show how closely the manipulation
of ethnicity and ethnic identity is related to the processes of accumulation
presided over by state agencies.

In order to get re-elected political parties need to appeal to a variety of
ethnic groups, cultures and identities at a variety of spatial scales. Each
political party claims to represent particular kinds of ethnic and cultural
constituencies in some way. States thus claim to incorporate and represent
this national cultural diversity in presenting themselves as the central source
of power and authority within national territory. Bayart's view is that
ethnicity acts as an agent of accumulation, allowing members of particular
ethnic groups to access opportunities for personal wealth and political
power.

In Kenya, corruption is a major outcome of this desire for opportunities
to create personal wealth and has no doubt gone some way to discrediting
the authority of the state in the eyes of many Kenyans. A further example
of the importance of ethnicity to the structuring of the state can be found
in Namibia which was under apartheid South Africa's control until 1990.
Up until then the state administration was heavily predicated upon
ethnic affiliation, but the government determined which ethnic affiliations
were to be recognised and which were not and this did not reflect the
true diversity of national cultural and political voices. After independence
in 1990, the new government led by the South West Africa People's

Organization (SWAPO) attempted a remapping of the country's 'internal political geography' on non-ethnic criteria (Simon 1996b) by establishing regional councils and promoting new regional identities. Although not entirely successful, an attempt had been made to at least address and respond to some of these issues.

Ethnic identities are formed at a variety of spatial scales. Ethnic communities can stretch across national borders despite the promotion of African boundaries as markers of ethnic and political difference. A good example of these complex geographies of ethnicity is the Tutsi community which stretches across Rwanda, Uganda and the DRC. In a similar sense, Mozambican migrants to Swaziland, for example, have forced a certain renegotiation of the terms of Swazi citizenship and the need for a more inclusive notion of national belonging, one that is not confined by state borders (McGregor 1994). As already discussed, there are important links between power and accumulation in African politics. The position of individuals in relation to the apparatus of the state influences their social status, the nature of their relationship to the economy and the nature of individual material power (Bayart 1993). Bayart's view of the state sees positions of power within the state as 'positions of predation' where those in power can monopolise 'legitimate' force to demand goods, cash and labour. This is often particularly evident in relation to the state's treatment of marginal groups according to Bayart. This can be illustrated through an examination of the relationship between the Bushmen, or San, in Botswana (Good 1999). Much has been made of the significant growth in Botswana's national economy since independence and this has clearly strengthened the position of the wealthiest groups in Tswana society. Arguably, however, it has worsened the plight of the San who have been marginalised as a direct result of the ascendancy of state elites who have sought greater wealth and political opportunity for some groups (especially themselves) while denying them to others. The San have in many cases been dispossessed of their land by some of these elites who often own many of the largest private ranches in the country (see, for example, Sporton *et al.* 1999). With reference to Botswana, Good (1999) has argued that 'the study of poverty is inherently a political problem', given the role of elites, and the exploitative practices that accompany their ascendancy. In his view, although Botswana has been portrayed as a 'shining light' of democracy in southern Africa, with regular elections, it remains very centralised in its political culture. In some respects Namibia and South Africa are characterised by less elitist and more participatory political cultures than Botswana.

Political manipulation of the state system by elites for their own economic ends can degenerate into what Allen (1999: 377) has called 'spoils politics systems':

> Under these conditions, patronage politics, involving the exchange of material benefits for political support, occurs only when actual competition for office occurs (mainly during the run up to elections) or to protect the ruling group from a short-term threat (e.g. from a general strike). State resources . . . are the subject of fierce winner-takes-all competition within the ruling group.

This sort of desperate politics in Africa is also sometimes characterized as a 'zero-sum game' whereby the loss of political power, and office, must be avoided at all costs. Political office can be used as a route to wealth and accumulation anywhere in the world, but the states of eastern and southern Africa are rendered much less secure and powerful than many of their Western counterparts in part because the *primary* route to the accumulation of wealth is through the holding of political office, rather than this being just one of several routes.

Civil society

The way in which political parties, parliaments and governments (especially their ideologies) have tended to dominate political analyses has perhaps suggested that African politics are much more state-centred than in reality. In fact, manipulation of the political process in the ways outlined above has not occurred without very serious opposition. Many observers now seek to analyse such opposition to the power of states through the concept of 'civil society' which Bayart argues should be defined in the following way:

> as society in its relation with the state . . . in so far as it is in confrontation with the state. . . . The notion of the civil society is thus an ambivalent, complex and dynamic relation between state and society.
>
> (Bayart, quoted in Chabal 1986)

Bayart highlights the ambivalent and dynamic nature of state–society relations in Africa. Civil society consists, then, not just of what is not part of the state but also of all those who have become disenfranchised and powerless (e.g. villagers, nomads, slum dwellers and all those who feel they do not have access to the state) – it is defined *in relation* to the politics of the state. A common factor is exclusion from the state; the relationship between state and civil society is thus defined by the state which determines the boundaries of political legitimacy. The politics of civil society are often not recognised or understood simply because they are not intelligible to the political language of the state (Kasfir 1998). However, although often without formal power structures, civil societies do have informal power and, individually, collectively or by avoidance, have been responding to post-colonial African states. Where political accountability has depth and power, legitimate civil society may be accommodated, but where it is shallow and where power is subject to challenge then the demands of civil society may be strongly contested.

Post-colonial order in many African societies was constructed around the assumption that nationalism was the only unifying force in civil society – political independence was a goal shared by many and nationalist politics tended to dominate the politics of civil society. When ethnic politics flashes up in post-colonial Africa it can thus be understood as part of the contest between state and civil society. States will also often try to co-opt sections of civil society to increase their legitimacy. Thus in many ways it is more useful to conceptualise politics in Africa as the relation between state and civil society rather than just as formal political structures. However, analysis of civil society should not be at the expense of class analysis for the two are interrelated.

The literature on civil societies in Africa does seem to be relevant in countries where political accountability lacks depth and power. In Zimbabwe, for example, the debate about civil society has raised questions about the role of the MDC or women's groups in pressuring Mugabe's government for democratic reform. In Kenya, the National Conventions Executive Council (NCEC), a radical lobbying group leading the struggle for constitutional reforms, has consciously projected itself as directing 'civil society', suggesting that the concept does have relevance for African politics (Southall 1999). Many commentators argue that 'civil societies' were also crucial to the democratic pressure placed on Kenneth Kaunda and President Banda in Zambia and Malawi, and to many of the political changes which have been occurring elsewhere in the region.

Civil society organisations are particularly important in countries where the predominant party system 'functions to firmly entrench political elites' (Good 1999: 556). In Zambia for example, the trade union movement was initially able to take an independent role which was crucial in the challenge to one-party rule and a peaceful transition of power from UNIP, the party which had ruled since independence (Clapham and Wiseman 1995). Kenneth Kaunda was replaced by Frederick Chiluba as President of Zambia after the elections in 1992. At first Chiluba, a former leader of the trade union movement in Zambia, was seen as a welcome change to the previous regime and a different kind of politician. Sadly, however, the new democratic political culture turned out to be of limited duration, for corruption has not improved (if anything it has intensified) and Kaunda was prevented from standing in the next round of elections on spurious grounds. The problems of the old regime thus did not disappear with a new state, as the once dynamic leaders of civil society who had led political opposition and debate, succumbed to the opportunities for self-enrichment once in power (Chan 1994; OneWorld 2000). After unseating a 27-year dictatorship, the ruling MMD (Movement for Multi-party Democracy) party is 'fast becoming the villain everyone loves to hate, widening popular disaffection with politics and politicians' (Clapham and Wiseman 1995).

Despite the increasing importance attached by political analysts to African civil society, it is also now being recognised that it is by no means a panacea for Africa's political problems. Some note the applicability of the concept of civil society outside Europe is questionable and that, contrary to the idealistic hopes pinned upon 'civil society' by many development organisations, it is highly contested, not always 'independent' from the state and often breeds anti-democratic behaviour (Mamdani 1996). A range of organisations which comprise the dynamic, complex and ambivalent relationship between state and society do, however, need to be recognised as important political forces. Many people view civil society as a distinct realm of political freedom, outside and separate from the state and party politics, even though it is arguably counterproductive to view 'state' and 'society' as entirely distinct and separate.

States and regional (dis)integration: imagining 'development' communities

> The world system is subject to a simultaneous process of globalization and loss of precise territorial definition, which may not lead to the eclipse of the state as an organ of power, but which is most surely leading to the development of transnational relations between societies.
>
> (Bayart *et al.* 1999: 9)

The theory and practice of 'development' vary enormously across Africa and it is extremely difficult to generalise about efforts to bring about a state of being 'developed' (Simon 1999). Some perspectives view development as a discourse, exploring the interconnected languages and representations involved while raising important questions about the failures of development projects and the limitations of development theorisations (Ferguson 1990; Escobar 1992, 1995; Watts 1993). Many recent writings about development have additionally sought to introduce post-modern and post-colonial perspectives into debates about development which, for example, explore the power of development as knowledge or how 'national development' plans can marginalise certain groups by defining the terms of public (political) participation (Crush 1995; Rahnema and Bawtree 1997). If knowledge is power (and power is also knowledge) then knowledges of development put in place certain kinds of relationships between different actors and agencies. Key concerns in these approaches have been first, the need to understand the contested nature of the idea of development and second, to focus on complex social identities like ethnicity and gender in order to incorporate an awareness of social and cultural difference in 'development'. Nationalist movements across the region have invented their own particular notions of what constitutes social, economic and political progress since independence which have had important implications for the citizens considered the subjects of these processes of social change and development. Development is thus invented and imagined at a national level (by politicians) through national plans and strategies which seek social change within national communities. However, the politicians' views of 'development' may differ considerably from those held by the communities themselves.

The Southern African Development Coordination Conference (SADCC) has been discussed as one important example of a regional strategy of 'imagining' an invented development community. SADCC first emerged in 1979/80 among the front-line states opposed to apartheid South Africa (Angola, Botswana, Mozambique, Tanzania, Zambia and, from 1980, Zimbabwe). A key objective was to reduce economic dependence on South Africa and seek common responses to South African interference in the affairs of member states. After the beginning of constitutional negotiations within South Africa to end apartheid at the beginning of the 1990s, shifting geopolitical conditions (including Namibia's addition to SADCC in 1990) led to a reassessment of the organisation's objectives (Simon and Johnston 1999) and the emergence of a successor, the Southern African

Development Community (SADC) in 1992. SADC's membership expanded in 1995 to include Mauritius and again in 1997 when the Democratic Republic of the Congo and the Seychelles were admitted. A year later three member states had become 'officially' involved in the conflict in the DRC (Angola, Zimbabwe and Namibia) without prior consultation with other SADC members and in violation of SADC's publicly stated position of neutrality (Simon and Johnston 1999). The 'security organ' of SADC, headed by Mugabe, played a role in facilitating the involvement of these three countries since there was a security agreement to support a SADC government if help was requested. The involvement of these countries in the DRC conflict, however, caused great strains within SADC and undermined the sense of a 'development community'. More materialistic issues have also played a part in inducing the help of SADC member states (see Box 9.3).

The cost of Zimbabwean involvement in the DRC war has been estimated at US$1–2 mn. *per day*, a fact which became a focal point for popular discontent in Zimbabwe and seemed to signify the isolation of some African states from the interests of their people as well as the failure of some post-colonial African states more generally. Each state has had its

---

Box 9.3 'Dirty diamonds'

Greed for diamonds and other 'lootable' commodities fuels civil wars.
(Collier 2000: 21)

Even the World Bank now recognises that the global precious gemstone industry has had an important role in perpetuating several African conflicts. Diamond-rich regions in Angola and the DRC have clearly 'fuelled' conflict in both countries over many years. The international diamond industry has increasingly been overshadowed by the issue of 'conflict diamonds' which have become the 'sinews of war, to feed and equip armies and rebel groups'. Conflict diamonds are 'diamonds that originate from areas under the control of forces that are in opposition to elected and internationally recognised governments, or are in any way connected to these groups'. Though this trade in gemstones has continued over many decades, only very recently have international diamond companies like De Beers begun to react by guaranteeing the origins of the diamonds it sells.

The active military presence of Zimbabwe and Namibia in supporting the Kabila regime when it came under attack from rebel forces backed by the Ugandans and Rwandans can be explained by a variety of factors but diamonds have evidently played a part. President Mugabe confirmed that Kabila offered Zimbabwe and Namibia a diamond mine each for their military help, though he denied that either he or Namibian President Sam Nujoma would personally gain from these 'gifts' (Nevin 2000: 21). The Zimbabwean army has also created companies to do business in the DRC, claiming that the Zimbabwean private sector had not taken advantage of new opportunities there. The profits from these business ventures go to generals and politicians. In an interview with Independent Newspapers in June 2000 Mugabe said:

> What Kabila has suggested to us, and we have worked on as payment for operations in the DRC, is that there are these mines, which the owners left some time ago. He has said 'let us be partners in regard to this mine and partners with Namibia in regard to that mine'. So there are two mines, but so far we haven't got a single diamond from them because we're still working on the paperwork.
>
> (President Robert Mugabe, cited in Nevin 2000: 21)

The Zimbabwean military's business involvement in the DRC was further publicised when a mining company, Oryx Diamonds, included a venture in the DRC as part of its reverse listing on the London stock exchange in May 2000. The Oryx prospectus showed that Osleg, the investment arm of the Zimbabwe Defence Force, would get 40 per cent of the profits earned by operating its concession at Mbuji-Mayi where reserves were calculated at US$1bn. This deal would have offered a financial lifeline to beleaguered governments in Kinshasa and Harare in the war against rebel movements backed by Uganda and Rwanda. The indirect involvement of the Zimbabwean Defence Forces with a company being floated on the London stock exchange highlights the global nature of the trade and the variety of interests that have become involved. Zimbabwe's material interests also evidently gave political and military elites a vested interest in remaining militarily active in the DRC in order to protect these interests from rebel forces. Even after their forces left in 2002, the networks established by Zimbabwe (and other protagonists in the war, such as Rwanda and Uganda) to profit from the DRC's mineral and timber wealth remained.

Another regional example of 'dirty diamonds' is provided by Unita in Angola which, after the end of the Cold War, turned increasingly to diamonds as a means of paying and feeding its troops, and buying arms. Unita amassed some US$3.7bn over a six-year period in the 1990s, much of it coming from the sale of diamonds from mines it had captured, which helped to maintain its military operations and prevent the establishment of democracy. De Beers always denied that it had knowingly bought diamonds from Unita, arguing that it was often unsure of the origins of the diamonds it was buying, particularly before the UN diamond sanctions of 1998. However, as Global Witness (2000: 9) points out:

> this is a complete abdication of corporate responsibility, and it further raises the question of whom exactly the De Beers staff, who were based in DRC along the Angolan border, thought they were paying for the diamonds that flooded across that border up until the fall of Mobutu in 1997.

In the DRC the strategic control of the diamond-producing areas was one of the key driving forces in the conflict and their control is crucial to any lasting peace settlement.

Sources: Global Witness (2000), *Guardian* (2000), Le Billion (2001), Nevin (2000), Parker *et al.* (2000), Regional Roundup (2002).

own reasons to enter this war and cultural and historical factors, as well as material ones, have played their parts. Simon and Johnston (1999: 7) have argued with respect to the various wars in the region at the end of the twentieth century that:

> Underlying the conflicts [in the DRC, Angola, Burundi and Rwanda] are inappropriate colonial borders and policies, often perpetuated or exacerbated after independence. In their respective ways, these four different countries illustrate the failures of the post-colonial African state established in the mould of the modern, European nation-state. The welfare and development of the population at large are not the real concerns of the governments or their rivals.

The significance of colonial borders and the policies of the colonial state are raised here as important developmental issues. Similar concerns are raised elsewhere in the Africanist literature. Ali Mazrui, for example, has written of the 'bondage of African boundaries' and Mamdani (1996) refers to the 'inherited impediments' of colonial history, such as the structure and form of the state apparatus. Such impediments can be illustrated by the case of Mozambique where, in the first decade of independence, Frelimo's socialist revolution relied heavily on a very bureaucratic and centralised state structure inherited from the Portuguese. This helped to distance the party from the rural areas and led to an excessive centralisation of governmental authority (Dinerman 1994) – neither factor was conducive to democracy and political stability. Frelimo was also confronted with the complicated task of nation-building within inherited colonial boundaries which included considerable ethnic diversity. Its attempt to build a national sense of identification by engaging in a denial of ethnicity, playing down ethnic differences and primarily characterising 'tribalism' as an evil legacy of the colonial era was thereby virtually doomed to failure (and of course further complicated by the emergence of a South African strategy of political destabilisation).

In post-apartheid South Africa, complex projects of nation-building are under way which also raise questions about how a national political party seeks to (re)imagine a national development community (Habib and Taylor 1999). The ANC is seeking to build a more inclusive multiracial society within the strictures imposed by the legacies of apartheid, an ideological system which was entirely based upon ethnic and cultural differentiation and segregation. Understandably this is proving to be an enormous and deeply complicated task. The case of Bophuthatswana is illustrative of the difficulties that are faced by attempts to create new national identities in South Africa. Bophuthatswana was defined as a 'homeland' or Bantustan under apartheid and some members of the elite accepted nominal 'independence' from South Africa. This spurious territorial entity 'disappeared' when apartheid ended, but the end of apartheid has opened up 'a Pandora's box of ethnic and regionalist claims' (Jones 1999: 509). The redrawing of provincial boundaries in South Africa led to a new province being formed, the North West province, which includes parts of the former Bophuthatswana. 'Fault lines' have since emerged around notions of 'Bophuthatswananess' and a conflict between ethnic, regional and national identities. This notion of national belonging, of being part of Bophuth-atswana, seemed at odds with the post-apartheid efforts to build a new *national* sense of inclusion, around provincial territories rather than Bantustans. Other attempts to deepen democratic stability, ensure political order and encourage national identification have been made in the 'New South Africa'. These include the restructuring of public media communications systems, for example (Barnett 1998), where national media agencies have sought to expand broadcasting in a variety of African languages and thereby to recognise and represent the multicultural nature of South African society.

Through SADC, the ANC government has taken on a new, unofficial, role as an arbiter of peace and regional security. President Mandela campaigned for the admission of the DRC to SADC in 1997 for example, arguing that this would be a way to guarantee regional stability (although this has not been borne out out by subsequent events, given the widening regional involvement in conflict in the DRC). As membership of SADC grows, the possibility of building closer cultural and political integration has become more complicated, however. Many South Africans remain 'unenthusiastic or uninterested' (Simon and Johnston 1999: 3) about the region's newly reinvented development community. Many state officials across the region find it hard to continually balance the different national and regional commitments derived from their overlapping membership of various regional integration agreements (McCarthy 1996; Sidaway and Gibb 1998). There are also important political and economic implications of 'surviving in South Africa's shadow' (Stock 1995) which divide the members of SADC in a number of ways. South Africa has had a crucial part to play in the formation of new regional spaces of identity since the end of apartheid. It is no longer the regional outcast but is not entirely free from the policy of 'military adventurism' (Rich 1994) employed by its apartheid predecessor. These issues have been particularly relevant in Lesotho (see Box 9.1) which is surrounded by the politico-economic power of South Africa. In September 1998, after the country's multiparty election results were disputed by the opposition party resulting in a coup, South Africa sent troops in the hope that the military would bring about a quick reversal of the coup but in fact public opinion turned against the intervention as its decision-making and operational aspects were widely questioned on the grounds that there were more pressing internal issues and that intervention was not the sole responsibility of South Africa.

SADC has been shaped simultaneously by political, economic and cultural processes, from which member states have gained certain forms of state legitimacy (Sidaway 1998, 2000). According to Sidaway, SADC represents an imagined regional community which continues to exist partly because it demonstrates that member states are 'real' and sovereign, even though (as in the case of Angola) their 'national' territories are often regularly under contention. Regional blocs like SADC have significance in terms of globalisation too, as they have become a way of constructing and bolstering the authority (or at least the image of authority) of particular states to operate in global markets or to respond to the challenges of globalisation, however these are defined. SADC thus also needs to be seen in the context of the emergence of a number of other transnational trading blocs around the world, such as in its relations with the European Union (see Chapter 10).

The idea of a 'development community' is important in understanding SADC's role in creating new forms of regional geopolitics and regional identity. Current SADC objectives include harmonisation and rationalisation of policies and strategies for 'sustainable development' in all areas (Makumbe 1999). This includes a SADC Trade Protocol to reduce trade tariffs in the region which came into effect in 2000. However the structure of the members' economies varies considerably and South Africa's GDP has often been larger than the combined GDP of all other SADC members. Thus the 'imagining' of this regional development community is complicated by the major economic and political differences between member states and the difficult challenge of harmonising objectives (e.g. see Chapter 10). On the other hand, according to neo-liberal analysts at the World Bank and IMF headquarters in Washington, Mozambique, Angola and Botswana were among the world's five fastest-growing economies in 2000 which might be taken to suggest one crucial, positive, common characteristic among some key SADC members. Such economic indices can be highly misleading, however. For example it is very difficult to compare Angola directly with Botswana, for their histories of political and economic change have been so very different, as are the developmental outcomes of economic growth. After three wars and decades of instability and conflict any sort of change in the profile of the Angolan economy can be made to look like 'progress', for example, but economic growth there rarely translates into real improvements in the livelihoods or welfare of the vast majority. By contrast, very significant improvements have been realised in Botswana. Mozambique, in particular, has been singled out as a 'success story' for its progress in 'rolling back' the state, privatisation and trade liberalisation – reforms demanded by the Bank – but a large part of this economic 'growth' was comprised of aid transfers from Western donors (Sidaway and Power 1998; see also Chapter 3).

The 'Washington consensus' on the need for neo-liberal reform of the state and the call for 'good governance' by a variety of aid and development organisations have clearly had an impact on the nature of politics and the allocation of economic resources across Africa (see Chapter 3). A generalised agenda of 'good governance' has been promoted by many national and international development organisations so as to improve efficiency and curtail corruption, although successes have often been more illusory than real. The view of 'development' imagined by many neo-liberal institutions has had enormous influence at both the national and regional level. The 'invention of development' in post-apartheid South Africa, for example, has taken place within this international context and has been conditioned by international institutions and coalitions of ideas in a wide variety of ways. Recent political transformations in South Africa have a high profile in the international development community (McEwan 2001), and many international development organisations have been keen to show that they are part of these changes and can contribute to them.

The next section looks at how democracy is one key area in which these organisations have sought to impose loan conditionalities and consequently to put forward particular views of 'good governance' in Africa.

Democratisation and conflict resolution in global political and economic context

'Development' and democracy are closely interconnected in Africa. Ake (1996) has even argued that development on the continent has never really 'failed', it just never really got started because the social and economic conditions have been lacking. Ake has also detailed the 'improbable' strategies or development policies adopted by a range of African states and demonstrated how these have often been poorly adapted to local conditions and unable to counter the political practices by which those in power extract monetary gains.

Perhaps the conditions which most disable democratic practices in southern and eastern Africa are those related to war. The ending of war, and demilitarisation and disarmament when peace is established, are key and interlocking challenges which determine the success of democratic government in Africa. For example, when the war in Angola resumed in 1992, the death toll reached 1,000 per day and in 1993 military expenditure represented some 20 per cent of GDP (Willett 1998). The campaign for democracy in Angola was thus directly related to ending the war and to demilitarisation and disarmament within a broader struggle over the meanings and interpretations of 'development'. For many observers within and outside Angola, the military budget has for too long received unjustified budgetary allocations, and services such as health and education have been sidelined. Angola has thus come under considerable pressure from the international financial institutions to increase its welfare and development expenditure. During the war years, however, the MPLA goverment could argue that significant defence expenditure was necessary to counter the threat of Unita, and that the available evidence suggested that peace talks with Unita were unlikely to yield any sustainable, positive outcomes. What is generally agreed though is that war in Angola has radically altered the nature of politics in the country and that entrenched corruption in the MPLA government has become almost 'a way of life' (Kapuscinski 1987). A major impediment to democratisation and attempts to monitor how state resources are allocated is that there is so little transparency in the way the country's vast oil wealth is actually spent (see also Chapter 3). In an unstable political environment such as this, corruption becomes a way of maximising resources and is often justified as necessary to the defence of national territory. The political context thus represents a 'key impediment to sustainable development' in Angola (Munslow 1999). Elsewhere in the region, particularly for countries involved in the conflict in the DRC, disarmament and the cessation of war remain crucial prerequisites to the establishment of democracy and to the development of better political institutions.

External pressures for democracy in southern and eastern Africa

Though the situation is dynamic, political independence in many parts of Africa often failed to deliver 'real' liberation and true democratic freedoms. Many observers now refer to a 'second independence' sweeping the continent, which began as the Cold War ended, where African voters seek liberation not from the colonial oppressor but from their post-colonial political leaders. The removal from office of President Kenneth Kaunda (Zambia) and President Hastings Kamuzu Banda (Malawi) were good examples of this trend. Both leaders were part of the national liberation struggles and had dominated politics in their respective countries during the post-colonial period with both presiding over single-party states. Kaunda and his UNIP (United National Independence Party) party were swept from office by an electoral defeat of considerable magnitude, and the highly personalised authoritarianism of President Banda, among the 'most dictatorial of dictators' (Wiseman 1995) was ended by a peaceful democratic election in 1994.

Multiparty politics have emerged in countries like Malawi and Zambia partly as a result of the loan conditionalities imposed by foreign aid agencies which seek to promote political and economic liberalisation. Assuming that development often fails due to the lack of 'good governance', this is now high on the agenda of the international donor community (Young 1999). These waves of democratisation are thus closely related to pressures from multilateral institutions like the World Bank and the IMF which have conditioned their lending to promote particular reforms of the state. These organisations have, however, been criticised for encouraging political approaches which 'lock society out' (Power 2000); they also often adopt an incredibly limited view of 'governance' in their definition of the state, prioritising, for example, the role of the state in wealth creation rather than in wealth distribution. This can have important implications for the political legitimacy of African states as they shift their attention towards the need to attract foreign investment and away from social redistribution and the problems created by inequality within their own borders.

In Kenya, President Moi was originally deeply hostile to democracy. He claimed it was part of a foreign ideology 'pedalled by some unpatriotic people with borrowed brains [whose advocates are] anarchists, rats and drug addicts' (Moi cited in Baker 1998). Nevertheless, under enormous pressure from Western governments (particularly the US) and aid agencies, Moi was eventually forced into multiparty elections in the early 1990s. However, although he managed to 'play along' with the demands for democratic reform and was for a while reinvented as a senior African social democrat, in reality he had successfully resisted any real radical reform of Kenyan political culture which remained deeply corrupt. Elsewhere in the region, despite the supposed strength of the tides of democratisation, other autocratic statesmen have also survived and are still in power. In Zimbabwe, Mugabe's determination to win the multiparty elections of 2000, for example, unleashed a period of violence and intimidation which undoubtedly helped to achieve his narrow victory. In Namibia, President

Sam Nujoma has managed to alter the constitution to allow himself a third term in the presidency. Thus it could be argued that 'democratic government' has often been rather narrowly defined (e.g. as simply having multiparty politics and regular elections) and can be manipulated by many politicians for some rather undemocratic ends. Much more meaningful definitions of democracy can be drawn up, however, that can provide a rather better set of objectives against which to measure political progress in the region. Ake (1996: 24) provides the following set of priorities:

> respecting the rule of law, working on the basis of negotiated consensus, broadening and deepening participation, cultivating tolerance and acceptance of the equality of all humankind, sharing the burdens and rewards of our common humanity with equity and building peace on justice.

The reason for the survival of some of these statesmen is directly linked to the precariousness of democratic political culture in countries like Zimbabwe and Kenya. In both it is only very recently that the range of political alternatives on offer to the electorate has really begun to open up, and it has proven very difficult for new parties to challenge the authority and respect of a political movement which brought about national liberation from colonial rule. A further significant hindrance to the development of truly democratic cultures in these countries, and elsewhere in the region, is that the material stakes are often high in elections. As discussed earlier in this chapter in relation to Allen's concept of a 'spoils political system', access to political power is often the main route to resources and wealth. This makes democratic culture exceedingly precarious for, as Diamond (1988: 69) suggests:

> Democracy requires moderation and restraint. It demands not only that people care about political competition but also that they not care too much; that their emotional and tangible stake in its outcome not be so great that they cannot contemplate defeat. Throughout much of Africa, the swollen state has turned politics into a zero-sum game in which everything of value is at stake in an election, and hence candidates, communities and parties feel compelled to win at any cost.

The role of the international donor community in regional politics is also exemplified by South Africa. The international financial institutions played a significant role in directing the policy choices of the ANC in the run-up to the 1994 elections in South Africa. Their choice of a market-oriented programme has, however, been criticised because it precludes, to a significant extent, the wealth redistribution which some would argue is fundamental to sustainable democracy in that extremely unequal, and highly politicised, society. Thus some have argued that the democratisation process in post-apartheid South Africa may be precarious until it is accompanied by more tangible forms of social and economic equity.

Another useful example of the intersections between politics and economic development is found in Ferguson's (1990) study of Lesotho which focused on the role of the World Bank in shaping national development trajectories. Ferguson (1990) described the Bank as an 'anti-politics machine'. He showed how key agents of the international development community depoliticised development in Lesotho by focusing on a limited range of 'technocratic issues', rather than considering wider issues of economic and political participation. National budgets, for example, are intensely political. Yet throughout this region they are often being decided by, or greatly influenced by, foreign economic and financial 'experts', who promote 'technical' interventions that ultimately disempower people.

The central importance of oil and diamonds to the continuation of conflict in Angola also illustrates something of the interconnections between politics and economics in the region. Illegal trade in diamonds sustained Unita and its purchases of arms and fuel for many years (see Box 9.3). In Angola it seems impossible to distinguish or deduce the political from the economic (or vice versa) and so it becomes important to grasp the contradictions between the two. The conflict was regularly characterised by the contradiction between the enormous profits being made from the sale of natural resources and the conditions of absolute poverty and instability which many Angolans were forced to endure. Huge geographical changes have been wrought by the conflict with massive migration to urban areas, especially Luanda, and many more people becoming refugees or 'internally displaced people' (IDPs) within Angola. People's livelihoods became desperately insecure, and hundreds of thousands of civilians were injured or killed. Each stage of the conflict in Angola has led to different forms of displacement and forced migration (Vines 1998); in 2001 it was estimated there were 4.36 million officially recognised IDPs and 421,000 refugees outside Angola of a national population of about 12 million. At Angolan independence in 1975 only about 15 per cent of the population was living in urban areas but this had grown to some 55 per cent by 1994. The population of Luanda has grown particularly rapidly and its infrastructure has been pushed beyond breaking point (Munslow 1999). These high levels of urbanisation may pose major challenges to any strategies of post-war development and reintegration of refugees, although after the wars in both Zimbabwe and Mozambique, once peace was established and the countryside became more secure, very large numbers of people who had been displaced to urban areas returned to their rural homes. As with a number of other complex situations of conflict in the region, Angola faces a range of longer-term reconstruction and rehabilitation issues (see Box 9.4) which are particularly acute in a country that had endured decades of almost continuous war.

Mineral wealth can also compromise democracy and good governance without directly fuelling open conflict. In the DRC for example, the Mobutu years were sustained by American and European support because of a desire to wield influence over the deployment of its strategic, and other valuable, minerals. Once Mobutu had gone there was an unseemly scramble by mining companies to secure concessions with the new regime in ways which were neither transparent nor conducive to honest government. The South African diamond company, De Beers, for example, negotiated an exclusive joint exploration deal for prospecting rights to 240,000 km^2 of known diamond areas. In 1997 Laurent Kabila accused De Beers of 'irregular business practices' during Mobutu's reign and also of trying to maintain its monopoly over DRC diamond production (Collins 1998).

Box 9.4   Some consequences of war and political instability: reconciliation and disarmament

Long periods of war and political instability in eastern and southern Africa have a range of post-war implications for internal governance and geopolitics. In South Africa a Truth and Reconciliation Commission (TRC) was established in 1995 to enable the country to come to terms with its divided past. The TRC recommended that R3bn be distributed to some 20,000 victims of apartheid, though many witnesses were not deemed eligible for relief. The conflict and genocide in Rwanda also raises complex and long-term questions about truth and reconciliation issues. By early 1998, reports of killings and 'disappearances' had become so routine across Rwanda that many seemed resigned to accept this violence as an inevitable part of their lives. The return of displaced peoples to post-genocide Rwanda put massive strain on social services by the late 1990s and women in particular have faced very specific challenges: '[w]hether sexually violated or not, Rwandan women of all groups and social strata saw their lives, their families, and their tenuous hold on economic security disrupted by the conflicts' (Newbury and Baldwin 2000: 3). In such circumstances, the psychological and sociological constraints on a resumption of 'normal' life, let alone development and improvement of livelihoods, can seem overwhelming.

In 1999 it was estimated that some 3.7 million people were then affected by the war in Angola with some 2 million people being internally displaced. Forced migration and displacement pose a number of complex challenges for states in the aftermath of conflict and instability, as large sections of the national population need to be reintegrated, in a variety of ways, into their societies. In addition, by the end of the 1990s about 95,000 people had been disabled by landmines in Angola and Mozambique alone, though this is probably a conservative estimate. The demobilisation of former soldiers and demilitarisation in countries formerly at war is another important part of post-war reconstruction. The presence of thousands of experienced ex-soldiers, who often still have access to guns, can pose serious problems for states after the end of conflict. Many years after the end of the Zimbabwean liberation war, for example, war veterans there were at the forefront of land invasions in 2000 and they have argued that their role in the country's liberation gives them very specific claims to state resources. Yet further problems stem from the conscription of children as soldiers, as in the case of Mozambique where many children suffered psychological traumatisation as a result of their involvement in the country's 16-year civil war. There have also been a large number of child soldiers involved in the conflicts in Angola and the DRC. In 1998 Kabila's regular army was said to include some 20,000 child soldiers, many of whom endured traumatic and psychologically disturbing conditions. UNICEF is one international agency that has been keen to develop action-oriented anti-war agendas which seek to reduce the impact of war on children, a key theme highlighted in its 1996 report on *The State of the World's Children*. Calls have been made for a universal ban on the production, stockpiling, sale, export and use of anti-personnel landmines led by organisations like UNICEF and the International Committee of the Red Cross (ICRC). The UN Security Council (whose permanent members dominate global arms sales) passed a resolution in August 2000 renewing its commitment to protecting children affected by war and to protect their rights during conflicts. This resolution also calls for special protection and assistance for refugees and internally displaced peoples, for those who have lost whole families through conflict or who have suffered some form of sexual abuse.

Sources: Amnesty International UK (1998), *Independent on Sunday* (6 December 1998), Júnior *et al.* (1996), Newbury and Baldwin (2000), Sangara (1998), Thompson (1999), UNHCR (1997), UNICEF (1996), UNICEF (2000), Vines (1998), Willett (1998).

Conclusions: Co-operation, corruption and complexity in African politics

[S]urely it would be very patronising and neo-colonialist for the world to say that Africa should not have any guns.

(Dowden 2000: 6)

'[T]he process of criminalization has become the dominant trait of a sub-continent in which the state has literally imploded under the combined effects of economic crisis, neo-liberal programmes, structural adjustment and the loss of legitimacy of political institutions.

(Bayart *et al.* 1999: 14)

This chapter has sought to demonstrate that it is necessary to think about how African politics is produced *geographically*, at a variety of spatial scales and by examining different forms of power and the changing nature of socio-spatial relationships. Conflict in countries like Angola, Zimbabwe and the DRC are fundamentally about what Ó Tuathail (1996) calls 'geo-power', or the relationships between political power and territory.[2] The war in the DRC illustrated the way in which some seven different countries became embroiled in a war ostensibly about DRC territory, but with a range of implications for participating states and their collective sense of regional unity and identity. SADC has also been discussed here as a useful example of the way in which geopolitics is important to the very imagination of 'development' and regional integration in southern Africa.

At a different scale, political boundaries *within* society are also of immense significance in determining whose voice is heard in national politics (Chabal 1992). Political boundaries define the limits of participation and thus who can be part of (and express opinions about) political processes. In Kenya, for example, there has been an active restriction of political spaces. As Kanyinga (1998: 36) has put it:

[2] The border conflict between Eritrea and Ethiopia, just beyond the region covered in this book, has been described as a 'geographer's nightmare' (BBC News 1999) and provides probably one of the clearest examples of the importance of boundaries and of the marking of political space in contemporary Africa.

Kenya's political history is replete with attempts by the state and political elites in general to hijack and co-opt or polarize popular forces. Thus, what initially began as a popular initiative is hijacked by 'forces from above' and its agenda diverted to routes of class and ethnicity. The potential of popular groups to undermine authoritarianism [therefore] has considerably declined.

Another example of the ways in which political participation can be limited in the region is illustrated in the Zimbabwean film *Flame*, a controversial film directed by Ingrid Sinclair in 1996. This traces the story of two women fighters in the Zimbabwean liberation struggle, exploring the ways in which they came to learn about and articulate their demands for political rights. The film was seized by the Zimbabwean police during editing on the grounds that it was subversive and pornographic and partly because it came uncomfortably close to the truth about the failings of ZANU-PF since 1980, illustrating the false promises of independence. Florence and Nyasha, the two principal characters in the film, are incorporated in very different ways into the *Chimurenga*, the Zimbabwean liberation struggle, leading to very different outcomes for each woman. Towards the end of the film both characters evaluate their chances of making a continued contribution to their society after independence but decide that this will always be limited by the fact that 'we are just women'. As they watch Zimbabwe's new leadership on television, they decide *'a luta continua'* (the struggle continues) – borrowing a Portuguese phrase from one of the Frelimo guerrillas of Mozambique with whom Nyasha falls in love. Zimbabwean liberation is thus depicted as incomplete since opportunities for women in post-colonial society have been limited and few women were given due recognition for their role in liberating their country from Rhodesian colonial power. The fact that this interpretation of events was considered unacceptable by Mugabe and ZANU-PF says a great deal about the way in which the party defines and shapes the political boundaries of participation in post-colonial society. Arguably, the political spaces mapped out by the state in Zimbabwe have been quite restricted and have sought to exclude some groups (such as homosexuals) while marginalising others (such as women).

Women are not the only group to be excluded by particular politicians and political institutions. Mugabe's homophobia has attempted to exclude gay men and women in particular from participating in Zimbabwean society. He has suggested that they behave 'worse than dogs and pigs' and do not act in accordance with African tradition (*Guardian*, 2 October 1999). Similarly pejorative statements have been made by President Moi in Kenya, and President Nujoma in Namibia. However, there has probably always been homosexuality in this part of Africa and it is not simply a by-product of colonial rule. As argued by the head of Gays and Lesbians in Zimbabwe (GALZ), much of the opposition is essentially political:

> Same-sex activity has always happened here. It's the organising around the identity of being gay that is new. That is what these presidents are objecting to. They see it as some kind of threat.
>
> (cited in *Guardian*, 2 October 1999: 19)

The way in which Mugabe's government has sought to deflect internal criticism of its post-colonial political strategies by playing to populist fears, misconceptions, and bigotry by outlawing gay Zimbabweans, further marginalising women and instigating land invasions by war veterans in 2000 is important here. Politically active men and women, whatever their sexuality, can represent a threat to the established order of Zimbabwean society and to the power of the state and its ability to define state–society relationships and the terms of citizenship or participation. The notion of such groups as forming part of a 'civil society' is useful (but not without its problems).

One element of civil society which has attracted a great deal of analytical attention recently in this respect is the sector of non-governmental organisations (NGOs). International organisations see elements of civil society, including NGOs, as a means of 'channelling opinion' into policy conduits in a coordinated fashion and of increasing people's confidence in involving themselves in political affairs. This is regarded as an important way of facilitating real democracy. NGOs theoretically have an important role to play in this respect since many try to strengthen the political interconnectedness between citizens and the state, particularly among marginalised and disempowered groups. Nevertheless, NGOs are not always less bureaucratic, corrupt and wasteful than the state. Nor are their members elected and accountability can be a problem. Aid directed at such civic organisations can also be diverted from its original purpose, just as with aid given to other institutions. Chabal and Daloz (1999: 24) thus argue that, 'far from strengthening civil society ... the role of NGOs ... [can] lead to the hijacking of genuinely needed development aid by the same old and well established elites'.

This concern with the role of elites, which is by no means new in African studies (Lloyd 1967; Cohen 1976; Hodder 1978), is crucial to any understanding of the shifting geographies of governance and politics in the region. As already discussed, in the 'new' South Africa the ANC has chosen to pursue conservative macro-economic policies (Marais 1997). The consequences of this political choice in terms of poverty reduction, ongoing violence and longer-term declines in the authority of the ANC and its democratic institutions may be quite profound. Between 1994 and 1999 only 4,000 of the 63,000 claims made to the Land Claims Commission had actually been processed (Allie 2000), indicating the magnitude of the task faced and the slow progress made towards resolution of complex problems like the land question. A more radical approach to development may well have sought to accelerate the process of land reform in South Africa. It was evident from the ANC's cautious statements on the Zimbabwean land invasions, which greatly disappointed Western governments, that they are aware of the dangerous tensions within their own society stemming from perceptions that redistribution is proceeding too slowly (African Business 2000; Tunbridge 2000; Wily 2000). The economic and social compromises that have been made by the ANC government in the making of a 'new' South Africa have been linked directly to the role of elites in the country. According to Bond (2000) there has been an 'elite transition' in which ANC elites have made political choices that have not confronted corporate power or sought real economic empowerment for the black majority. Many ANC officials cite, as a justification for this stance, the negative economic example of Mozambique where, after the adoption of Marxism-Leninism by Frelimo in 1975, most of the settlers (with their privileged educations and access to capital) fled the country, *en masse*, and foreign investment was actively discouraged.

In other countries the process of privatisation, which is strongly encouraged by the international financial institutions, has also been linked to elite self-interest. The state enterprise privatisation programmes pursued since the early 1990s in Mozambique and Zambia have been described as being some of the most successful in Africa (Campbell *et al.*, cited in Craig 2000) and as examples for other 'developing countries' to follow. However, in both countries privatisation has been regarded by some political elites as an opportunity for personal accumulation. Although this need not be attended by corruption, the empirical evidence is that elites often do use their knowledge and power to acquire illegitimate stakes in such privatisations.

In the 1990s, several African politicians (e.g. Nelson Mandela, Thabo Mbeki and Robert Mugabe) made reference to the beginnings of an 'African renaissance', a phrase which became part of a variety of discourses about Africa and its future. Many African leaders held the view that a rebirth of African politics could lead to the construction of new and more inclusive political spaces which empowered the electorate to participate in politics. Politicians from outside Africa, such as the former US President Bill Clinton and his Secretary of State Madeleine Allbright, also referred to a perceptible African renaissance as a rebirth and renewal of African politics. The late Julius Nyerere was more cautious about the use of the term, arguing that it was merely a distant hope, but nevertheless he felt that there was room for optimism in that the tide had turned in favour of 'trends to have elected governments who are committed to developing the continent, not looters and gangster movements intent on stripping their countries to the bone' (Nyerere, cited in Cliffe 2000: 2). A much bleaker view of this 'renaissance' was offered by playwright Wole Soyinka, however, in a speech in Cape Town in 1999. While recognising that there had been a cultural and literary awakening in Africa which was linked to broader political events, Soyinka pointed out that there was still far too much conflict, killing and impoverishment on the continent and that 'among the slaughter, there can be no African renaissance' (*Mail and Guardian* 1999).

Similarly it could be argued that any rebirth and renewal of African political affairs need to begin with the problematic of geo-power which has been discussed in this chapter: the entrapment of African politics in the bondage of outdated colonial boundaries. A 'remapping' of African political geographies (beginning with national, local and regional political boundaries) is therefore necessary if political cultures are to become more inclusive and are to reflect dynamic contemporary cultural, political and social realities.

An example of such a remapping has been put forward by Gakwandi (1996) who urges the need for a move towards a new political map of Africa and suggests that much of eastern and southern Africa could become a single region named *Mozambia*. Although it is unlikely that the massive cultural, social and economic diversity of the region could be subsumed into a single political region (or even a simple map), it is becoming increasingly important to rethink the importance of regions in Africa, of regional integration and the possibility of regional co-operation. At the end of the 1990s, for example, Oxfam proposed a wider strategy for reconstruction in the Great Lakes region, with the EU, UN, World Bank, IMF, NGOs and civil societies coming together to 'take advantage of the end of the Mobutu era' (Oxfam 1998: 1). Sadly this vision proved to be completely at odds with subsequent events. The high level of debt in the DRC also remains a major stumbling block to the building of alternative structures of regional government.

Arguably, the international development community has focused unduly on the issue of state corruption in its governance agenda in Africa, when exploring alternative forms of governance or regional integration might have been more valuable. Corruption is not easy to define and, as has been argued here, is often an extension of the very processes of economic liberalisation promoted by international donors (which promote private accumulation). Corruption is thus directly linked to the political elites that dominate African politics in that it partly explains their enthusiasm for privatisation. Political corruption has been linked to authoritarianism as an explanation for developmental failure (Szeftel 1998) and offers a simplistic 'causal' explanation which has been popular with many international organisations. International financial institutions like the World Bank and the IMF seem to prefer simplistic explanations in which countries have either 'good' or 'bad' governance. As a result they struggle to comprehend the complexity, uncertainty and unpredictability of African politics.

Useful websites

www.africaanalysis.com   Africa Analysis.

www.africaintelligence.com   Africa Intelligence. Available also in French.

www.africanews.org   Africa News Online.

www.africaonline.com   Africa Online.

www.allafrica.com   AllAfrica.

www.bbc.co.uk/worldservice   British Broadcasting Corporation (BBC).

www.csf.colorado.edu/ipe/africa.html   IPE Resources Related to Africa. Offers extensive data sources for news and analysis on Southern Africa.

www.eldis.org   Eldis Development Gateway.

www.idasa.org.za   Institute for Democracy in South Africa (IDASA). Aims to promote democracy and a culture of tolerance.

www.in2zw.com/mdc   Movement for Democratic Change (MDC) (Zimbabwe).

www.marekinc.com/NCN.html Central Africa Watch/New Congo Net (NCN). Formerly known as Zaire Watch, this site aims to monitor the transition to democracy and also includes link to MetroNet Afrique, which provides business briefs on Africa.

www.mg.co.za/mg The Mail and Guardian (South Africa).

www.misanet.org Media Institute for South Africa (MISA). Focuses primarily on the need to promote free, independent and pluralistic media. Looks at freedom of political expression in Southern Africa.

www.nationaudio.com.News/DailyNation/Today Daily Nation (Kenya).

www.oau-oua.org Organisation of African Unity (OAU).

www.oneworld.org/odi One World Organisation (UK).

www.truth.org.za Truth and Reconciliation Commission (South Africa).

www.uganda.co.ug/govern Government of Uganda.

www.zamnet.zm/zamnet/post The Post (Zambia).

References

África Hoje (May 2000) Pobreza em 65 per cent dos lares angolanos: 45.

African Business (July/August 2000) Land grab jitters in Namibia, Business Briefs: 7.

Ajulu, R. (1998) Kenya's democracy experiment: the 1997 elections, *Review of African Political Economy* **76**: 275–85.

Ake, C. (1996) *Democracy and Development in Africa.* Brooks, Washington.

Alexander, J. (1997) The local state in post-war Mozambique: political practice and ideas about authority, *Africa* **67**(1): 1–26.

Allen, C. (1999) Warfare, endemic violence and state collapse in Africa, *Review of African Political Economy* **81**: 367–84.

Allen, C. and **Burgess, J.** (1999) Julius Kambarage Nyerere, *Review of African Political Economy* **26**(81): 315–16.

Allie, M. (2000) Justice at a snail's pace, *BBC Focus on Africa*, July–September: 2.

Amnesty International (AI) UK (1998) *Rwanda: The hidden violence.* AI, London.

Baker, B. (1998) The class of 1990: how have the autocratic leaders of sub-Saharan Africa fared under democratisation? *Third World Quarterly* **19**(1): 115–27.

Barnett, C. (1998) The contradictions of broadcasting reform in post-apartheid South Africa, *Review of African Political Economy* **78**: 551–70.

Bayart, J.F. (1993) *The State in Africa: The politics of the belly*. Longman, London.

Bayart, J.F., Ellis, S. and **Hibou, B.** (1999) From kleptocracy to the felonious state? in Bayart, J., Ellis, S. and Hibou, B. (eds) (1999) *The Criminalization of the State in Africa*. James Currey, Oxford: 1–31.

BBC Focus on Africa (2000) Angola/Namibia: friends in need, *BBC Focus on Africa*, April–June, **11**(2): 8, British Broadcasting Corporation, London.

BBC News (1999) Border a geographer's nightmare, BBC Online News [www.news.bbc.co.uk accessed 23 July 1999].

Bond, P. (2000) *Elite Transition: From apartheid to neoliberalism in South Africa*. Pluto, London.

Braden, K.E. and **Shelley, F.M.** (2000) *Engaging Geopolitics*. Longman, London.

Chabal, P. (1992) *Power in Africa: An essay in political interpretation*. Macmillan, London.

Chabal, P. (ed.) (1986) *Political Domination in Africa: Reflections on the limits of power*. Cambridge University Press, Cambridge.

Chabal, P. and **Daloz, J.P.** (1999) *Africa Works: Disorder as political instrument*. James Currey, Oxford.

Chan, S. (1994) The diplomatic styles of Zambia and Zimbabwe, in Rich, P. (ed.) *The Dynamics of Change in Southern Africa*. St Martins Press, London: 218–33.

Clapham, C. (1996) *Africa and the International System: The politics of state survival*. Cambridge University Press, Cambridge.

Clapham, C. and **Wiseman, J.** (1995) in Wiseman, J. (ed.) *Democracy and Political Change in Sub-Saharan Africa*. Routledge, London.

Cliffe, L. (2000) African renaissance?, Plenary discussion paper presented at 'Africa: Capturing the Future', Conference at the University of Leeds, 28–29 April, Leeds University Centre for African Studies (LUCAS).

Cohen, A. (1976) *Two-dimensional Man: An essay on the anthropology of power and symbolism in complex society*. University of California Press, Berkeley.

Collier, P. (2000) *Economic Causes of Civil Conflict and Their Implications for Policy*. World Bank, Washington.

Collins, C.J.L. (1998) Congo/ex-Zaire: through the looking glass, *Review of African Political Economy* **75**: 112–23.

Craig, J. (2000) Evaluating privatisation in Zambia: a tale of two processes, paper presented at the ROAPE conference, Leeds University Centre for African Studies (LUCAS), 28–30 April.

Crush, J. (1995) 'Imagining development' in Crush, J. (ed.) *Power of Development*. Routledge, London: 1–26.

Diamond, L. (1988) Democracy in developing countries, in Diamond, L., Linz, J. and Lipset, S. (eds) *Africa*: Vol. 2. Lynne Rienner, Boulder, Colo.; Adamantine, London.

Dinerman, A. (1994) In search of Mozambique: the imaginings of Christian Geffray, in *La Cause des Armes à Mozambique – Anthropologie d'une guerre civile*, *Journal of Southern African Studies* 20(4): 569–87.

Dorman, S.R. (2000) Zimbabwe: change now? *The World Today* 56(4): 25–7.

Dowden, R. (2000) The hopeless continent, *The Economist*, 13–19 May: 6.

Du Pisani, A. (1994) Namibia: impressions of independence, in Rich, P. (ed.) *The Dynamics of Change in Southern Africa*, St Martins Press, London: 199–217.

Englebert, P. (1997) The contemporary African state: neither African nor state, *Third World Quarterly* 18(4): 767–75.

Escobar, A. (1992) Reflection on 'development': grassroots approaches and alternative politics in the Third World, *Alternatives* 10: 377–400.

Escobar, A. (1995) *Encountering Development: The making and unmaking of the Third World*. Princeton University Press, Princeton.

Ferguson, J. (1990) *The Anti-Politics Machine: Development, depoliticisation and bureaucratic power in Lesotho*. Cambridge University Press, Cambridge.

Frimpong-Ansah, J.H. (1991) *The Vampire State in Africa: The political economy of decline in Ghana*. Africa World Press, Trenton, NJ.

Gakwandi, A.S. (1996) Towards a new political map of Africa, in Abdul-Raheem, T. (ed.) *Pan-Africanism: Politics, economics and social change in the twenty-first century*. Pluto, London.

Global Witness (2000) *Conflict Diamonds: Possibilities for the identification, certification and control of diamonds*. Global Witness, London.

Goldsmith, A. (2000) Sizing up the African state, *Journal of Modern African Studies* 38(1): 1–20.

Good, K. (1999) The state and extreme poverty in Botswana: the San and destitutes, *Journal of Modern African Studies* 37(2): 185–205.

Griffiths, I. (1993) *The Atlas of African Affairs*. Routledge, London.

Guardian (2000) The generals profit as the people pay with their lives, 20 January: 13.

Guardian (1999) Debt? War? Gays are the real problem say African leaders, 2 October: 19.

Habib, A. and **Taylor, R.** (1999) Parliamentary opposition and democratic consolidation in South Africa, *Review of African Political Economy* 80: 261–7.

Hanlon, J. (1984) *Mozambique: The revolution under fire*. Zed Press, London.

Hanlon, J. (1986) *Beggar Your Neighbours: Apartheid power in Southern Africa*, Indiana University Press, Bloomington.

Hartung, W.D. and **Moix, B.** (2000) Cold War legacies, *BBC Focus on Africa*, April–June **22**(2): 20–1.

Haughton, J. (1998) *The Reconstruction of a War-torn Economy: The next steps in the Democratic Republic of Congo*. Technical Paper, Harvard Institute for International Development, Cambridge, Mass.

Hodder, B.W. (1978) *Africa Today: A short introduction to African affairs*. Methuen, London.

Human Rights Watch (2000) *Leave None to Tell the Story: Genocide in Rwanda*. Human Rights Watch, Washington.

Hyden, G. and **Bratton, M.** (eds) (1992) *Governance and Politics in Africa*. Lynne Rienner, London.

Independent on Sunday (1998) March of the child soldiers, 6 December: 15.

Jarosz, L. (1992) Constructing the dark continent: metaphor as geographic representation of Africa, *Geografiska Annaler (B)*, **74**(2): 105–15.

Jones, P.S. (1999) From nationhood to regionalism to the north west province: Bophuthatswananess and the birth of the new South Africa, *African Affairs* **98**: 509–34.

Júnior, B.E., Riedesser, P., Walter, J., Adam, H. and **Steudtner, P.** (1996) *Criancas, Guerra e Perseguição – Reconstruindo a Esperança*. Tipografia ABC, Instituto Nacional de Livro e do Disco (INLD), Maputo.

Kamrava, M. (1999) *Cultural Politics in the Third World*. UCL Press, London.

Kanyinga, K. (1998) Contestation over political space: the state and demobilisation of party politics in Kenya. Centre for Development Research (CDR) Copenhagen, Working paper CDR 98.12, November.

Kapuscinski, R. (1987) *Another Day of Life*. Penguin, London.

Kasfir, N. (ed.) (1998) *Civil Society and Democracy in Africa: Critical perspectives*. Frank Cass, London.

Le Billion, P. (2001) Thriving on war: the Angolan conflict and private business, *Review of African Political Economy*: special issue on Patrimonialism and Petro-Diamond Capitalism: Peace, Geopolitics and the Economics of War in Angola **28**(90): 625–35.

Lloyd, P.C. (1967) *Africa in Social Change*. Penguin, Harmondsworth.

Luke, T. (1993) Discourses of disintegration, texts of transformation: re-reading realism in the new world order, *Alternatives* **18**: 229–58.

McCarthy, C. (1996) Regional integration: part of the solution or part of the problem? in Ellis, S. (ed.) *Africa Now: People, policies, institutions*. James Currey, Oxford: 211–31.

McEwan, C. (2001) Gender and citizenship in South Africa: learning from South Africa, *Agenda* **47**: 47–59.

McGregor, J. (1994) 'People without fathers'. Mozambicans in Swaziland 1893–1993: nationality, citizenship and entitlements. *Journal of Southern African Studies* **20**(4): 545–67.

McNulty, M. (1999) The collapse of Zaire: implosion, revolution or external sabotage? *Journal of Modern African Studies* **37**(1): 53–82.

Mail and Guardian (1999) Speech by Wole Soyinka, Cape Town, 3 September.

Makumbe, J. (1999) Anti-corruption efforts in the SADC, *Southern African Political and Economic Monthly* December: 45–7.

Mamdani, M. (1996) *Citizen and Subject: Contemporary Africa and the legacy of late colonialism.* James Currey, London.

Marais, H. (1997) *South Africa: Limits to change? Transforming a divided society.* Zed, London.

Matlosa, K. (1998) Democracy and conflict in post-apartheid southern Africa: dilemmas of social change in small states, *International Affairs* **74**(2): 319–37.

Mazrui, A. (1993) The bondage of boundaries: why Africa's maps will be redrawn, *The Economist*, 11 September: 150–1.

Minter, W. (1994) *Apartheid's Contras: An inquiry into the roots of war in Angola and Mozambique.* Zed, London.

Munslow, B. (1999) Angola: the politics of unsustainable development, *Third World Quarterly* **20**(3): 551–68.

Nevin, T. (2000) Dirty diamonds? *African Business* July/August: 21–2.

Newbury, C. and **Baldwin, H.** (2000) Aftermath: women in postgenocide Rwanda. Center for Development Information and Evaluation, United States Agency for International Development, Working Paper 303, July, Washington.

Nugent, P. and **Asiwaju, A.I.** (1996) Introduction: the paradox of African boundaries, in Nugent, P. and Asiwaju, A.I. (eds) *African Boundaries: Barriers, conduits and opportunities.* Pinter, London: 1–18.

Ó Tuathail, G. (1996) *Critical Geopolitics: The politics of writing global space.* Routledge, London.

OneWorld (August 2000) Greed and corruption bring Zambia to its knees, [www.oneworld.net/editors letters/aug2000.html].

Oxfam (1998) *The Importance of Engagement: A strategy for reconstruction in the Great Lakes Region.* Oxfam, Oxford.

Parker, A., Sangera, S. and **Guerrera, F.** (2000) *Fatal Transactions: An investigation into the illicit diamond trade.* Financial Times, London [www.ids.ac.uk].

Power, M. (2000) The short cut to international development: representing Africa in New Britain, *Area* **32**: 91–100.

Rahnema, M. and **Bawtree, V.** (1997) (eds) *The Post-Development Reader.* Zed, London.

Regional Roundup (2002) Information from the SADC Press: Democratic Republic of Congo, 29 October [www.iht.com/articles/75123].

Rich, P.B. (1994) South Africa and the politics of regional integration in Southern Africa in the post-apartheid era, in Rich, P. (ed.) *The Dynamics of Change in Southern Africa.* St Martin's Press, London: 32–51.

Rodrianja, S. (1996) Nationalism, ethnicity and democracy, in Ellis, S. (ed.) *Africa Now: People, policies and institutions*; James Currey, London: 20–41.

Rostron, B. (2000) Africa 2000: lines on a map, *BBC Focus on Africa* **11**(1): 56–8.

Salmond, J. (1997) Swaziland: of trade unions and transformation, *Southern Africa Report* **12**(3): 5–8.

Sangara, B.A. (1998) Congo, in Hampton, J. (ed.) *Internally Displaced People: A global survey.* Earthscan, London: 62–5.

Sejanamane, M. (1996) Peace and security in Southern Africa: the Lesotho crisis and regional intervention, in Mandaza, I. (ed.) *Peace and Security in Southern Africa.* SAPES Books, Harare.

Shearer, D. (1999) Africa's great war, *Survival* **41**(2): 89–106.

Sidaway, J.D. (1998) The (geo)politics of regional integration: the example of the Southern African Development Community, *Environment and Planning D: Society and Space* **16**: 547–76.

Sidaway, J.D. (2000) Imagined regional communities: undecidable geographies, in Cook, I., Crouch, D., Naylor, S. and Ryan, J.R. (eds) *Cultural Turns/Geographical Turns.* Prentice Hall, London: 234–58.

Sidaway, J.D. and **Gibb, R.** (1998) SADC, COMESA, SACU: contradictory formats for regional integration, in Simon, D. (ed.) *South Africa in Southern Africa: Reconfiguring the region.* James Currey, Oxford.

Sidaway, J.D. and **Power, M.** (1998) 'Sex and violence on the wild frontiers': the aftermath of state socialism in Mozambique, in Pickles, J. and Smith, A. (eds) *Theorising Transition: The political economy of post-communist transformations.* Routledge, London.

Simon, D. (1996a) Strategic territory and territorial strategy: the reintegration of Walvis Bay into Namibia, *Political Geography* **15**(2): 193–219.

Simon, D. (1996b) What's in a map? Regional restructuring and the state in independent Namibia, *Regional Development Dialogue* **17**(2): 1–31.

Simon, D. (1998a) Angola: things fall apart again? *South African Journal of International Affairs* **6**(1): 67–72.

Simon, D. (1998b) Angola: the peace is not yet fully won, *Review of African Political Economy* **77**: 495–553.

Simon, D. (1999) Rethinking development, in Simon, D. and Närman, A. (eds) *Development as Theory and Practice: Current perspectives on development and development co-operation*. Longman, Harlow, DARG Regional Development Series 1.

Simon, D. (2001) The bitter harvest of war: continuing social and humanitarian dislocation in Angola, *Review of African Political Economy*: special issue on Patrimonialism and Petro-Diamond Capitalism: Peace, Geopolitics and the Economics of War in Angola **28**(90): 503–20.

Simon, D. and **Johnston, A.** (1999) *The Southern African Development Community: Regional integration in ferment*. RIIA, South African Study Group, Briefing Paper No. 8, London.

Slater, D. (1993) The geo-political imagination and the enframing of development theory, *Transactions of the Institute of British Geographers* **18**(4): 419–37.

Smith, B.C. (1996) *Understanding Third World Politics: Theories of political change and development*. Macmillan, London.

Southall, R. (1999) Re-forming the state? Kleptocracy and the political transition in Kenya, *Review of African Political Economy* **79**: 93–108.

Sporton, D., Thomas, D. and **Morrison, J.** (1999) Outcomes of social and environmental change in the Kalahari of Botswana: the role of migration, *Journal of Southern African Studies* **25**(3): 441–60.

Stock, R. (1995) The Politics and economics of surviving in South Africa's shadow, in Stock, R. *Africa South of the Sahara: A geographical interpretation*. Guildford Press, London: 397–410.

Strassaldo, R. (1989) Border studies: the state of the art in Europe, in Asiwaju, A.I. and Adenyi, P.O. (eds) *Borderlands in Africa: A multidisciplinary and comparative focus on Nigeria and West Africa*. University of Lagos Press, Lagos.

Szeftel, M. (1998) Misunderstanding African politics: corruption and the governance agenda, *Review of African Political Economy* **76**: 221–40.

Thompson, C.B. (1999) Beyond civil society: child soldiers as citizens in Mozambique, *Review of African Political Economy* **80**: 191–206.

Tunbridge, L. (2000) The domino effect, *BBC Focus on Africa* July–September: 22–3.

UNHCR (1997) *The State of the World's Refugees: A humanitarian agenda*. UNHCR, Oxford University Press, Oxford.

UNHCR (2000) *The State of the World's Refugees: Fifty years of humanitarian action*. UNHCR, Oxford University Press, Oxford.

UNICEF (1996) *The State of the World's Children*. UNICEF, New York.

UNICEF (2000) UNICEF hails new Security Council resolution on children and war [www.unicef.org/newsline.00pr61.html].

Vines, A. (1998) Mozambique and Angola, in Hampton, J. (ed.) *Internally Displaced Peoples: A global survey*. Earthscan, London: 89–94.

Watts, M. (1993) The geography of post-colonial Africa; space, place and development in sub-Saharan Africa (1960–1993), *Singapore Journal of Tropical Geography* **14**(2): 173–90.

West, H. and Kloeck-Jenson, S. (1999) Betwixt and between: traditional authority and democratic decentralization in post-war Mozambique, *African Affairs* **98**(4): 455–84.

Willett, S. (1998) Demilitarisation, disarmament and development in Southern Africa, *Review of African Political Economy* **77**: 409–30.

Wily, L.A. (2000) Land tenure reform and the balance of power in Eastern and Southern Africa, *Natural Resource Perspectives* **58** (June) [www.odi.org.uk/nrp/58.html].

Winter, J. (2000) Up for grabs, *BBC Focus on Africa* July–September: 18–19.

Wiseman, J.A. (ed.) (1995) *Democracy and Political Change in Sub-Saharan Africa*. Routledge, London.

Young, T. (1999) The state and politics in Africa, *Journal of Southern African Studies* **25**(1): 149–54.

CHAPTER 10

International and regional trade in eastern and southern Africa

Richard Gibb

The trading characteristics of any region can be examined from a variety of scales, ranging from the global and international to the regional and national. Trade is of central significance to the economies of the countries of southern and eastern Africa. Both intra-regional and international patterns of trade, and the conditions under which such trade occurs, have experienced some important shifts over the past 20 years as structural adjustment within the region, and the wider implications of economic globalisation more generally, have made their impacts. Above all, trade has been liberalised, reducing or removing the protection from global competition that many governments previously afforded specific sectors of their economies. Global agreements are also gradually removing the special status, often derived from colonial legacies, that particular countries enjoyed with respect to access to markets in Europe or particular European countries. This chapter is based on an empirical examination of the region's trading characteristics. It will first explore the position and place of eastern and southern African trade in the world economy. How peripheral is this region of Africa? Second, the trading relationship between eastern and southern Africa and the European Union (EU), which is the region's most important trading partner, is examined in order to explore further the international characteristics of the region's trade. This is followed by an analysis of regional trading transactions and an evaluation of various schemes designed to promote regional integration via trade. Through an examination of international and regional trade, this chapter will highlight not only those phenomena common (if not necessarily unique) to the region of eastern and southern Africa, but also factors that can be used to subdivide the region.

Peripherality in the world economy

Africa is a marginal player in global trade. According to the World Trade Organisation (WTO) (1997), the whole continent accounts for just 2.1 per cent of all merchandise exports and 2.4 per cent of merchandise imports. Furthermore, Africa's importance to the world economy, which has never been very great, has been falling in recent years. Between 1980 and 1995, Africa's GDP fell by over 7 per cent in real terms (Harvey 1997). The total value of Africa's trade in 1995/96 (imports and exports) stood at US$222bn. This is equivalent to just one middle-ranking European country. For example, Africa's trade is significantly smaller than the US$300bn traded by the Netherlands. This comparison is even more striking because the Netherlands has long regarded its economy as being too small to be economically viable or independent.

As a region, eastern and southern Africa accounts for approximately 50 per cent of Africa's trade. Again, by way of comparison, this is similar in size to the trading profile of Austria. In other words, the collective strength of the 16 states of eastern and southern Africa, with a combined population of 269 million in 2000, is equal to just one of Europe's smaller economies with a population of just over 8 million. This level of international trade reflects the limitations imposed by the size and structure of the region's economy. Table 10.1 lists the gross domestic product of eastern and southern African countries. As shown, their collective GDP amounted to $178.9bn in 2001 (excluding the DRC), of which South Africa alone accounted for 63 per cent. The region's total GDP was then approximately equal to Austria's and only 13 per cent of the United Kingdom's (although conversion of GDP to PPP$ somewhat improves the comparison, giving the region a total GDP equivalent to Austria combined with the Netherlands). Much of eastern and southern Africa's GDP is represented by subsistence output and, as a consequence, has no monetary or trading value.

From a global perspective, therefore, the region of eastern and southern Africa represents a relatively peripheral component of the world economy. Furthermore, the data above refer only to trade in merchandise and not trade in services. Services represent the fastest growing component of the world economy; an activity in which Africa is grossly underrepresented (WTO 1997). As far as foreign direct investments are concerned, Africa is almost totally neglected. Moreover, the very small amount that is invested tends to go to mining enclaves and not manufacturing (Harvey 1997). While the broad aggregate data mask the importance of countries, like South Africa, and commodities, such as gold, diamonds and coal, that are more fully integrated into the world economy, it nonetheless reveals why, as far as the major economic powers are concerned, and perhaps most importantly international capital, eastern and southern Africa is perceived to be detached from the world trading system. Harvey (1997: 1) comments:

> Africa's role in the world economy derives from its economic import-
> ance to the rest of the world: in international trade and as a destination
> for international investments. Africa's importance defined in this way is
> currently very small, and has declined over a very long period.

Table 10.1 Economic and trade indices for eastern and southern Africa

	Population 2000 (millions)	GDP 2001 (US$bn)	GDP 2001 (PPP US$bn)	GDP AAGR 1990–2000 (%)	Exports to EU as % total exports	Member of SADC	Member of SACU	Member of EAC	Member of COMESA
Angola	13.1	9.5	25.2[a]	1.3	33	✓			✓
Botswana	1.5	5.1	13.2	4.7	80	✓	✓		
Burundi	6.4	0.7	4.1	−2.6	48				✓
DRC	50.9	–	–	−5.1	64	✓			✓
Kenya	30.7	10.4	31.7	2.1	47			✓	✓
Lesotho	2.0	0.8	4.3	4.1	35	✓	✓		
Malawi	11.3	1.8	6.6	3.8	47	✓			✓
Mozambique	18.3	3.6	20.1[a]	6.4	24	✓			
Namibia	1.8	3.2	11.9[a]	4.1	62	✓	✓		
Rwanda	7.6	1.7	8.8	−0.2	57				✓
South Africa	43.3	113.3	413.6[a]	2.0	41	✓	✓		
Swaziland	0.9	1.3	4.9	3.3	41	✓	✓		✓
Tanzania	35.1	9.1	18.8	2.9	46	✓		✓	
Uganda	23.3	5.7	28.6[a]	7.0	75			✓	✓
Zambia	10.4	3.6	8.4	0.5	30	✓			✓
Zimbabwe	12.6	9.1	30.8	2.5	69	✓			✓
Regional total	269.2	178.9[b]	631.0[b]						
South Africa (%)	16%	63%	66%						
Netherlands	16.0	375.0	420.7						
Austria	8.1	188.7	223.6						
United Kingdom	60.0	1,406.3	1,463.0						

[a] Based on regression.
[b] Regional total excludes DRC for which data not available.
Sources: World Bank, World Development Indicators Database for GDP data; UNDP (2002) for population.; trade data EU (1997a), SADC (1999).

In eastern and southern Africa, South Africa is the country most fully integrated into the world economy. In terms of its GDP, trade and foreign direct investment (FDI), South Africa dominates completely not only eastern and southern Africa but also sub-Saharan Africa and, in many respects, the whole of Africa. However, as the previous data indicate, even South Africa is a comparatively small and weak player when viewed from a global perspective. This rather depressing picture of eastern and southern Africa's role in the world economy is compounded by its prospects for recovery. It has been well argued that recovery must be based on export growth (Daniels 1996; Harvey 1997). However, with a few rare exceptions, Africa's comparative advantage rests in raw materials and cheap, low-paid, unskilled and often ununionised labour. Technological developments are making both of these resources less attractive.

The European connection

It may at first appear somewhat bizarre, if not completely Eurocentric, to place an analysis of eastern and southern Africa's trading links with Europe before an analysis of intra-regional trade and attempts to promote regional integration. However, it is appropriate to accord the European link such priority because it remains vitally important for almost all the countries of the region. This importance is a result of eastern and southern Africa's dependence on the EU market for approximately 50 per cent of export earnings (see Table 10.1). Indeed for the majority of countries in eastern and southern Africa, the EU is the single most important trading partner, significantly more important than the USA, Japan or even neighbouring countries. Furthermore, the co-operation agreements linking the EU with eastern and southern Africa constitute a tangible and formal linkage between this region and the world economy.

One factor common to all the states of eastern and southern Africa is their membership of the Cotonou Agreement, a trade and development co-operation agreement linking the EU with 77 African, Caribbean and Pacific (ACP) states (see Box 10.1). Cotonou will, over a transitional period lasting eight years, replace the trading arrangements established by the Lomé Convention, the EU's comprehensive developmental trade-and-aid agreement from 1975 to 2000. In order to understand eastern and southern Africa's trading relationship with the EU, it is therefore necessary to examine both Lomé, the trade regime of which will be maintained temporarily up to 2008, and Cotonou.

All the states of Africa, with the exception of those North African states bordering the Mediterranean and western Sahara, are members of Cotonou. However, South Africa, which joined the ACP group in April 1998, does not benefit from Cotonou trade preferences and has negotiated a separate free trade agreement with the EU. South Africa, because of its relatively sophisticated economy as well as being classified as 'developed' by the WTO, is restricted to 'political' membership of Cotonou. Thus, while all the countries of eastern and southern Africa are Cotonou members, South Africa alone is excluded from non-reciprocal trading

Box 10.1   Lomé and Cotonou

The Lomé Convention was an agreement linking the 15 member states of the EU to the 71-member African, Caribbean and Pacific group of states (ACP states). The countries belonging to the ACP under Lomé, 48 in Africa, 15 in the Caribbean and eight in the Pacific, were the ex-colonies of the UK, France, Portugal and the Netherlands. These states are amongst the poorest countries in the world. When the Cotonou Agreement (signed in Cotonou, the capital of Benin, on 23 June 2000) replaced Lomé, six new members from the Pacific were added to the ACP group.

Cotonou therefore links one of the world's largest and most powerful economies, the EU, to a group dominated by Least Developed Countries. It is certainly not a relationship of equals. Officially, both Lomé and Cotonou are described as 'partnership' agreements based on the idea of 'mutual inter-dependence'. In reality, key decisions affecting the future of the EU-ACP relationship are taken by the EU.

The Cotonou Agreement proposes to replace Lomé with a series of free trade areas, an idea which has both its supporters and critics. At present, however, it is the only option on the negotiating table, and Economic Partner-ship Agreements linking the EU with groups of ACP states look likely sometime in 2008. Supporters argue that free trade areas will provide a stimulus for change in the ACP states, forcing them to integrate in to the world economy. Critics argue that free trade areas will force open the economies of the ACP states to increasingly fierce competition, leading to higher unemployment and reduced standards of living. Many aid agencies are predicting that free trade areas will damage developing countries. They think that European multina-tionals rather than the ACP's poor will be the principal beneficiaries.

Source: Gibb (2000).

privileges. Cotonou therefore serves both to unite and to divide the region.

The level of access that third countries have to the EU market, which is the world's largest trading bloc (Heidensohn 1995), can be critical to their economic and social well-being, particularly for lesser developed coun-tries (LDCs) which depend on the EU market for exports. The EU has a number of preferential trading arrangements with developing countries that grant varying levels of access to the Union's market (Figure 10.1). At the very top of this hierarchy are the 'Everything But Arms' (EBA) agree-ment, allowing non-reciprocal duty-free access for all LDCs, and the ACP signatories to Cotonou/Lomé. The ACP group enjoy the lowest tariffs and most generous quota allowances. Lower down the hierarchy of privileged access are two categories of the generalised system of preferences (GSP) which are designed to assist developing countries, which are not signatories to Cotonou, export manufactured goods and some processed agricultural products. At the very bottom of the hierarchy is the (somewhat misnamed)

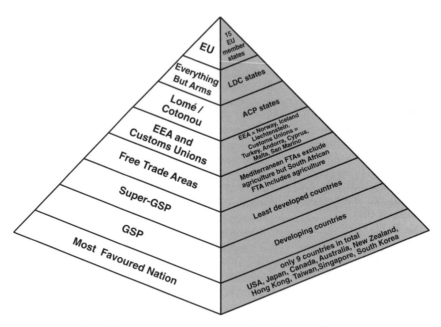

Figure 10.1   The EU's pyramid of preferential trading privileges

most favoured nation (MFN) status, a status bestowed on just nine of the world's most developed countries.

One of the distinguishing features of Lomé was the discriminatory character of its preferential trading agreement. Thus, Lomé discriminated positively in favour of European ex-colonies, a status common to all the countries of eastern and southern Africa, at the expense of other developing countries excluded from the Convention but at a similar level of development, such as Bangladesh, Honduras or Guatemala. For the countries of eastern and southern Africa that had full Lomé membership, the Convention's most important attribute was its non-reciprocity, allowing exports duty-free access to the EU market while at the same time allowing the countries of the region to maintain tariff barriers against European goods.

The European Commission (1999a) estimate that in 1999, 92 per cent of all products originating in the ACP group (excluding South Africa) entered the EU duty free. If agricultural products subject to quotas are included, then this figure rises to over 99 per cent. Essentially, all industrial products had a tariff exemption under Lomé and most agricultural exports benefited from quota regimes or commodity protocols. Thus, compared to the MFN protocol, the preferential margin granted to the ACP group was equal to approximately 3.6 per cent of the value of ACP exports, which in 1997 amounted to ECU734mn. Moreover, the commodity protocols covering beef, sugar, bananas and rum generated a further ECU1.6bn in exports.

These overall figures for the value of the Lomé Convention trade regime conceal considerable disparities between different ACP regions and countries. The EU's trade data on Lomé preferences subdivide the region of eastern and southern Africa into the existing regional groupings of the East African Community (EAC) and the Southern African Development Community (SADC) (see Figure 10.1). The EAC comprises Kenya, Tanzania and Uganda and SADC comprises Angola, Botswana, the DRC, Lesotho, Malawi, Mauritius, Mozambique, Namibia, Seychelles, South Africa, Swaziland, Tanzania, Zambia and Zimbabwe. However, South Africa's data are excluded from the EU's Lomé data because of its exclusion from Lomé trade preferences. Tanzania's membership of both regional groupings means that it is included in both SADC and EAC trade data. Neither Burundi nor Rwanda are included, a factor which singles them out from the rest of the region in this respect.

As Table 10.2 illustrates, different regions of the ACP group do not benefit equally from the value of trade preferences (excluding protocols). SADC enjoys the highest preferential margin (compared to the rates applied under GSP) of any ACP region (4.4 per cent) and the EAC (4.3 per cent) has the second highest (European Commission 1999b). This places eastern and southern Africa in a very privileged position as far as its access to the EU market is concerned. As Figure 10.1 illustrates, ACP states have a privileged position in the hierarchy of preferential access granted to the EU market, and within the ACP group the region of eastern and southern Africa benefits from the greatest preferences. Both SADC and the EAC benefit from a large proportion of their exports enjoying a significant preferential margin, 53 and 34 per cent respectively (the ACP average being 29 per cent).

It is difficult to evaluate accurately the impact of Lomé trade preferences. Trading performance is dependent on a whole range of factors that are local, national, regional and international in character. However, despite being accorded wide-ranging preferences, the ACP's trading performance with the EU has been less than impressive. There is little dispute that Lomé, with a few notable exceptions, failed either to promote ACP

Table 10.2 Preferential margin of EU trade preferences compared to GSP (%)

Country/regional grouping	On all imports	On imports not liberalised under GSP
Sub-Saharan Africa	2.1	10.0
SADC	**4.4**	**11.0**
EAC	**4.3**	**11.7**
UEMOA	3.6	10.8
CEMAC	0.3	6.1
Nigeria	0.2	3.9
Caribbean	3.9	6.8
Pacific	2.5	5.9
Other ACP LDC	1.6	8.3

Source: European Commission (1999a).

integration into the world economy or to enable ACP states to maintain their market share in the EU. According to Eurostat data (1996), the share of ACP exports in the EU market fell from 6.7 per cent in 1976 to just 3.4 per cent in 1995. At the same time the ACP states, and particularly sub-Saharan Africa, remain dependent on EU trade. In ACP Africa, dependency on trade with the EU is higher (41.21 per cent) than for the ACP Caribbean (18.18 per cent) and ACP Pacific (27.27 per cent). The figure for eastern and southern Africa is in fact higher than the African average, standing at 50.8 per cent in 1997. Subdividing eastern and southern Africa, the EU is the major trading partner of SADC, accounting for 37 per cent of total imports and 50.7 per cent of SADC exports. It is also the major trading partner for the EAC, with 42 per cent of imports and 56 per cent of exports.

Did the fact that SADC and the EAC enjoyed the highest preferential margins result in them being able to exploit more effectively Lomé trade preferences? In order to evaluate which ACP regions better exploited Lomé preferences, the export trends over the period 1988–97 have been analysed (Table 10.3). As the table illustrates, the overall increase in the value of all ACP exports between 1988 and 1997 was negligible, representing a rise of just 3.6 per cent in nine years. The most significant exception to this underperformance was the EAC, which boosted the value of its exports by 41 per cent. SADC, on the other hand, despite benefiting from the most generous trade preferences, experienced negative growth of –5.4 per cent. For those products enjoying a significant preferential margin (3 per cent or more), all ACP regions, with the exception of CEMAC (Communauté Économique et Monétaire de l'Afrique Centrale), saw an expansion in the value of their exports. The two regions with the highest growth rates were the EAC (135 per cent) and SADC (84 per cent). However, as will be shown below, the benefits of these preferences are concentrated on just a few key products and have not in the majority of cases promoted diversification.

Table 10.3 EU trade with the ACP, 1988–97

Country/regional grouping	Growth of all imports (%)	Growth in products with preferential margin over 3% (%)	Preferential imports as % of total imports
Total ACP	3.6	61.9	28.8
Sub-Saharan Africa	–1.6	57.2	26.6
SADC	**–5.4**	**83.6**	**53.0**
EAC	**40.9**	**134.5**	**34.3**
UMEOA	4.3	21.6	39.3
CEMAC	–17.3	–29.5	5.5
Caribbean	25.0	17.6	65.8
Pacific	–17.8	46.4	16.3
Other ACP LDC	–5.0	80.6	25.5
Other PED	75.7	77.7	

Source: European Commission (1999a).

Commodity protocols

Attached to the Lomé Convention were four commodity protocols covering beef, sugar, rum and bananas. Since 1975, these protocols have extended some duty-free access and fixed quota regimes to so-called 'traditional' ACP suppliers. The value of the protocols is significant and, for many states, represents the most valuable element of the Lomé Convention, guaranteeing both employment and hard currency revenue.

Two protocols are relevant to this study on eastern and southern Africa; beef/veal and sugar. In fact the beef and veal protocol excludes all other ACP regions and is limited to just six states within eastern and southern Africa (Table 10.4). The price paid for beef originating from eastern and southern Africa is linked to the artificially high prices established by the EU's Common Agricultural Policy (CAP). The protocol therefore generates substantial foreign exchange revenues. For example, in 1996 the USA market price for a tonne of beef was ECU1,780 against ECU2,700 in the European market. The six beneficiaries of the beef/veal protocol share a total quota of 52,100 t. Botswana has the largest quota (18,916 t or 36 per cent of the total quota) with Kenya allocated just 142 t (0.2 per cent of the total quota).

The protocol on beef and veal has never been completely utilised (Table 10.5). In 1997, just under 50 per cent of the quota was filled compared to 80 per cent in 1995. Botswana, Namibia and Zimbabwe have been the

Table 10.4 The beef/veal and sugar commodity protocols

Country/regional grouping	Beef/veal	Sugar
East and southern Africa	Botswana Kenya Madagascar Namibia Swaziland Zimbabwe	Kenya Malawi Mauritius Madagascar Swaziland Tanzania Uganda Zambia Zimbabwe
Rest of Africa		Congo Côte d'Ivoire
Caribbean		Barbados Belize Guyana Jamaica St Kitts Suriname Trinidad and Tobago
Pacific		Fiji

Source: Dunlop (1999).

303

Table 10.5 Quota allocations of the beef protocol, 1996/97

Country	Quota (t)	Utilisation in 1997	% Utilisation
Botswana	18,916	10,670	56.4
Namibia	13,000	6,026	46.4
Zimbabwe	9,100	7,825	86.0
Swaziland	3,363	225	6.7
Kenya	142	0	0.0
Total	52,100	25,181	48.3

Source: European Commission (1999a).

most active in exploiting their quota share. For Botswana, this resulted in annual payments of over ECU24mn. between 1990 and 1994. The falling utilisation rate may be a result of environmental problems, particularly drought, and increasingly stringent veterinary standards imposed by the EU.

The sugar protocol is different from the beef/veal protocol in that all ACP regions are involved and, because the EU needs raw cane sugar for its refineries, it is more generous to ACP producers. Table 10.4 lists the nine regional states that benefit from the sugar protocol. Collectively, the region accounts for over 50 per cent of the total ACP quota by volume. According to Dunlop (1999), it is estimated that the sugar industry employs, directly and indirectly, over 635,000 people in the 16 ACP states allocated quotas. In Swaziland, over 60 per cent of the country's total GDP is derived from the sugar industry. Collectively, the seven states in the region gained ECU452mn. from the sugar protocol, with Mauritius alone benefiting by ECU299mn.

Clearly, both the beef/veal and sugar protocols are important to the region of eastern and southern Africa. The real value of the protocols is more extensive than employment and foreign exchange. For example, the sugar industry in Zimbabwe is also responsible for a whole range of social services, including housing, health and education, and in Mauritius the proceeds of the sugar protocol have been used to diversify into manufacturing. However, the commodity protocols also have their critics. In particular, concerns have been expressed about the high levels of dependence generated by the protocols. The argument here is that the protocols have created a situation in which competition is stifled and, as a consequence, attempts to promote diversification are rare. In certain countries the dependence on a particular product can make economies vulnerable to both environmental conditions and the management of the protocols. This is particularly the case in countries like Swaziland and Mauritius which depend on the income generated from a protocol.

Commodity composition of trade

The commodity composition of eastern and southern African–EU trade mirrors a classic developing–developed country relationship, with the EU exporting value-added manufacturing goods and services, and eastern and southern Africa exporting primarily primary and unprocessed raw materials. This trading profile is common to all the states of eastern and southern Africa, including South Africa. The following analysis, by way of example, examines the commodity composition of SADC (including South Africa), Rwanda and Burundi.

The largest proportion of SADC exports to the EU comprises primary products, particularly minerals. Indeed, SADC's exports to the EU are concentrated on a few key products comprised of relatively unprocessed commodities and dominated by the importance of gold. In 1998 the top four commodities exported were gold (12 per cent), diamonds (7 per cent), coal (7 per cent) and sugar (4 per cent). Of the 44 most important commodity exports to Europe, all but 10 were either raw materials or agricultural products. A similar commodity composition of exports to the EU exists throughout the region. In Burundi, coffee accounts for 87 per cent of all exports to the EU. In Rwanda, gold (47 per cent) and coffee (43 per cent) are the only two commodities exported on any scale. In Botswana, diamonds constitute 70 per cent of all exports to the EU and beef 30 per cent. Although the commodity composition of South Africa's exports is more diversified, it is nonetheless dominated by a few primary products; gold (17 per cent), coal (10 per cent), diamonds (6 per cent) platinum (4 per cent) and agricultural produce (approximately 11 per cent).

In contrast, the largest proportion of eastern and southern Africa's imports from the EU comprises high-value capital goods in the form of machinery and equipment. Only 4 of the 43 most important items imported by SADC from the EU were primary products. The single most important items imported in 1998 were radio transmission apparatus (5 per cent), electrical appliances (4 per cent), motor vehicles (4 per cent) and motor vehicle parts (4 per cent).

In summary, four key points about the eastern and southern Africa–EU trading relationship require emphasis. First, the trading relationship is far more important to eastern and southern Africa than it is to Europe. The EU is the region's most important trading partner, taking 48 per cent of all regional exports. For eastern and southern Africa, the maintenance of a favourable long-term trading relationship with the EU is of critical importance to the economic well-being of the region. Not surprisingly, given that the EU is the world's largest trading bloc (Gibb 1998), eastern and southern African trade is relatively unimportant in comparison with the EU's principal trading partners, comprising only 2 per cent of all EU merchandise trade in 1997.

Second, Cotonou/Lomé gives eastern and southern Africa privileged access to the European market, even when compared to other ACP regions. While the region has benefited considerably from those commodities with a significant preferential margin, in particular textiles, footwear and some agricultural produce, in other products it has fared less well.

Third, Cotonou/Lomé provides an extremely valuable and, perhaps most importantly, reliable source of income via the beef/veal and sugar commodity protocols. The beef/veal protocol is restricted to just six states within eastern and southern Africa, and for countries like Namibia it provides a significant percentage of government revenue. The sugar protocol alone provides eastern and southern Africa with an almost guaranteed income of ECU452mn. (1996/97 prices). Finally, the single most important characteristic of the eastern and southern Africa–EU trading relationship is its unequal nature. For eastern and southern Africa the EU market is critical to the economic, political and social security of the region. For the EU, eastern and southern Africa is a relatively unimportant but useful market.

The post-Lomé era

In September 1998, the ACP and EU states started negotiating a successor agreement to replace the Lomé Convention. While the ACP states were keen to preserve the non-reciprocal trading preferences accorded to them by Lomé, the EU was adamant that the existing trade regime had to be renegotiated (Solignac-Lecomte 1998, 1999). The European Commission (1997b, c) advanced three principal reasons why the Lomé Convention needed to end. First, it failed either to promote ACP integration into the world economy or enable ACP states to maintain their market share in the EU market. Although, as data presented in this chapter indicate, the EAC did in fact increase its exports to the EU market in the period 1988–97 by some 41 per cent, the majority of ACP regions, including SADC, experienced minimal or negative growth. As for promoting the integration of ACP states into the world economy, the data at the start of this chapter illustrate that the opposite has happened, with Africa becoming significantly more marginalised over the past decade. Lomé, by providing secure market access and special commodity protocol arrangements, may have acted to increase eastern and southern Africa's dependency on the EU. There have, however, been success stories in the region. Examples include Zimbabwe diversifying into textiles, Swaziland into furniture, and Kenya and Zimbabwe have developed a lucrative horticulture business. However, as Solignac-Lecomte (1999) points out, these success stories are limited to countries that enjoy special privileges or commodities with a substantial margin over either MFN or GSP rates. For example, it is possible for Zimbabwe to export horticultural produce like fresh peas to Europe, despite significant air freight costs, because this produce is subject to a zero tariff regime when it enters the EU market. In contrast, the same produce grown in East-central Europe is subject to CAP taxes and, despite enjoying much lower transport costs, is unable to compete with produce entering the EU market duty free.

Second, the Commission argued that the status quo of continued preferences is no longer viable as a result of pressures arising from the WTO-sponsored multilateral trading system (see Box 10.2). The value of the ACP group's preferences (excluding commodity protocols) has been eroded with successive GATT tariff rounds and will be reduced still further once the Uruguay Round is implemented in full and the new WTO Millennium Round of trade talks agrees on further liberalising measures. The EU's common external tariff barrier has been reduced from approximately 10 per cent in the 1970s to somewhere between 3 and 4 per cent following

Box 10.2   Debates about the GATT/WTO

In 1995, the World Trade Organisation replaced the General Agreement on Tariffs and Trade. The primary objective of the WTO is to promote 'open, fair and undistorted competition' (WTO, 1997). It aims to do this by liberalising trade and opening up markets through a reduction of barriers to trade. The benefits of this system, and free trade more generally, to developing countries is contested on many fronts. The social and political dangers of liberalising economies that lack competitive industries is widely recognised. For many countries, like Zambia and Zimbabwe, liberalisation has resulted in rapid de-industrialisation, job losses and political instability. Many academics and NGOs argue that the WTO is a multilateral form of governance dominated by the economic superpowers, in particular the USA and EU, and is used by them to govern the world economy in a way that benefits their economies. For example, the EU has resisted the liberalisation of agriculture, that would be to the benefit of many developing countries, because of the uncompetitive state of the European agricultural industry. However, multilateral free trade and its application to developing countries has many powerful advocates. The International Monetary Fund, World Bank, USA and EU, supported by international capital, which by any measure is an extremely powerful alliance, all support the WTO's objective of promoting multilateral free trade. These institutions and governments support the liberalisation of developing country economies on the grounds that previous protectionist policies failed to generate growth and integrating these countries into the world economy offers the best long-term solution to marginalisation.

Sources: WTO (1997), Gibb (2000).

the implementation of the Uruguay Round. The value of the ACP group's duty-free access to the EU has therefore been eroded significantly. In the industrial sector, Cotonou provides few benefits above those now available under the MFN protocol. Under MFN, 84 per cent of all ACP exports would be duty free in any case, with the remaining 16 per cent being taxed at an average of 8 per cent (European Commission 1999b). Furthermore, following the implementation of the Uruguay Round, the value of trade preferences is set to fall further. For example, one of the most significant Lomé preferences was an exemption to the Multifibre Agreement (MFA) that governs and restricts the export of most textiles and clothing. As a result of the Uruguay Round, the MFA will eventually be phased out and ACP textile exports to the EU will face intense competition from Southeast Asia and elsewhere. Within eastern and southern Africa, Zimbabwe, Zambia and Malawi are likely to be affected detrimentally by the demise of the MFA.[1]

---

[1] It is worth noting that Mauritius, also a member of SADC, is likely to be particularly badly affected, although this volume is generally confined to mainland eastern and southern Africa.

Third and finally, the European Commission favoured an end to Lomé because it believed the trade regime to be inconsistent with the values and rules of the multilateral trading system. By providing preferential access to European ex-colonies, yet excluding other developing countries at similar levels of development, Lomé was seen to violate both the WTO's MFN principle and a 1979 Enabling Clause designed to support LDCs. The WTO's MFN principle is founded on the idea that no country should confer any special privilege on another or discriminate against it. International trade must therefore be conducted on the basis of non-discrimination.

The Lomé preferential trading regime had already come into direct conflict with the WTO over the issue of European banana imports. In theory, the same arguments used to denounce the banana protocol could be used against the beef/veal and sugar protocols and, ultimately, the whole Lomé trading regime. The EU was therefore adamant that any post-Lomé agreement has to be compatible with WTO principles.

(Regional) economic partnership agreements

The EU's proposals for a post-Lomé trading agreement, as outlined in the Cotonou Agreement, are based on two, potentially conflicting, differentiating criteria: developmental and regional. Most importantly, the EU is dividing the ACP bloc according to development status, categorising states as either developing or lesser developing countries (LDCs) according to their categorisation by the World Bank. At the same time, one of the EU's principal objectives is to promote regional integration among groups of ACP countries. ACP regionalism is seen as a mechanism to promote the cause of trade liberalisation within ACP regions and, in turn, the integration of ACP countries into the world economy (Solignac-Lecomte 1999). How then does this proposal affect eastern and southern Africa?

If the countries included in this volume are divided according to development status (see Figure 10.2), 11 would be classified as LDCs, 4 as developing and South Africa as either a 'country in transition' (EU categorisation) or developed (WTO categorisation). The EU's strategy is to offer all developing countries the option of joining an Economic Partnership Agreement (EPA), and LDCs that are unwilling or unable to enter into an EPA, the option of super-GSP that would be non-discriminatory and therefore open to all LDCs.

In January 1999 the EU amended its super-GSP to match Lomé preferences, with the important exception of commodity protocols. Hence Lomé trade preferences were de facto extended on a non-reciprocal basis to all LDCs and were, as a consequence, perceived to be WTO-compatible. For the 11 LDCs of eastern and southern Africa that meant an end to their preferential treatment compared to other non-ACP LDCs. In February 2001 the EU introduced a new GSP regulation establishing duty-free access for all exports, including agricultural produce, originating in LDCs, with the exception of arms and munitions (Goodison 2002). The 'Everything But Arms' initiative covers all LDCs.

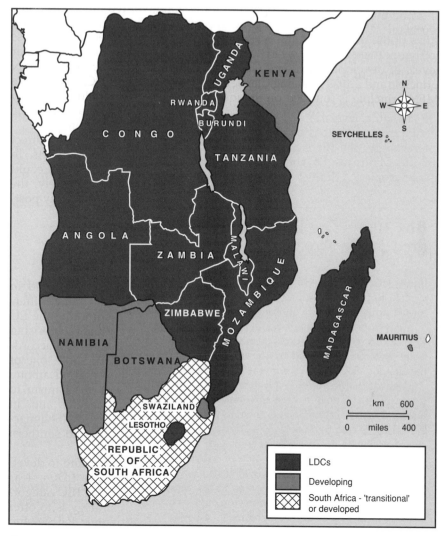

Figure 10.2 Eastern and southern African countries, development status in 2000

The EU's strategy is to offer all ACP developing countries the option of joining an EPA by creating a series of free trade areas (FTAs) centred on the basic principle of reciprocity. This presents a dilemma for eastern and southern Africa. The region contains a mixture of developing and less developed countries, and any attempt to separate the region using development criteria has the potential to be economically and politically divisive. In 1999, the EU published a series of six consultancy studies on the impact of the EPA proposal on different ACP regions. Much along the lines of previous EU thinking, the reports divided eastern and southern Africa into SADC (without South Africa because of its Trade, Development and Co-operation Agreement with the EU, see Box 10.3) and the EAC. The report on SADC, known as the Imani study (European Commission 1999b), argues that in the long term SADC would be a 'natural' regional partner for the EU. However, in the short term it concludes that 'SADC has not yet reached an advanced level of economic integration' and the 'present institutional structure is not conducive to negotiating trade agreements with other regions'. The biggest threat to SADC of an EPA relates not to the changing levels of access SADC exports have to the EU, but to the significantly enhanced access EU exporters will gain to SADC markets. Interestingly, the three SADC countries with the highest tariffs against the EU, Mauritius, Seychelles and Zimbabwe, are (or were) all classified as 'developing'.[2] Zimbabwe was, however, reclassified as an LDC by the

---

Box 10.3   The European Union–South Africa Free Trade Area

Immediately after South Africa's first democratic election in April 1994, it applied to become a full member of the Lomé Convention. This request was turned down by the EU on the grounds that the South African economy was 'too developed'. Although South Africa subsequently joined Lomé in June 1998, it did so as a 'political' member only and was not accorded any of the trade preferences enjoyed by the other 70 Lomé members. In October 1999, the EU and South Africa signed the Trade Development and Co-operation Agreement that includes provisions leading, after a period of transition, to a free trade area. The free trade agreement took over four years to negotiate and included many acrimonious exchanges between European and South African officials. The free trade area is, in theory, 'asymmetrical' in both timing and coverage. While the EU will liberalise up to 95% of its imports from South Africa, South Africa will open 86% of EU imports to duty free trade. Tariff dismantlement on the EU side will take place over a 10-year period, while South Africa will have a transition period of 12 years. However, the free trade deal has been criticised for protecting too many uncompetitive European industries. In particular, the agreement prevents 40% of South Africa's agricultural produce from entering the EU market at a duty free rate.

The free trade agreement will be extended to include the other members of the Southern African Customs Union (SACU). This is ironic for two reasons: first, Botswana, Lesotho, Namibia and Swaziland were never invited to participate in the negotiating process and are therefore having their terms of trade altered radically without formal consultation; second, South Africa was prevented from joining Lomé on the grounds that its economy was different to other ACP states, particularly those in southern Africa, yet these same states are now expected to join South Africa's free trade agreement.

[2.] This point is supportive of the argument made in Chapter 3 that trade barriers can, as in the NICs, be used as a tool for economic development, contrary to the free trade prescriptions of the IFIs.

World Bank in 2001, due to a dramatic decline in its economy. An EPA would have the greatest proportional fiscal impact on the Seychelles, with a 30 per cent reduction in government revenue following a 70 per cent reduction in customs revenues as a result of tariffs against European goods falling to practically zero. Similarly, Mauritius and Zimbabwe will experience a fall in government revenue of 27.9 and 18 per cent respectively.

The Imani study (European Commission 1999b) recommended an EPA scenario based on: first, extending the already agreed EU–South Africa FTA (which is an EPA in all but name) to the member states of the Southern African Customs Union (SACU – Botswana, Lesotho, Swaziland and Namibia (see Figure 10.3)); second, concluding a separate EPA with the three remaining non-LDCs in SADC (Mauritius, Seychelles and Zimbabwe (at that point still classified as 'developing' rather than 'least developed'); and third, leaving six LDCs outside of an EPA with access to the EU via EBA. The possibility of a SADC-wide EPA is not envisaged before 2010. Within southern Africa, the European Commission therefore proposes replacing Lomé with three different EU–southern Africa preferential trading agreements. As observed by McQueen (1999: 7), the EPA proposal 'would also weaken incentives to investors to locate in the SADC region since the regional market would be more easily served from the EU, the common denominator in these increasingly complex agreements'.

The EAC currently comprises two LDCs (Tanzania and Uganda) and one developing country (Kenya). The EU commissioned a report on its future trading relationship with the EAC and, notwithstanding the LDC status of two of the three EAC members, this calls for a EAC–EU EPA to be established (European Commission 1999b). The report notes the considerable progress made by the EAC towards internal trade liberalisation with the prospect of zero tariffs by 2005.

An EPA would guarantee the continued benefit of preferential access to the EU market. However Tanzania and Uganda, because they are LDCs, would not suffer a loss of preferences under a non-EPA scenario. Kenya, therefore, is the only country that stands to gain. However, as already shown for the SADC case, the greatest potential impact of an EPA is not EAC access to the EU market but, vice versa, that EU exporters gain better access to the EAC markets. Under an EPA, imports into the EAC from the EU would be given the same tariff treatment as imports from other EAC countries: zero tariffs on almost all products except for sensitive items such as agricultural produce, textiles, tobacco, soap and vehicles. This has potentially damaging consequences because of *trade deflection*, where EAC importers switch from less efficient EAC producers to more efficient EU producers as a result of the introduction of an EPA. This would result, in the short term at least, in decreasing regional trade, a loss of income and security, and, perhaps most importantly, further unemployment. In addition, there could be large costs arising from *trade diversion*, with more efficient non-EU producers being replaced by EU imports benefiting from preferential access to the EAC market. For both Tanzania and Uganda, the net overall impact of an EPA is likely to be reduced living standards. For all EAC countries, an EPA would reduce significantly both tariff and tax revenues. Because the EU is the EAC's largest trading partner, zero tariffs

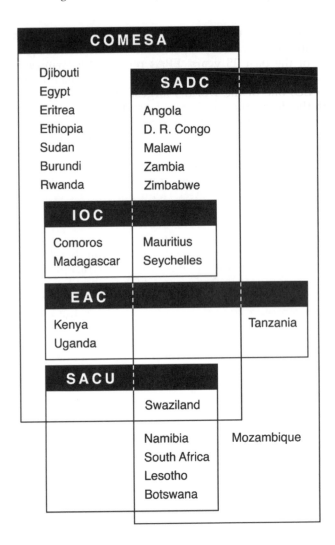

Figure 10.3   Regional organisations and memberships

translate into significant tariff revenue losses. This, coupled to trade diversion and deflection, could result in a substantial revenue loss for Tanzania (–20 per cent), Uganda (–16 per cent) and Kenya (–12 per cent). Furthermore, given that an EPA would disproportionately benefit Kenya, there is the potential for regional tensions to arise.

The EU's post-Lomé Cotonou strategy for eastern and southern Africa therefore represents a radical departure from the non-reciprocal trading agreements of the past 25 years. EPAs represent a form of free trading agreement that will, after a period of transition, create a single market linking the EU, one of the world's largest and most sophisticated economies, with the developing and some least developed countries of eastern and southern Africa. Although LDCs have little to gain, and perhaps much to lose, from joining an EPA, 3 out of the 10 LDCs in the region may join with neighbouring states to form an FTA with the EU. The most significant negative impacts of EPAs relate to the prospect of enhanced EU penetration of eastern and southern African markets and the loss of customs revenues following the introduction of duty-free tariffs. For some countries in this region this could result in really substantial economic losses. Furthermore, EPAs may lead to more efficient EU producers forcing eastern and southern African industries out of production. De-industrialisation is already a serious problem as a result of trade liberalisation under SAPs (see Chapter 3), and further industrial decline is bound to lead to a rise in unemployment (which is already one of the most serious problems facing many of the region's economies) and yet further losses in government revenue. As observed by Clements (cited in Panos 1998: n.p.):

> The current proposals for free trade areas would transform Lomé into a battering ram for free trade, forcing infant industries of the developing countries in the ACP into unfair competition with the industrialised economies of Europe.

The essence of the Cotonou trade agreement is that Lomé non-reciprocal trade preferences will be maintained up to January 2008 in order to allow for the negotiation of WTO-compatible EPAs. Negotiations to establish the Cotonou trade regime commenced in September 2002. However, as observed by Tekere (2000), the EU and the ACP states 'do not agree on the content and substance of the trade arrangements following the transitional (eight year) arrangement'.

The future of the commodity protocols under EPAs is also uncertain. The EU mandate is extremely vague and states that 'the banana, beef and sugar protocols will be reviewed in the context of the negotiation of economic partnership agreements'. The greatest single threat presented by the EU's post-Lomé strategy is to the cause of regional integration, both political and economic. By fragmenting the import and tariff regimes of eastern and southern African countries, the Cotonou EPA model will undermine attempts to promote regional integration. As observed by Solignac-Lecomte (1999: 4):

> A major drawback stressed by '(R)EPA-critics' is that it may actually do more harm than good to the process of regional integration itself, by fragmenting the regimes applying to imports from the EU and by impeding the adoption of external policies.

Clearly, external economic analyses by the EU serve to undermine any supposed unity for the countries of eastern and southern Africa. The next section of this chapter examines the state of trade within eastern and southern Africa and the level of economic integration that exists within the region's various trading blocs.

Intra-regional trade

Trading patterns in eastern and southern Africa are determined by three principal factors: first, the dominant size of the South African economy (see Table 10.1); second, the structure of production which is competitive rather than complementary (most countries produce a similar range of products and compete in similar markets); and third, the regional organisations designed to promote trade are institutionally weak and suffer from overlapping memberships and inter-organisational competition (see Figure 10.3).

Before examining these factors, it is first useful to review empirically the state of intra-regional trade. However, this task is made difficult by the lack of reliable and comprehensive trade data for the majority of eastern and southern African countries (African Development Bank 1993; Simon and Johnston 1999). Moreover, as observed by Maliyamkono and Bagachwa (1990), unrecorded trade is estimated to be anything from 15 to 50 per cent of official trade. The most comprehensive data on trade flows within eastern and southern Africa are provided by a 1993 study by the African Development Bank (1993) that refers to the period before South Africa joined SADC. Although many trade studies have been undertaken since then, most rely on the African Development Bank data for intra-regional trade. As Simon and Johnston (1999) note, there appears to be no substantive research into the net impact of increased trade with South Africa in the period after 1994.

Up to 1994, the level of intra-regional trade within eastern and southern Africa, excluding South Africa, was very low indeed. Most studies cite a figure of five per cent or less (Mayer and Thomas 1997). This lack of intra-regional trade is a result of the lack of complementarity between national economies. Notwithstanding the geopolitical division of the region in the period up to 1994, split between apartheid South Africa and its attempts to destabilise the region and the rest of eastern and southern Africa pursuing policies designed to isolate Pretoria, South Africa was the region's dominant trading partner. Large parts of southern Africa, which created the Southern African Development Coordination Conference (SADCC) as a means of isolating South Africa (see Box 10.4), remained almost totally dependent on South Africa. However, the EAC was not dependent on South Africa, being too far away to have been drawn into South Africa's attempts to create dependency through its ruthless policy of regional destabilisation (Hanlon 1986).

South Africa's accession to SADC in 1994 had a profound impact on intra-regional trade (Mayer and Thomas 1997). Indeed South Africa is, as far as trade is concerned, the region's common denominator with strong links to most of the region's economies, with the exception of Burundi, Rwanda, Uganda and, to a lesser extent, Kenya, Angola and the DRC. South Africa's membership of SADC has raised the level of intra-regional

---

Box 10.4   The Southern African Development Co-ordination Conference

In August 1992, the Southern African Development Co-ordination Conference (SADCC) was replaced by SADC (see Figure 10.2). The original aims of SADCC were to reduce the region's dependence on apartheid South Africa and to promote regional integration. SADCC explicitly rejected traditional integration and customs union theory in recognition of the likelihood that free trade would enhance regional inequalities. The cornerstone of SADCC's integrative strategy was 'project coordination' with a particular emphasis on large-scale infrastructure projects. Evaluating the success or otherwise of SADCC has generated considerable debate. On the one hand there are those who argue that SADCC was successful in *limiting* the region's dependence on apartheid South Africa and promoting and preserving regional solidarity in the face of South African destabilistaion. Empirical evidence for this argument can be sought in the rapid increase in intra-SADC trade following South African membership of SADC in 1994. On the other hand, critics of SADC highlight its failure to *reduce* the region's dependence on South Africa throughout the apartheid period and its weak record in promoting meaningful moves towards regional co-operation or integration. Empirical evidence for this argument can be sought in the low levels of intra-SADCC regional trade and the fact that the region's dependence on South Africa actually increased from the mid-1980s.

Sources: Simon and Johnston, 1999.

---

trade from below 5 per cent in the early 1990s to an estimated 22 per cent by 1999, largely, at first, by the incorporation of SACU (see below) trade into SADC trade statistics (SADC 1999; Simon and Johnston 1999). Although this remains relatively low compared to intra-regional trade in the EU (60 per cent), the North American Free Trade Area (45 per cent) and Mercosur (30 per cent), annual growth rates are high. It is estimated also that once (or perhaps, if) all the SADC Free Trade Area measures agreed under the 1996/97 trade protocol for the regional organisation are implemented, intra-SADC trade could increase to 35 per cent of total regional trade (Simon 2001).

The vast majority of intra-regional trade in eastern and southern Africa is between South Africa and other countries of the region, although Zimbabwe and Kenya also have important regional trading links. Given South Africa's dominance of intra-regional trade it is useful to structure this examination of intra-regional trade through an analysis of South Africa's trade with the region. South Africa's trade can be divided into three distinct categories; trade with the other member states of the Southern African Customs Union (SACU), SADC and trade with the EAC plus Burundi and Rwanda.

The Southern African Customs Union

The origins of SACU date back to 1891, making it one of the oldest customs unions in the world. It is also one of the most imbalanced (Gibb 1997; Simon and Johnston 1999). The SACU agreement has been re-negotiated on several occasions, in 1910, 1969 and, most recently, in 2002. SACU member states have a common external tariff (CET) barrier against the rest of the world, the free movement of goods, capital, services (but not labour) and, with the exception of Botswana, a common currency through the Multilateral Monetary Area (MMA). The level of integration represented by SACU, both economic and institutional, is therefore relatively sophisticated.

SACU's longevity, in sharp contrast to the many other attempts in Africa to promote regional integration, is a result of an agreement whereby South Africa gains a captive market for its internationally uncompetitive manufacturing goods and, in return, Botswana, Lesotho, Namibia and Swaziland (BLNS) receive a disproportionately large share of SACU's revenue pool. SACU has had a profound impact on trade patterns in the region. The SACU tariff structure has in the past been designed by South Africa to protect South African industry and interests from more competitive overseas operations. The CET therefore acts as a considerable incentive to intra-SACU trade. South Africa's trade with SACU is larger than its trade with either Asia or North America. That trade is, however, asymmetrical in favour of South Africa. In 1994/95, 61 and 58 per cent of South Africa's African exports and imports were conducted with BLNS, respectively. However, the absolute size of these trade flows varies considerably. South Africa's exports to the rest of Africa stood at R22,420m., almost four times greater than the R5,697mn. imported. Consequently, South Africa ran a trade surplus with SACU of over R10bn in 1994/95.

There are, of course, substantial benefits accorded to the BLNS states from SACU membership. The most important benefit is the very significant contribution made to state revenue from the SACU customs revenue pool. The proportion of the revenue pool allocated to BLNS has been increasing over time, particularly throughout the 1980s and 1990s. In 1995/96, SACU receipts accounted for between 16 and 51 per cent of government revenue in individual SACU countries. Table 10.6, which provides a summary of the trading profile of Botswana, illustrates this high level of integration among SACU states. In 1997, 72 per cent of Botswana's imports were sourced from within the Customs Union (overwhelmingly South Africa), compared to just 12 per cent from Europe. Conversely, 80 per cent of Botswana's exports (mainly diamonds by value) were destined for Europe compared to only 14 per cent for the Customs Union. A very similar pattern is evident in the trading profiles of Swaziland and Namibia and, in the case of Lesotho, South Africa dominates both exports and imports.

SACU therefore represents a sophisticated form of regional integration and is at the very core of intra-regional trade within eastern and southern Africa.

Table 10.6   Botswana's trading profile, 1995–97 ('000 pula)

Area/country	1995	1996	1997[a]
Exports to:			
Common customs area	1,276,948	1,489,690	1,106,172
Zimbabwe	181,601	250,518	282,377
Other Africa	49,257	50,511	63,942
United Kingdom	2,222,973	4,424,115	4,574,287
Other Europe	2,147,337	1,827,156	1,554,927
United States of America	52,352	77,875	72,698
Total	5,941,470	8,141,832	7,670,290
Imports from:			
Common customs area	3,924,950	4,474,242	4,267,173
Zimbabwe	293,028	329,320	276,456
Other Africa	17,817	23,306	27,219
United Kingdom	134,688	147,026	129,469
Other Europe	319,250	234,859	371,184
South Korea	377,499	250,055	637,263
United States of America	107,410	73,781	64,903
All other	132,431	196,211	188,272
Total	5,307,073	5,728,800	5,961,939

[a] 1997 data are up to September.
Source: SADC (2000).

The Southern African Development Community

In 1994/95, South Africa's exports to SADC (including BLNS) accounted for 93 per cent of all its African exports (Table 10.7). Non-SACU SADC countries accounted for 32 per cent of this total. The most striking feature of South Africa/non-SACU SADC trade is the high ratio of exports to imports, standing at approximately 7.5 : 1 in 1995. South Africa derives many benefits from this trade in terms of both generating a healthy trading surplus and providing a destination for manufactured high-value exports (Mayer and Thomas 1997): exactly the sort of 'non-traditional' exports that all developing countries are keen to foster. It is estimated that non-SACU SADC countries account for approximately 25 per cent of South Africa's exports of motor vehicles, 40 per cent of plastics, 24 per cent of chemicals and 30 per cent of mechanical appliances (Mayer and Thomas

Table 10.7   South Africa's Africa trade, 1994/95

Country/regional grouping	% Exports	% Imports
Botswana	18.6	9.9
Namibia	18.1	24.8
Swaziland	12.6	20.1
SACU (subtotal)	61.5	58.5
Non-SACU SADC		
Zimbabwe	11.0	17.9
Mozambique	6.3	1.6
Zambia	5.2	3.4
Malawi	2.8	3.2
Mauritius	2.4	0.3
DRC	1.6	6.4
Angola	1.4	0.3
Tanzania	0.8	0.3
Seychelles	0.4	0.1
Non-SACU SADC (subtotal)	31.9	33.5
SADC	93.4	92.0
Kenya	3.0	0.5
Burundi	0.1	Negligible
Uganda	0.1	Negligible
Rwanda	0.05	
EAC not incl. Tanzania (subtotal)	3.25	0.5
East and southern Africa	96.7	92.5
Total rand (millions)	22,419.8	5,697.1

Source: SADC (2000).

1997, citing Davies 1996). SADC, including SACU, is therefore a very important market for the South African economy. In contrast to their poor performance in the markets of Europe and other developed economies, where they are generally non-competitive, South Africa's manufactured goods are in demand in southern African countries. In part, this demand has arisen from historical links and geographic proximity, with South African exporters having an intimate knowledge of the needs and problems of the 'local' market. Mayer and Thomas (1997: 26) observe:

> In contrast to the poor trade potential amongst southern African countries, there is a great deal of complementarity between the South African economy and its neighbours. In some ways the SADC countries trade with the Northern countries is similar to their trade with South Africa. The African market is clearly very important for South Africa.

Following South Africa's first democratic election in 1994, exports to SADC increased substantially; between 1994 and 1995, they increased by 59 per cent and in 1996 by a further 50 per cent again (SADC 1999). Moreover, since 1996, it has been reported widely that South African exports to SADC have continued to experience significant growth (Simon 2001; *Business Day*

1999). There has not, however, been a corresponding increase in the level of South African imports from SADC. South Africa's imports from the region remained largely stable and may have actually declined in relative terms. With the exception of Zimbabwe, South Africa's imports from SADC are concentrated on agricultural commodities, in particular coffee, fish, tea, cocoa and beef. Moreover, South Africa's imports of these commodities from the rest of the world greatly exceed the value of the imports from African sources.

One of the most influential factors explaining South Africa's export growth to SADC, alongside the ending of apartheid, is the impact of various tariff liberalisation programmes in SADC countries. South Africa, because of its diversified production structure and supportive policy environment, was in a position to exploit the better access offered to non-SACU SADC countries. The structural adjustment programmes (SAPs) adopted by most of the region's economies throughout the 1980s, at the behest of the IMF and World Bank, opened up previously closed markets to South African exports. Clearly, the asymmetrical nature of this trading relationship is in the long term unsustainable. Illegal migration into South Africa, already large, will in all likelihood grow significantly if the imbalance continues. Furthermore, if non-SACU SADC countries fail to prosper there is the potential for negative economic and political spillover that will ultimately damage the South African economy. For example, the political tension and economic crisis experienced by Zimbabwe in 2000 led to South African–Zimbabwe trade, which had previously expanded at a rapid rate, declining by over 7 per cent in 1999.

The pattern of trading transactions outlined above developed in the absence of any official SADC trading arrangement. However the position is now being formalised, albeit very slowly. The SADC Treaty (1992) calls for the development of policies aimed at the progressive elimination of the barriers to the free movement of capital, labour, goods and services, but the Treaty does not impose economic obligations on member states. Economic policies, together with most other policies, are developed via protocols which amend the Treaty. In August 1996, 12 SADC members (all but Angola) signed a 'Protocol on Trade' which committed signatories to establish a free trade area. In broad terms, the Protocol calls for the establishment of a free trade area eight years from the 'entry into force' of the Protocol. For the Protocol to 'enter into force' it had to be ratified by at least a two-thirds majority, which was achieved in May 2000. When implemented, the FTA should promote the gradual elimination of all barriers, both tariff and non-tariff, to intra-regional trade. The objectives of the SADC Trade Protocol are set out in Article 2:

1. To further liberalise intra-regional trade in goods and services on the basis of fair, mutually equitable and beneficial trade arrangements, complemented by protocols in other areas.
2. To ensure efficient production within SADC reflecting the current and dynamic comparative advantages of its members.
3. To contribute towards the improvement of the climate for domestic, cross-border and foreign investment.

4. To enhance economic development, diversification and industrialisation of the region.
5. To establish a free trade area in the SADC region.

(SADC 1996).

Recognising the potential for the more developed economies, particularly South Africa, to benefit at the expense of the least developed, the Protocol calls for an element of asymmetry and differentiation to be built into the FTA. Thus, SACU states will 'frontload' their tariff reduction programme, reducing tariffs on a greater number of products more quickly than other SADC states. Mauritius and Zimbabwe will 'midload' and the rest of SADC will 'endload', with a slower pace of tariff reduction and less coverage. Asymmetric trading agreements are clearly absolutely crucial in FTAs where there are severe imbalances. However, whether there is sufficient asymmetry in the arrangements outlined above to redress the increasingly imbalanced intra-regional trading transactions between South Africa and SADC is open to question.

The EAC, Rwanda and Burundi

The EAC was originally established by treaty in 1967 and dissolved in 1978. While a wide variety of factors contributed to the EAC's demise, Kenya's large and persistent trading surplus with Uganda and Tanzania was critical to the breakdown of the agreement. In 1978, Tanzania took the decision to 'seal' its border with Kenya in order, in part, to foster closer economic and political relations with SADCC. However, by 1992, the then President of Tanzania, Julius Nyerere, described his decision to withdraw from the EAC as his 'biggest mistake' (AfDB 1993: 20). Since 1992 various attempts have been made to try to resuscitate the EAC with limited success. As Figure 10.3 illustrates, all the member states of the EAC are also members of the Common Market for Eastern and Southern Africa (COMESA) and Tanzania also belongs to SADC.

South Africa's trade with the EAC has been limited until recently (Table 10.7). In 1994/95, only 3.2 per cent of South Africa's African exports went to the EAC. Exports from the EAC to South Africa were negligible. Thus one feature of the EAC which clearly distinguished it from southern Africa was South Africa's lack of historical dominance in EAC markets. However this situation began to change quite significantly in the last few years of the twentieth century as corporate South Africa made determined moves into the East African market. By 1998 South Africa was Tanzania's fourth largest trading partner, ahead of Kenya, and according to a survey it was perceived by South African firms as their most promising investment market (Simon 2001). There has also been increasing direct investment in East Africa. South African Breweries has been a particularly notable actor, and is now highly influential in the Kenyan and Tanzanian markets. Tourism- and telecommunications-related investment has also been important (Simon 2001).

The level of intra-EAC trade is limited by a lack of complementarity and the dominance of primary exports. At the end of the 1990s, Kenya's and Tanzania's exports were dominated by primary products (90 per cent) and Uganda had negligible manufacturing exports (in 1991 it had a recorded zero per cent). Intra-EAC trade is estimated to be below 5 per cent of total trade and dominated by Kenya.

Rwanda and Burundi are members neither of SADC nor the EAC. They are, however, members of COMESA (see Figure 10.3), a regional organisation whose principal objective is to promote intra-regional free trade. However, since gaining independence in 1962, both countries have suffered many serious ethnic confrontations which have spilt over into bloody conflict. The civil war in Rwanda, which lasted from 1994 to 1998, culminated in the genocide of 800,000 people. Since then, the return of over 2 million refugees has exacerbated the already dire economic and political conditions of the country. In Burundi, where the first democratic elections were held in June 1993, only to be followed by a bloody *coup d'état* in September of the same year, there is a serious economic crisis and a structural trade deficit problem. Following the coup, not only did Western countries freeze aid but neighbouring states imposed an economic embargo on the country. Lasting peace remains elusive in both countries and their economies remain extremely weak and largely detached from intra-regional trade.

Informal and 'unrecorded' trade

Informal trade, as a component of intra-regional trade within eastern and southern Africa, is extremely significant. Not unsurprisingly, however, there are no reliable estimates of the value of this trade. There is also a problem defining informal trade. Ellis and MacGaffey (1996) prefer the term 'underground trade' when describing all forms of commercial transactions, both legal and illegal, that cross international boundaries but are unrecorded in official data. Informal trade does not therefore have to be illegal and, although often associated with small one-person enterprises, can involve large-scale operations. While informal trading is not necessarily illegal it often involves 'a wide range of illegality' (Ellis and MacGaffey 1996: 20). The degree of illegality can range from trade which is partially legal, such as the movement of foodstuffs that are concealed as a means of gaining commercial advantage, to trade in cocaine and heroine. Unrecorded trade may also be legal, such as the small-scale movement of foodstuffs and clothing.

Informal trade is, by definition, difficult both to measure and detect (Alessandrini and Dallago 1987). According to the AfDB (1993) the trend towards informal trade has been lessened as a result of liberalisation programmes and import support measures improving the availability of basic consumer and household goods. In theory, structural adjustment should precipitate a move away from the public and informal sectors to the formal sector as private sector activities and entrepreneurial culture are encouraged by the reform process. However, Aryeetey (1996) argues that the expected transition from informal to formal sector activity has failed to materialise and that unrecorded trade represents a dynamic and increasingly important component of intra-regional trade. According to Aryeetey (1996: 121), the principal reasons behind the failure of SAPs to promote more formal sector activity are 'structural impediments to the transmission of macro-economic incentives'.

A wide variety of commodities are involved in informal trade. Foodstuffs, and in particular maize, are often cited as the principal commodity types that dominate informal intra-regional trade. The AfDB note that 1 million bags of maize, representing over 100,000 t, is transported illegally from Zambia to surrounding countries on an annual basis. Unrecorded maize also moves from Mozambique to Malawi. Periodic and specific shortages, like the fuel shortage in Zimbabwe in 2000 which precipitated a clandestine cross-border movement of petrol, can also add significantly to informal intra-regional trade. Within southern Africa, one of the most cited illegal movements is that of stolen South African luxury vehicles and vehicle parts going to Mozambique and Zambia, and often even further.

One of the most detailed studies into informal trade within eastern and southern Africa by Maliyamkono and Bagachwa (1990, cited in AfDB 1993: 27) concluded: 'The value of unregistered exports is visibly enormous. . . . In the absence of official estimates or records that is probably all that can be said.' However, the level, coverage and importance of informal intra-regional trade should not be exaggerated. The AfDB study (1993: 27) concludes that 'such flows add 1 to 2 per cent to total flows'.

Informal and unrecorded trade is not restricted to intra-regional trading transactions. The intercontinental trade in ivory and drugs has been well documented (Parker and Mohamed 1983; Ellis 1994) and the 'illegal' and unrecorded export of minerals, notably gold, diamonds and precious gemstones, is a growing problem (Kibble 1998; Bayart *et al.* 1998; Chabal 1996; Munslow 1999; ACTSA 2000). This situation is made worse in certain states – states which are to all intents and purposes dysfunctional – such as Angola and the DRC, where large parts of the country are, or have been until recently, held in 'rebel' hands dependent, in part, on the export of high value minerals (see Chapter 9). There is also evidence to suggest that the major operators of intercontinental unrecorded trade have close connections to holders of high political office (Ellis and MacGaffey 1996). Even in situations where this is not the case, as in South Africa, the unrecorded and illegal movement of precious minerals can be significant. South Africa's Chamber of Mines estimate that criminal syndicates operating at the international level are costing their industry approximately R2.3bn ($295m.) a year, equal to about a third of operating profits (*Financial Times* 2001). The scale of unrecorded illegal gold trade in South Africa is therefore significant, running at 35t a year, equal to 8 per cent of total annual production.

Conclusion

As noted at the start of this chapter, an analysis of trade characteristics and patterns can be used both to identify commonalities within a region and differentiation. From an international perspective, the countries of eastern and southern Africa do have much in common. First, all of them, including South Africa, are marginal players in global trade. This puts the region at a distinct disadvantage in terms of its trading relationships with the rest of the world. Eastern and southern Africa's importance to the world economy is currently very small and its influence in determining the character and regulation of the world economy, through institutions like the WTO and IMF, is limited. Viewed from a world-systems or dependency perspective, this region is without doubt 'peripheral' and 'dependent'. Second, nearly all the countries of eastern and southern Africa share a common dependence on exporting a disproportionately large share of their exports to the EU, ranging from almost 80 per cent in the case of Botswana, Mauritius and Uganda to 25 per cent in Mozambique. This trade is almost uniformly unbalanced, with the EU exporting high-value manufactured goods and eastern and southern Africa exporting low-value added primary products. This trading profile reinforces the region's peripherality in the world economy. Third, all the countries of eastern and southern Africa are signatories to the Cotonou Agreement and share a common difficulty in negotiating new trading arrangements with the EU (although in this they do not differ from other ACP countries). The existing Lomé trade preferences will almost certainly not be extended beyond 2008. All the countries in the region therefore face a common problem of how to manage the EU's post-Lomé strategy which aims to differentiate between countries deemed to be 'developing' and those deemed 'less developed'. The critical question arising from the prospect of a multiplicity of EU–eastern and southern African trading regimes is the impact this differentiation will have on regional integration and, in particular, trade and investment flows.

The intra-regional trading characteristics of eastern and southern Africa are more difficult to interpret. The level of intra-regional trade, although starting from a low base, is experiencing considerable growth. In the 10-year period 1986–96, intra-SADC trade as a percentage of members' world trade grew from 10 to 20 per cent. Although all SADC countries expanded their exports, the best performers were South Africa, Zimbabwe, Botswana, Namibia, Swaziland and Zambia, which collectively account for 95 per cent of intra-SADC trade. However, the vast majority of this trade is with South Africa and non-South African intra-SADC trade is still limited. Trade among the other countries of eastern and southern Africa remains low, around the 5 per cent level. In the case of Burundi and Rwanda it is probably less than 2 per cent.

Patterns of intra-regional trade thus appear to highlight contrasts within the region of eastern and southern Africa and to define three distinct, but by no means separate, sub-regions. First, there is a core area focused on South Africa and including the other members of SACU, Zimbabwe and Zambia (and perhaps Mozambique). Second, there is an area comprising other SADC members, Uganda and Kenya, where limited levels of intra-regional trade reflect the structural rigidities of their competitive, as opposed to complementary, economies. Finally, there is an area covering Rwanda and Burundi which continues to suffer from the impact of economic sanctions imposed in the mid-1990s and appears remarkably detached from the region in terms of trading links. This chapter has therefore revealed a complex geography of trade which sometimes unites and sometimes separates the economies of eastern and southern Africa.

References

ACTSA [Action for Southern Africa] (2000) *ACTSA News.* Spring.

African Development Bank (1993) *Economic Integration in Southern Africa,* 3 vols. Biddles, England.

Alessandrini, S. and **Dallago, B.** (eds) (1987) *The Unofficial Economy.* Gower, London.

Aryeetey, F. (1996) Formal and informal economic activities, in Ellis, S. (ed.) *Africa Now.* James Currey, London.

Bayart, J.F., Ellis, S. and **Hibou, B.** (1998) *The Criminalisation of the State in Africa.* James Currey, Oxford; Indiana University Press, Indianapolis.

Business Day (1999) Regional trade aid mooted, 18 May.

Chabal, P. (1996) The curse of war in Angola and Mozambique, *Africa Insight* **26**(1):

Daniels, J. (1996) Southern Africa in a global context: forever in the slow lane, *South African Journal of International Affairs* **4**(1): 33–61.

Davies, R. (1996) Promoting regional integration in Southern Africa: prospects and problems from a Southern African prospective, *African Security Review* **5**(5): 27–38.

Dunlop, A. (1999) *What Future for Lomé's Commodity Protocols?* ECDPM Discussion Paper 5, Maastricht, ECDPM.

Ellis, S. (1994) Of elephants and men: politics and nature conservation in South Africa, *Journal of Southern African Studies* **20**(1): 53–70.

Ellis, S. (1996) *Africa Now: People, policies and institutions.* James Currey, London.

Ellis, S. and **MacGaffey, J.** (1996) Research on sub-Saharan Africa's unrecorded international trade: some methodological and conceptual problems, *African Studies Review* **39**(2): 19–41.

European Commission (1997a) *Green Paper on Relations between the European Union and the ACP Countries on the Eve of the 21st Century.* Office for Official Publications of the European Communities, Luxembourg.

European Commission (1997b) *The Lomé Trade Regime.* Commission of the European Communities, Europa server [europa.eu.int.en/comm/dg08/ps1/trade.htm].

European Commission (1997c) *Guidelines for the Negotiations of New Co-operation Agreements with the African, Caribbean and Pacific (ACP) Countries.*

Communication from the Commission to the Council and the European Parliament, ISSN 1012-2184, Office for Official Publications of the European Communities, Luxembourg.

European Commission (1999a) *An Analysis of Trends in the Lomé IV Trade Regime and the Consequences of Retaining it.* Europa server (DGVIII) [europa.eu.int (Document CE/TFN/GCE3/09-EN)].

European Commission (1999b) *Consequences for the ACP Countries of Applying the General System of Preferences (GSP).* Joint analysis by EU and ACP experts for negotiating Group 3, Economic and Trade Co-operation, Europa Server (DGVIII) [europa.eu.int (Document CE/TFN/GCE3/29-EU)].

Eurostat (1996) *Trade Statistics of the European Union.* Centre for Official Publications of the European Union, Luxembourg.

Financial Times (2001) South Africa mines spread their net to Jinaper gold Mieres, 7 February: 12.

Gibb, R.A. (1997) Regional integration in post-apartheid Southern Africa: the case of renegotiating the Southern African Customs Union, *Journal of Southern African Studies* **23**(1): 67–86.

Gibb, R.A. (1998) 'Europe in the world economy' in Pinder, D. (ed.) *The New Europe: Economy, society and environment.* Wiley, London.

Gibb, R.A. (2000) Post-Lomé: the European Union and the south, *Third World Quarterly* **21**(3): 437–81.

Goodison, P. (2002) *Implications of the Reform of the EU Sugar Regime for Southern African Countries, Part II: Reform of the EU Sugar Regime: Issues arising in EU–Southern Africa sugar relations.* European Research Office, Brussels.

Hanlon, J. (1986) *Beggar Your Neighbours: Apartheid power in Southern Africa.* James Currey, London.

Harvey, C. (1997) *The Role of Africa in the Global Economy: The contribution of regional co-operation, with particular reference to Southern Africa.* BIDPA Working Paper No. 11, Gaborone.

Heidensohn, K. (1995) *Europe and World Trade.* Pinter, London.

Kibble, S. (1998) *Drugs and Development in South Africa: How Europe could help.* CIIR, London.

Mail and Guardian [Johannesburg] (1995) September 1–7: 12–13.

Maliyamkono, T. and **Bagachwa, M.** (1990) *The Second Economy in Tanzania.* James Currey, London.

Mayer, M. and **Thomas, R.** (1997) Trade integration in the Southern Africa Development Community, in Kritzinger-van Niekerk, L. (ed.) *Towards Strengthening Multisectoral Linkages in SADC.* Development Bank of Southern Africa, Development Paper No. 33, Halfway House, Midrand.

McQueen, M. (1999) *The Impact Studies on the Effects of REPAs between the ACP and the EU.* ECDPM Discussion Paper No. 3, ECDPM, Maastricht.

Munslow, B. (1999) Angola: the politics of unsustainable development, *Third World Quarterly*, special issue on Corruption, **20**(3): 551–68.

Panos (1998) *Trading in Future EU–ACP Relations. Putting commerce before co-operation?* Panos Briefing Paper No. 31 [www.ips.org].

Parker, I. and **Mohamed, A.** (1983) *Ivory Crises.* Chatto and Windus, London.

SADC (1992) *Treaty of the Southern African Development Community.* SADC, Gaborone.

SADC (1996) *Protocol on Trade in SADC.* SADC, Gaborone.

SADC (1999) *Industry and Trade: Co-ordinated by Tanzania* [www.sadcreview.com].

SADC (2000) *Official SADC Trade, Industry and Investment Review*, millennium edition. SADC, Gaborone.

Simon, D. (2001) Trading spaces: imagining and positioning the 'new' South Africa within the regional and global economies, *International Affairs* **77**(2): 377–405.

Simon, D. and **Johnston, A.** (1999) *The Southern African Development Community: Regional integration in ferment.* RIIA, Southern African Study Group. Briefing Paper No. 8, London.

Solignac-Lecomte, H. (1998) *Renegotiating Lomé: Would ACP–EU free trade agreements be a stimulus for change?* European Centre for Development Policy Management, Maastricht.

Solignac-Lecomte, H. (1999) The impact of the EU–South Africa agreement on Lomé, paper presented to a conference on 'Assessing the EU–SA Agreement', South African Institute of International Affairs, 2–3 September.

Tekere, M. (2000) Cotonou Agreement – opportunities and challenges for EU–Southern Africa, paper presented at the ACTSA Conference (Action for Southern Africa), 13 October, London.

Treaty on European Union (1992) *The Unseen Treaty: The treaty on European Union.* Nelson and Pollard, Oxford.

UNDP (2002) Human Development Report. Oxford University Press, New York.

WTO (1995) *The Results of the Uruguay Round of Multilateral Trade Negotiations: The legal texts.* World Trade Organisation, Geneva.

WTO (1997) *Annual Report*, vols I and II. World Trade Organisation, Geneva.

CHAPTER 11

Regional urbanisation and urban livelihoods in the context of globalisation

Deborah Potts

Introduction

Urban growth patterns, rural–urban migration, the provision and adequacy of urban services and infrastructure, and the nature of urban employment and urban households' livelihoods have all experienced significant change in eastern and southern African urban settlement systems since the 'independence' and 'modernisation' decade of the 1960s. In some cases the changes have been dramatic: urban poverty is much worse and the difficulties of a risky urban existence have affected net in-migration rates to large centres; the urban economy has informalised and labour force participation rates within households have increased as people struggle to maintain necesssary expenditure on their most basic needs; the urban environment has deteriorated. Globalisation has had profound effects on urban centres and urban people, and the balance of influences has generally been negative. African cities have felt the chill of the winds of globalisation, but are essentially so marginal to the current global corporate agenda that they rarely participate in any associated economic development (see Simon 1992, 1997; Satterthwaite 2002; Jauch 2002; Beall 2002; Jenkins *et al.* 2002).

A key factor in the nature of urbanisation in the region (as anywhere in the world) has been the general trends in economic development within countries. Where and when national economies have declined and per capita incomes have fallen (see Chapters 5 and 10, this volume), urban economies have also faltered and suffered. While this essential relationship underpins any analysis of urbanisation in the region, other more nuanced factors which have been more variable in their geographical impact have also been of significance. One factor has been the trends in development theory and practice that have been adopted by (and often foisted on) countries in the region.[1] The vagaries of modernisation theory, of basic needs approaches and, from the 1980s, structural adjustment programmes (SAPs) and trade liberalisation have each in turn profoundly influenced urban growth patterns, urban service provision and the sectoral composition of urban economies with its vital corollaries: urban employment and income patterns (see Simon and Närman 1999, for a discussion of trends in development theory and aid/conditionality policies). The attitudes of development economists towards the role of urban centres in national economic development have been significant in this respect. In the early post-independence period, when modernisation theory held sway, there was broad acceptance that industrialisation held the key to improved productivity, and hence 'development', and that this entailed government intervention. The activities, employment and investment associated with this approach favoured urban centres. The neo-liberal economic conditionality of SAPs, on the other hand, contained an important assumption that urban bias in African economic development policies had been very damaging to national economies and recommended (and, if possible, imposed) allocative shifts in resources towards the rural sector (particularly exports) and against urban-based 'formal' production, jobs and investment. The impact of these policies, which were strongly influenced by market-oriented ideologies (see Chapter 3, this volume), on urban economies was profound: rapid de-industrialisation was effected, urban-based public and private sector retrenchments occurred, and private sector provision of services was promoted which often excluded the majority of the urban population (although in some areas, since public sector provision had become so inadequate, in effect this made little difference to the poor). Quite extraordinary adaptations were made by urban residents to cope with the effects of these impacts which often rendered previous livelihood strategies completely non-viable.

By the end of the 1990s, however, there was yet another change in development theorists' conceptualisations of the role of urban centres. Urban areas are yet again being projected as central to national economic transformation, albeit within a still mainly market-oriented policy framework (see, for example the World Bank's 1999 *World Development Report*) and urban poverty is no longer dismissed. As yet, however, there is little evidence that the sectoral trends of the last 20 years can be reversed in the cities of eastern and southern Africa.

These shifts in the urban policy environment over time provide the backdrop to this chapter on urbanisation in eastern and southern Africa. Urbanisation is too broad a topic to provide comprehensive coverage of all its facets. Here the focus is mainly on processes within cities, rather than debates about urban systems and hierarchies. First, the nature and composition of urban growth rates over time are analysed. Second, changes in the way urban people make their livings are examined. Finally, as an example of trends in urban service provision, the urban housing sector is discussed.

[1.] The following sections on the urban policy environment and urban population growth draw upon a chapter in D. Bryceson and D. Potts (eds) *African Urban Economies* (forthcoming).

Urban growth in eastern and southern Africa

Many cities and towns in the region are not growing nearly as fast as they did in the 1960s and 1970s. A number of reasons can be advanced to explain this. Some of them are basic characteristics of urbanisation processes, although they are seemingly often forgotten by regional urban analysts. One important factor is that a given number of net in-migrants to centres with small populations translates into high rates of annual population growth, but yields a much smaller growth rate if added to a much larger urban centre. Also, in countries with extremely low levels of urbanisation, the pool of potential rural–urban migrants in relation to current urban residents is huge: thus as urban levels increase (e.g. to over 50 per cent in today's South Africa), even if the propensity to migrate to town from rural areas remains the same, the rate of growth yielded by in-migration drops significantly. This means that as urban centres get bigger, and as the population becomes more urbanised, the rates of growth naturally tend to drop.[2] This may seem too obvious to state but it is believed to be worthwhile simply because there has been a degree of mythologising about the 'stubborness' of sub-Saharan African urban growth rates to refuse to 'abate'. In fact, in the absence of war and/or drought (sadly not absent enough in eastern and southern Africa), many of the reasonably reliable data which are available (which are far too few) indicate that a downward trend in urban growth is already established (Potts 1995, 1997).

This trend in the region is not, however, only due to statistical inevitability. Many countries still have very low *levels* of urbanisation (see Table 11.1) and many centres are still small, so there remains significant scope for very rapid urbanisation; indeed by world standards only Kinshasa and the larger towns of South Africa are 'large cities'. Another set of factors is related to significant changes in the rate and nature of in-migration (see Box 11.1): essentially falling rates of in-migration combined with increased rates of return migration (see Potts 1995). These have reflected urban economic decline and have also led to reductions in urban growth rates, in certain cases quite dramatically.

Urban growth has two demographic components – net in-migration and natural increase (NI). Through the 1960s, 1970s and 1980s a rule-of-thumb annual rate of NI for most of the countries in the region (except South Africa) was 3 per cent, or even higher (see Chapter 2). Although fertility was generally lower in urban areas, so was mortality, and by the 1980s the sex ratios in urban populations in much of eastern Africa (with the notable exception of Nairobi) were roughly balanced. Since age groups who were at their most fertile also tended to be over-represented in African urban populations, due to selective migration patterns, lower fertility rates need not translate directly into lower crude birth rates compared to rural areas. Thus NI was high in urban areas and, as base populations rose, steadily increased its contribution to annual growth. By the 1980s,

---

[2] An obvious illustration of this is the hypothetical situation where the level of urbanisation has reached 100 per cent. Annual urban growth from then on (excluding international migration) will equal natural increase only.

Table 11.1 Urbanisation level (%), 1960s–2000s, southern and eastern Africa[a]

Country	1960s	1970s	1980s	1990s	2000s
Angola	11[60]	14[70]			*60–70*[01b]
Botswana		9[71]	18[81]	46[91] (24%)	52[03c]
Burundi		4[79]			
DRC		22[70]	32[84]		
Kenya	8[62]	15[79]	18[89]		
Lesotho	5[66]	11[76]	14[86]		
Malawi	5[66]	8[77]	11[87]	14[98]	
Mozambique			13[80]	29[97]	
Namibia		25[70]		28[91]	
Rwanda		5[78]			
South Africa	47[60]			54[96]	
Swaziland	11[66]	15[76]	23[86]	23[97]	
Tanzania		13[78]	18[88]		
Uganda	8[69]		9[80]	11[91]	
Zambia	21[63]	29[69]	40[80]	42[90]	
Zimbabwe	19[69]		26[82]	31[92]	

[a] Data derived from census data, date indicated by superscript, unless otherwise noted. Estimates in italics.
[b] Estimate given by Ann Condy, DfID Luanda Urban Poverty Programme, at Chatham House British-Angola Forum seminar, 14.6.01.
[c] Estimate from Botswana census projections (UN projections for 1995 were 60%!). It should be noted that the apparent, sudden leap in urbanisation level in Botswana in the 1990s is largely definitional. The country has historically been characterised by nucleated settlements which were essentially agricultural in economic (if not political) terms. Many of these large agro-villages had grown to a size sufficient to be included in the 'urban' population by the 1990s (see Table 11.2). However, functionally their status remains ambiguous: using a *functional* definition for urbanisation, the census reported an alternative level in 1991 of only 24% (as shown in brackets).

therefore, NI made up about half of the annual growth in many cities' populations in eastern, and parts of southern, Africa. Thus the share of *net* in-migration, which had been the primary generator of urban growth in the early post-independence era, had fallen even before, as outlined above, net in-migration itself began to decline. There were important exceptions, however: for example, urban settlements where populations were swelled by rural people fleeing war (e.g. in Mozambique, Angola and Rhodesia in the 1970s), or centres whose development was particularly fostered by government policies. Examples of the latter would include new capital cities in Botswana (Gaborone) and Malawi (Lilongwe). The new mining towns of Botswana (e.g. Selibe-Phikwe, Orapa) also burgeoned in this period. Urban patterns in the white minority regimes had their own peculiar features in this era which are explored in a separate section below.

The impact of NI on eastern and southern African urbanisation will have changed again in the 1990s, and earlier in some cases, as national rates of NI have begun to decline. Fertility decline is now established

Box 11.1   Declining net in-migration to cities in conditions of economic decline

There is evidence from some African census data that urbanisation in the 1980s and 1990s had not been occurring as rapidly as was assumed. Table 11.2 shows, for example, that a number of urban centres in Uganda and Zambia were growing at rates not much greater than, or even less than, the national rate of population growth over certain periods.

In Uganda the period of much slower than expected urban growth occurred in the 1970s, when political disruption in that country was at its height. Here Mbale, Tororo and Entebbe appear to have experienced net outmigration, and Jinja even registered an absolute fall in its population. Even the capital city in Uganda, Kampala, grew at a rate not much above that of the national population (3.2 per cent compared to 2.8 per cent, suggesting net in-migration accounted for only about one-eighth of its growth). Urban growth in Uganda picked up again in the 1980s, as political stability was restored. However, the impact of SAPs in that country may have reduced the growth from net in-migration again (although it should be noted that this is purely speculative as, at the time of writing, no data for urban growth in the 1990s were available). The impact of AIDS will undoubtedly have had a dramatic effect on the contribution of natural increase within the city to its growth during the 1990s.

In Zambia the evidence for very restricted urban expansion, and of net outmigration, is very clear. During both the 1970s and 1980s it is apparent that the Copperbelt towns, which had experienced significant employment-led growth in the 1960s, were no longer attracting and keeping migrants: Kitwe, Chingola, Mufulira and Luanshya all grew at rates lower than natural increase in the 1970s, and by the 1980s Kabwe could be added to the list. Of the mining towns, only Ndola was experiencing net in-migration by the 1980s. Lusaka, the capital city, on the other hand, maintained a rapid growth rate through the 1970s and 1980s, with in-migration accounting for about one-half of population increase since the end of the 1960s (and much more than that in the immediate post-independence period). In part this may have reflected in-migration from the Copperbelt towns. Overall in Zambia the level of urbanisation had reached 40 per cent in 1980, but increased to only 42 per cent in 1990 (despite the fact that many authorities assumed that Zambia's population was 50 per cent urban by the 1980s). Preliminary data from the 2000 census indicate that even Lusaka's growth slowed to less than the rate of national natural increase during the 1990s and that the Copperbelt towns continued to stagnate or decline. By the end of the 1990s deaths from AIDS will certainly have been contributing to this significant change. Thus over a period of 40 years Zambia has been transformed from possibly the fastest urbanising country in the region, to one where urbanisation (in the narrow demographic sense of an upwards trend in the urban share of the national population) is now stagnant.

The evidence for increasing rates of return migration in Zambia which the census data indicated has been fleshed out by a number of qualitative studies. In 1988 Johan Pottier published an anthropological study of Mambwe villages in Zambia, tellingly titled *Migrants No More*. He found that by the late 1970s, the economic downturn in the Zambian Copperbelt was leading to former migrants returning to their areas of origin. However, as his research in the small town of Mbala (the former Abercorn) showed, although many of these returnees expressed an affection for their rural village 'homes', they did not necessarily feel they had to return to rural life – it was enough to be near to those 'homes'. Indeed many of them did not want to return to a village life-style, and were concerned that they would not be 'too welcome in the villages they left as young labourers' (Pottier 1988: 43). Their return migration stream was thus increasing the growth rate of the small town relative to the large Copperbelt centres. Pottier's findings that some return migrants who had spent a long time in large urban centres were reluctant or even fearful to participate in village life and agricultural production were confirmed by Ferguson's more recent study of the Copperbelt and the plight of people forced out of urban livelihoods and into rural ones through lack of economic alternatives. Other recent research on the Copperbelt has also shown that many former urban residents have moved out of the cities, driven by economic circumstances. For all the reasons cited above, and to stay near to urban services, they are often settling in the forest reserves near the Copperbelt towns and supporting themselves first by charcoal production and then by agriculture.

Growth rates at the upper end of Tanzania's urban hierarchy during the last intercensal period are also worthy of note. First, the capital, Dar es Salaam, grew in the 1980s at only half the extraordinary rate it achieved in the late 1960s and 1970s, although this nevertheless remained quite rapid (4.8 per cent per year). One official projection had been that its population would be over 1.9 million by 1990, but it was only 1.2 million at the 1988 census. Second, some of the other major urban areas (mostly district capitals) grew faster than Dar, and some more slowly (including two, Tanga and Tabora, which grew not much faster than the rate of natural increase). Third, the *national* urban growth rate, at 10.7 per cent, is incompatible with the growth recorded in the main centres due to definitional issues in the smaller centres (see Box 11.2).

When large African cities prove to be growing more slowly than projected, there is sometimes some disbelief and it also often takes many years for the 'new' figures to be realised by analysts (and to filter through to UN and World Bank tables). For example in 1992 it was reported for Dar that 'in recent years' it had been growing at over 9 per cent (the rate recorded in the 1960s and 1970s) and that this 'growth has perplexed policy makers because it is happening even as data show faster deterioration in living conditions in the city than the national average' (Mtatifikolo 1992: 230). As noted, in fact, the growth rate had halved by the 1980s. Other major centres which have been found to be growing more slowly than 'expected' include Maputo, Bulawayo, Harare, Nairobi and Mombasa. In Harare and Nairobi's cases their growth has still incorporated significant net in-migration (although natural increase has been more important), but this has not been *as* high as the authorities' projections assumed.

Sources: CSO (2002), Ferguson (1999), Kulaba (1989), Mtatifikolo (1992), Palmer (2001), Pottier (1988), Potts (1995).

in the southern part of the region (i.e. South Africa, Namibia, Lesotho, Botswana, Swaziland, Zimbabwe) as well as Kenya, and there is evidence of small falls in most of the other countries (see Chapter 2; Potts and Marks 2001). Again these are most marked in urban areas. By the latter part of the 1990s, very significant increases in mortality rates due to AIDS in almost all the countries under study have made further inroads into NI rates. In the original regional 'core' area of AIDS – Uganda, Rwanda, eastern Zaire, north-western Tanzania – heightened mortality rates would have had their impact somewhat earlier (Barnett and Blaikie 1992). Urban areas will have been particularly affected as they generally have higher rates of HIV-positive people (although it is very common for urban people to go to rural areas in the latter stages of AIDS, to be nursed by rural relatives). In Zimbabwe by mid-2001, according to some estimates, national NI was zero, essentially due to the impact of AIDS (there remained a small excess of births over deaths but this was countered by emigration). If these estimates were correct it must mean that the demographic motor of NI in the main urban areas there had gone into reverse (although registrations of urban births and deaths might fail to depict the shift given the predilection for urban residents' deaths from AIDS to occur in rural areas).

There are therefore four sets of factors operating to reduce urban growth rates in eastern and southern Africa in the last 20 years or so (in the absence of war or other calamitous events). First, an inevitable statistical decline as cities and urbanisation levels grow; second, a fall in net in-migration rates accompanying economic decline and structural adjustment; third, a fall in fertility rates; and, fourth, a rise in mortality rates. The latter two demographic factors will affect the rate at which the urbanisation *level* rises only if their impact is differentially experienced between rural and urban areas – but as noted already, there are good reasons to expect this now to be the case and that urban NI rates in the 1990s have fallen below those in rural areas.

What is the combined impact of these factors on the growth of individual cities in the region? In most cases it is not yet possible to be definitive because reliable, recent data on urban populations are unavailable. Published data by international organisations, although often cited, are of little use for the 1990s unless based on national census data; usually they are simply projections based on outdated assumptions, and there is good reason to believe that for many countries and cities the figures are misleading (Potts 1995; Sattherthwaite 1996). In most cases, the impact of NI changes will only have been significant towards the end of the 1990s, so even were reliable data available to compare city size in, say 1988, and 1998, the average annual rate of growth might be masking quite major changes over the decade. A compilation of data on urban growth rates from the 1960s for every country in the region is found in Table 11.2. Where possible, census data have been used, supplemented by occasional estimates. The table shows there have been some very significant shifts in average annual growth rates for many major centres, with the trend downwards. Intermediate and smaller towns are, however, often experiencing strong growth (in part because of their lower base). In several cases growth rates for individual centres have been recorded over specific

Table 11.2 Urban and national populations and annual average growth rates, 1960s–2000s

Country/urban centre	1960s		1970s		1980s		1990s		2000s
	Population ('000s)	*AAGR % 1960s*	*Population ('000s)*	*AAGR % 1970s*	*Population ('000s)*	*AAGR % 1980s*	*Population ('000s)*	*AAGR % 1990s*	*Population ('000s)*
Angola	4,830^{60}	0.84$^{60–70}$	5,250^{70}	7.1$^{70–80}$	940^{80}[2]	7.8$^{80–90}$	2,000^{90}	2.7$^{70–01}$	12,000^{01}
Luanda	225[1]	7.8	475^{70}	12.9$^{70–88}$	800^{88}[3]	–		3.8–6.5$^{90–01}$	3,000–4,000
Huambo			90						
Lobito			88						
Botswana	515^{64}	1.8$^{64–71}$	585^{71}	4.8$^{71–81}$	941^{81}	3.5$^{81–91}$	1,327^{91}		
Total urban	21	8.2	54	10.7	166	18	606		
Gaborone	4	25	18	12.5	60	8.4	134		
Francistown	10	11.1	19	4.7	30	7.7	65		
Lobatse	8	7.8	13	4.0	19	3.2	26		
Selibe-Phikwe		–	5	20	30	3.0	40		
Serowe							30		
Kanye							31		
Molepolole							37		
Burundi	3,210^{65}	1.6	4,028^{79}		NC	2.6$^{79–90}$	5,356^{90}		
Bujumbura							235		
DRC	12,734^{55}	3.6$^{55–70}$	21,638^{70}	2.3$^{70–84}$	29,671^{84}	See 1970s			
Total urban					9,495				
Kinshasa	400^{60}[4]	10.6$^{59–67}$[5]	1,680^{75}	4.9$^{67–76}$	2,664	5.4$^{76–84}$			
Kenya	10,943^{69}	3.4$^{62–69}$	16,141^{79}	4.0$^{69–79}$	21,444^{89}	3.4$^{79–89}$	28,687^{99}	3.0$^{89–99}$	
Total urban	1,053	6.6	2,300	8.1	3,700	4.9	–	–	
Nairobi	509	9.7	828	5.0	1,325	4.8	2,143	4.9$^{89–99}$	
Mombasa	247	4.7	341	3.3	462	3.1	665	3.7	
Nakuru	47	3.1	93	7.1	164	5.8	231	3.5	
Eldoret	18	-1.1	51	10.7	112	8.2	197	5.8	
Kisumu	32	4.7	153	16.9	193	2.4	323	5.3	
Machakos	6	5.5	84	30.0	116	3.3	143	2.1	
Meru	4	4.4	72	33.5	95	2.8	126	2.9	
Nyeri	10	3.5	36	13.7	91	9.8	101	1.0	
Thika	18	4.0	41	8.9	58	3.4	107	6.4	

Table 11.2 (cont'd)

Country/urban centre	Population ('000s) 1960s	AAGR % 1960s	Population ('000s) 1970s	AAGR % 1970s	Population ('000s) 1980s	AAGR % 1980s	Population ('000s) 1990s	AAGR % 1990s	Population ('000s) 2000s
Lesotho	852[66]	(see 1970s)	1,217[76]	3.6[66-76]	1,605[86]	2.6[76-86]	NC		
Total urban	72		136	6.4	222	4.9			
Maseru	28		65	6.6	109	6.9			
Teyateyaneng	7		9	2.5	13	4.1			
Mafeteng	6		8	3.7	12	3.9			
Malawi	4,040[66]	(see 1970s)	5,547[77]	2.9[66-77]	7,989[87]	3.7[77-87]	9,934[98]	2.0[87-98]	
Blantyre	109		219	6.3	333	4.2	480	3.3	
Lilongwe	19		99	15.8	223	8.5	435	6.1	
Mzuzu	8		16	5.8	44	10.6	87	6.2	
Zomba	20		24	1.9	43	5.8	64	3.6	
Total urban	193		471	8.4	850	6.1	1,400	4.7	
Mozambique	6,604[60]	2.1[60-70]	8,169[70]	2.7[70-80]	12,130[80]	1.7[80-97]	16,075[97]	see 1980s	
Total urban			–		1,539	6.6	4,594		
Maputo			355	7.8	755	1.6	989		
Matola			–		–		441		
Beira			114	7.3	231	3.5	413		
Nampula			126	2.1	156	4.2	315		
Chimoio			–		74	5.3	178		
Quelimane			72	-1.5	62	5.5	153		
Nacala			–		80	4.3	164		
Tete			53	-1.2	47	4.8	105		
Xai-Xai			64	-3.5	45	5.0	103		
Lichinga			37	1.0	41	4.7	89		
Pemba			–		43	4.3	88		
Inhambane			27	7.4	55	-0.1	54		

	1960/66	1960–70 (%)	1970/76	1970s (%)	1980s	1980s–91 (%)	1991/97	1990s (%)	1995/96
Namibia			746[70]	3.0[70–81]	1,031[81]	3.1[81–91]	1,402[91]	5.4[91–95]	
Total urban				4.1	268	3.6	383		
Windhoek					96	4.4	147		182[95][6]
Walvis Bay							30		
Oshakati							22		
Rehoboth							21		
Rwanda			3,573[70]	3.8[70–78]	4,832[78]	3.1[78–91]	7,149[91]		
Kigali			–		116		–		
South Africa[7]	15,988[60]	3.1[60–70]	21,794[70]	2.8[70–85]	33,121[85][8]			1.9[85–96]	40,584[96]
Total urban									21,782
Cape Town	MA[9] 941	3.2	MA 1,290	2.9[70–80]	MA 1,712[80]	2.8[80–91]	MA 2,327[91]	3.6[91–96]	MA 2,773
Durban	MA 714	4.3	MA 1,091	2.6[70–80]	MA 1,412[80]	2.4[80–91]	MA 1,836	3.2	MA 2,149
Johannesburg	MA 2,830	3.0	MA 3,788	2.7[70–80]	MA 1,609[80]	2.3[80–91]	MA 6,343	2.8	MA 7,272
Pretoria			563		a 823[85]	4.6[85–91]	a 1,080	0.4	a 1,104
Soweto					522[85]	2.3[85–91]	597	–	a 1,098
Port Elizabeth	MA 355	3.1	MA 481	3.0	MA 649[80]	2.4[80–91]	MA 840	2.7	MA 958
Pietermaritzburg					a 192[85]	3.0[85–91]	a 229	10.5	a 378
Vereeniging					a 541	6.2[85–91]	a 774	-14.8	a 347
Bloemfontein	148	2.1	182		a 233	7.0[85–91]	a 350	-0.9	a 334
Botshabelo					96	3.5[85–91]	118	–	a 178
East London			125		a 194	5.7[85–91]	a 270	-4.7	a 212
Mdantsane							–	–	a 183
Newcastle					a 153	5.7[85–91]	a 213	0.6	a 220
Welkom					a 320	5.0[85–91]	a 428	-13.9	a 203
West Rand			428		a 647	5.1[85–91]	a 870	–	–
Swaziland	375[66]	2.8[66–76]	495[76]	3.3[76–86]	681[86]	See 1990s	930[97]	2.9[86–97]	
Mbabane	14	5.3	23	5.0	38		–[10]	–	
Manzini	16	6.0	29	4.7	46		–	–	
Total urban	40	9.7	101	4.3	155		214	3.0	

Table 11.2 (cont'd)

Country/urban centre	Population ('000s) 1960s	AAGR % 1960s	Population ('000s) 1970s	AAGR % 1970s	Population ('000s) 1980s	AAGR % 1980s	Population ('000s) 1990s	AAGR % 1990s	Population ('000s) 2000s
Tanzania									
Total urban	12,313[67]	See 1970s	17,513[78]	3.3[67-78]	23,174[88]	2.8[78-88]	NC		
—			–	11.4		10.7			
Dar es Salaam	273		757	9.7	1,205[11]	4.8			
Mwanza	35		111	11.1	172	4.5			
Tanga	61		103	4.9	137	2.9			
Mbeya	12		77	17.9	131	5.5			
Morogoro	25		62	8.5	118	6.6			
Arusha	32		55	5.0	103	6.4			
Moshi	27		52	6.2	97	6.3			
Tabora	21		67	11.2	93	3.2			
Dodoma	24		46	6.2	83	6.2			
Iringa	22		57	9.2	85	4.0			
Mtwara	20		49	8.2	77	4.7			
Kigoma	21		50	8.0	74	4.0			
Uganda									
Total urban	9,535[69]	NC	NC	2.8[69-80]	12,636[80]	2.5[80-91]	16,672[91]		
—	747			3.6	1,100	5.0	1,890		
Kampala	330			3.0	459	4.4	774		
Jinja	48			-0.6	45	3.4	65		
Mbale	24			1.6	28	6.1	54		
Masaka	13			7.6	29	5.0	50		
Entebbe	21			0.0	21	6.6	43		
Zambia									
Total urban	3,490[63]	2.5[63-69]	4,057[69][12]	3.0[69-80]	5,662[80]	3.2[80-90]	7,383[90]	3.4[90-00]	10,286[00][13]
—	715	8.9	1,192	5.8	2,259	3.7	2,905		
Lusaka	121	13.8	262	6.5	536	6.1	982	2.8	1,289[00][14]
Ndola	93	9.5	160	4.0	251	4.0	376		
Kitwe	123	8.4	200	2.6	266	2.4	338		
Chingola	60	9.6	103	2.1	131	2.5	168		
Kabwe	46	6.3	66	6.6	136	2.0	167		
Mufulira	81	5.0	108	2.1	135	1.2	153		
Luanshya	75	4.2	96	1.3	111	2.8	146		

Zimbabwe	$3,857^{62}$	4.1^{62-69}	$5,099^{69}$	3.1^{69-82}	$7,546^{82}$	3.3^{82-92}	$10,412^{92}$
Total urban	802	3.1	990	5.4	1,962	5.0	3,188
Harare	310	2.3	364	4.6	656	6.2	1,189
Bulawayo	210	1.7	236	4.4	414	4.1	621
Chitungwiza	0	–	15	20.7	173	4.7	275
Gweru	38	2.6	46	4.2	79	4.7	128
Mutare	43	-0.2	42	4.0	70	6.5	131
KweKwe	21	6.0	31	3.3	48	4.6	75
Kadoma	19	3.8	25	4.6	45	4.1	68
Hwange	20	-0.1	20	5.2	39	0.8	42
Masvingo	10	1.3	11	7.9	31	5.3	52
Chinhoyi	8	7.9	13	4.7	24	6.0	43
Marondera	7	6.5	11	4.9	20	7.2	40

Source: Population figures derived from census data unless otherwise stated.

Notes: [a] Estimates in italics; census or estimate date and AAGR period shown in superscript; country figures in same column from same source unless otherwise indicated; NC = no census in that decade.
[b] – no data or growth rate uncalculable as urban populations not comparable.

1 Johnson (1970).
2 This estimate and those for 1990s and 2001 given at 'Rebuilding civil society in Angola', Royal Commonwealth Society meeting at Chatham House, 24.5.01.
3 Estimate from *Africa Confidential* (1997).
4 Estimate from Piermay (1997).
5 All AAGRs for Kinshasa are estimates taken from Piermay (1997).
6 1995 household survey, Windhoek, reported in UNCHS city profiles.
7 Data for urban settlements depicted 'a' are for specific urban agglomerations (not equivalent to Metropolitan Areas which are much larger and may combine such agglomerations) and are taken from Brinkhoff (2002).
8 National population for 1985 here *includes* the four homelands normally excluded from apartheid South Africa's census statistics after the date of their so-called 'independence' and has been adjusted for a significant undercount (van Zyl 1988). The 1991 census excluded these areas and its estimate of the national population is thus not shown here.
9 Figures denoted MA are for Metropolitan Areas and are derived from Crankshaw and Parnell (2002).
10 Swaziland's 1997 census report does not disaggregate urban populations by specific settlement.
11 The urban populations for 1988 for Tanzania are derived from the urban district data in the final 1988 regional census tables. These are broken down into 'urban' and 'rural' populations and the rural populations have been excluded from the urban figures provided here.
12 1969 census data have been used as a surrogate for 1970.
13 Data from CSO (2002).
14 Estimated by author from provincial data in preliminary census report. Copperbelt Province AAGR for 1990–2000 calculated as negligible (<0.1%).

periods which appear aberrant and can only be understood with reference to specific local conditions and policies – examples would include, in Malawi, Lilongwe's exceptional growth in the 1970s and, in Botswana, Gaborone's growth at the same time: in both cases, as discussed earlier, the cause was designation as a new capital city (Potts 1985a, b). Other individual circumstances are explained in the table footnotes which are central to interpretation of the table.

Another fundamental factor in urban growth rates is the redrawing of urban boundaries as settlements grow and extend beyond their original 'space'. Where this incorporates reasonably fairly the essentially 'urban' population in *functional* terms, it need not affect urban analysis. However when the population *enumerated* as urban, due to such redesignations, suddenly does not tally with the functionally urban population (i.e. because the boundaries have been cast far too wide), the possibility exists for fundamental misunderstandings about urban growth in that town or, worse, in the nation as a whole. This is of significance when analysing reported growth rates in Tanzania and Kenya in the 1980s and 1990s, for example, whose cases are examined in Box 11.2.

Urban growth under influx controls and war conditions: the legacies of white minority rule in southern Africa

The broad chronology of growth trends outlined above does not apply well to the urban centres of those countries in the southern part of the region which were affected by various types of influx control, imposed by white minority regimes in an effort to maintain political supremacy and control flows of black labour. These controls were formally removed in 1980 in Zimbabwe, in 1977–78 in Namibia and in 1986 in South Africa, but in the 1960s and 1970s their impact was profound and these countries did not experience the high rates of urbanisation typical of the region in those decades. One effect was to skew African urban sex ratios in favour of men which constrained the role of natural increase as a generator of urban growth[3] in these countries compared to those further north. Influx controls also succeeded for some time in maintaining a net in-migration rate well below that which would have occurred in their absence (see Box 11.3) – indeed in direct contrast to the other countries in the region the African population of South Africa did not 'urbanise' (i.e. the urban share of the population did not increase) in the 1960s and 1970s and neither did the Rhodesian African population in the 1960s. On the other hand, certain policies encouraged very specific types of rapid 'urbanisation' at these times. For example, some of the 'border' settlements in South Africa to which housing and industrial developments were deliberately diverted in order to foster apartheid, and the notorious 'rural slums', experienced rapid growth (see Box 11.3).

[3.] Although in all three countries much looser controls in earlier decades, particularly on women's migration, had already led to the establishment of a relatively permanent, family-based African labour force in cities which had its own natural growth impetus.

Box 11.2   Urban boundaries and misleading population growth rates

The urban growth rates recorded by the various censuses in Kenya and Tanzania (see Table 11.2) exhibit some curious features, with certain centres (or whole swathes of the urban hierarchy) seeming to grow suddenly and unexpectedly. In most cases these features are explained by significant, and functionally confusing, boundary changes whereby many rural dwellers have been enumerated as urban.

The 1967–78 growth rates for Mbeya and Mwanza in Tanzania, and Machakos and Meru in Kenya, for example, illustrate this problem. If only a limited number of intermediate and small centres are thus affected this should not have much impact on the national urban growth rate for that period and, as can be seen in the case of Machakos and Meru, in the next intercensal period the growth rates may be much lower than might be expected because the base figure was overinflated.

However, if boundary changes affect a significant proportion of national urban centres, the distortions can be more serious. This situation pertained in Tanzania in the 1980s and Kenya in the 1990s. Tanzania's national urban growth rate of 10.7 per cent per year for the last intercensal period does not tally with the growth experienced in Dar es Salaam and the other main urban centres at all. This either implied spectacular growth at the lower end of the urban hierarchy, or definitional issues, or both. In fact very many of the enumerated urban subdistricts, at the lower end of the hierarchy in particular, contained rural areas, thus greatly inflating their population. An attempt to adjust the data for this factor in a sample of regions came up with an average annual growth rate for small towns of 5,000–10,000 people of 6.4 per cent, and for the next level in the hierarchy of 10,000–35,000 people of 7.0 per cent. This suggests that urban growth *was* more vigorous in these smaller centres than the larger ones, but not at the spectacular rates necessary to yield a national rate of over 10 per cent. It is clearly also plausible that villagisation under *ujamaa* had created a whole tranche of settlements that became definitionally 'urban' through their size alone in this intercensal period. However, it seems safe to say that the 10 per cent national rate of urban growth for the 1980s is an exaggeration.

Even greater distortions are evident in Kenya's 1999 census, which also appeared to suggest that its urban population had grown at over 10 per cent per year. Here a very large number of new areas (43 in total), usually further down the urban hierarchy than the main centres shown in Table 11.2, was suddenly incorporated into the 'urban' population (i.e. centres over 20,000). However, several of these are really densely settled regions which may include a number of small towns or even villages, and many rural people. For example, Tala/Kangundo in Eastern Province appears in 1999 as a town council with 180,000 urban residents, Vihiga in Western Province is suddenly listed as a municipality with 110,000 people, and Bomet, in Rift Valley, appears as a town council with 84,000. None figured in any previous census. Several other centres also grew to implausibly high numbers from tiny settlements according to the 1989 census. The listing of such regions as one urban 'settlement' apparently stems from their having one local government to serve them. Vihiga, for example, is basically a region.

Therefore, any UN or World Bank publications which use these data to calculate Kenya's total urban population in 1998 and its intercensal growth rate will be giving a totally false picture of the country's uban dynamics in the 1990s.

Based on: Holm (1992), J. Nealis (pers. comm. 19.9.2001), United Republic of Tanzania (1988).

Box 11.3 The impact of influx controls on urban growth rates and settlement types in Rhodesia and South Africa

Black (African) urban population of South Africa: apartheid era ('000s)

Year	Total African urban population	% of total African population	Number in 'white' South Africa	% African population in 'white' urban South Africa	Number in homelands' urban areas (a)
1946	1,689	24.3	1,689	**24.3**	–
1960	3,471	31.8	3,460	**31.7**	11
1970	5,070	33.0	4,475	**29.1**	595
1980	6,870	33.0	5,324	**25.6**	1,546

Note: [a] These figures *include* the African population in urban settlements in the so-called independent states (Transkei, Venda, Bophuthatswana, Ciskei). Most estimates for 1980 tend to exclude these people.

Urbanization level of black (African) population of Rhodesia: 1961–77 (% living in towns over 10,000)

Year	% Africans urbanised
1961–62	14.5
1969	13.9
1977	15.8

The tables above illustrate the extraordinary effectiveness of racist apartheid policies aimed at restricting in-migration to the major urban centres in national space designated as 'white' in white minority South Africa and Rhodesia. In South Africa, as influx controls and associated policies pushing African urban housing and employment out of the metropolitan areas became increasingly draconian, the African population designated as urban in successive censuses stagnated between 1960 and 1980. The proportion of the African population living in 'white' urban South Africa actually fell so that by 1980 it was little more than it had been just after the war. In Rhodesia, African urbanisation occurred entirely within 'white' space, and here it also stagnated from 1960 until the mid- to late 1970s.

These figures are all the more startling when compared with processes of urbanisation in the rest of southern and East Africa over the same periods. Elsewhere, this was when the most rapid urban growth rates were recorded, and urbanisation levels were rising rapidly. In the absence of apartheid policies there is no doubt that the rate of urbanisation of the African population in 'white' South Africa and Rhodesia would have been positive and high – thus apartheid did succeed in bucking the trend and in partially achieving one of its objectives – to discourage African urbanisation outside of the homelands (although numerically the African urban population in 'white' space continued to increase).

However, the economic forces within South Africa which promoted outmigration from rural areas could not be denied. A host of forces including dire poverty in the homelands, lack of investment in African agriculture, land shortages, evictions from white farms as mechanisation progressed, and apart-heid removals of so-called 'surplus people' or 'black spots' (inconveniently located homeland fragments or other African farming areas) from 'white' space meant there were millions of people who, in normal circumstances, would have headed for employment opportunities in the big towns. Apartheid decentralisation and housing policies succeeded in redirecting some of these to residential towns within the homelands, and work across the 'border' in border industry areas and nearby 'white' urban settlements or in growth points within the homelands. This accounts for the very rapid increase in the homelands' 'de-signated' urban population shown in the table. However, in addition to those enumerated as urban by the censuses, there were many more who ended up (forcibly or voluntarily) in what Colin Murray termed 'rural slums': large, sprawling settlements at quasi-urban densities but without normal urban economic functions. Residents largely had to rely on commuting, migrant remittances, pensions and charity. If these people are included as part of the African *urban* population, the urbanisation level by the 1980s increases signifi-cantly over the figures given in the table above. One estimate made by Charles Simkins was that 8.6 million Africans were 'urbanised' in 1980, of whom 3 million were in the homelands, resulting in an African urbanisation level of 41.6 per cent; de V Graaff argued for an even higher estimate of 10.7 million (about 4 million in the homelands) and an urbanisation level of 51.8 per cent.

Whether incorporating such settlements into urban figures when analysing South African urban processes during the 1960s and 1970s is helpful is diffi-cult to say, for it disguises the efficacy of apartheid in controlling functional African urbanisation in 'white' South Africa (which was its intention), and it is debatable whether anything that is not 'rural' must always therefore be 'urban'. By the twenty-first century, however, many of these settlements, some of which had hundreds of thousands of residents, had inevitably developed their own dynamics and, although fairly stagnant in population terms, are part of the difficult legacy of apartheid faced by the new South Africa.

Sources: de V Graaff (1987), Murray (1987), Potts and Mutambirwa (1990), RSA census data, Simkins (1981).

In Namibia, settlements in the northern part of the country where the African population is most concentrated also grew rapidly in the 1970s and 1980s because of the insecurity created by the military activities of South African forces and the liberation movement, SWAPO. Similarly, towards the end of the 1970s in Rhodesia urban growth was swollen by displaced people escaping the effects of the liberation war there (although when this ended in 1980 it is clear that most of the displaced people returned to rural areas). The effect on urbanisation of war has been even more profound in Mozambique and Angola. In both countries liberation wars against the Portuguese, ending with independence in 1975, were followed by wars against anti-government, apartheid-backed, forces. The civil wars have waxed and waned: Mozambique's ended in 1992 although security in rural areas remained compromised for some years, but in Angola the war continued with only brief respites right through until 2002. Millions of people have been internally displaced in both countries and a significant proportion ended up in the relative security of the towns, especially Maputo and Luanda.

In further contrast to the trends in the countries further north where, by the 1980s, net in-migration rates were often reducing under the economic conditions of structural adjustment, in South Africa and Zimbabwe this decade witnessed heightened rural–urban migration as institutionalised controls were abolished. In terms of economic geography, these migrant flows can be conceptualised in terms of some inevitable rebalancing in the distribution of the population from extremely poor, under-resourced rural areas to much wealthier urban areas where there were many opportunities to make a living, even if there were not enough 'formal jobs'. In the absence of influx controls these flows would have been spread more evenly over time, as they were in the rest of the region, rather than being concentrated in the 1980s. The evidence of urban growth in South Africa at this time was highly visible as massive informal settlements developed in and around the major metropolitan areas.

Census enumeration in the cities in South Africa in the 1980s was fraught with difficulties as there was so much political resistance to the state, and so many good reasons to avoid enumeration. Accurate assessments of growth rates and the size of metropolitan areas, individual townships and informal settlements were therefore hard, although estimates abounded. It appears that some of these estimates were too generous. There was often political capital to be made (on both sides) from exaggerating the size of the 'problem', and city authorities may err towards high estimates in the hope of capturing more resources from the state or donors. A specific survey of Durban in 1995, for example, found that its population had been significantly overestimated; after careful surveys the metropolitan area was estimated at about 2.3 million (Urban Strategy Department 1995), compared to a 1990 estimate by the local Built Environment Support Group of 3.5 million.

South Africa's first post-apartheid census in 1996 was therefore an important marker, theoretically making it possible to assess urbanisation trends during the dying years of apartheid. It found that the level of urbanisation was somewhat higher than anticipated, at 54 per cent, making South Africa one of the few nations in the region with more than half its people in urban areas, and the only one where that level is related meaningfully to its economic development patterns. However, now that the urban situation in the country has had time to normalise, further surges of in-migration are unlikely. Given South Africa's low levels of fertility

(2.9 children per woman in 1998), and its tragically high HIV rate, urban NI rates are bound to be low. Overall, then, urban growth rates in South Africa are now likely to be very considerably lower than they were in the 1980s. Indeed, the Greater Johannesburg Metropolitan Area was estimated to be growing in 2001 at only 0.7 per cent per year, more slowly than the national population (Parnell and Beall 2001). Whether this represents deurbanisation is hard to judge as there are difficult boundary issues involved, but the 1996 census suggests that the white population had reduced by about a quarter since 1991 (although underenumeration in 1996 may also have been a factor). Some of the former 'rural slums', a unique feature of South Africa's settlement landscape (see Box 11.2), are declining or stagnating. For example, Botshabelo in the Free State was estimated to have zero growth by the mid-1990s (Krige 1996) and Phuthaditjaba, a sprawling rural slum which included the capital of the former homeland of QwaQwa, decreased by 5.3 per cent from 1991 to 1996 (Crankshaw and Parnell 2002).

Boundary issues make analysis of recent (i.e. 1990s on) South African urban growth rates complex in nearly all major urban areas. In addition, the 1996 census naturally incorporated the four homelands which had been deemed 'independent' under apartheid and thus excluded from the later apartheid censuses. As a result the nominal intercensal growth rate of the 'national' population between 1991 and 1996, at 5.5 per cent per year, is an artifice[4] (the growth rate of 1.9 per cent per year shown in Table 11.2 for 1985–96 is more representative as these homeland populations are incorporated into the statistics). The same goes for the nominal urban growth rate, a seeming 4.4 per cent for that period, but this was partly caused by the integration into the statistics of numerous urban settlements in the former Transkei, Ciskei, Venda and Bophuthatswana (Crankshaw and Parnell 2002). The data shown in Table 11.2 for 1980, 1985, 1991 and 1996 are based mainly on metropolitan area boundaries for the largest cities. For some other urban areas, census estimates for specific urban *agglomerations* are used and indicate some wild fluctuations in growth. Figures relating to more narrowly defined urban areas for 1985 and 1991 yield rather different growth rates and also indicate some interesting variations in growth in settlements *within* urban agglomerations or metropolitan areas. These data are analysed in Box 11.4. To complicate matters further, the Municipal Demarcation Board has subsequently used the 1996 census to demarcate *new* metropolitan areas and their populations for six 'metros': Cape Town (1996 population: 2.563 million); Durban (1.738 million); Ekurhuleni (East Rand) (1.448 million); Johannesburg (2.638 million); Nelson Mandela Port Elizabeth (0.969 million) and Pretoria (1.144 million).[5] For the Witwatersrand as a whole the demarcated population was 4.995 million. Tracking urban growth rates in South Africa's large urban areas in a meaningful way has thus become exceptionally difficult.

Although South Africa suffered very slow economic growth through the 1990s and urban unemployment is high, it has not suffered the extreme

---

[4.] Growth rates calculated from data contained in Crankshaw and Parnell (2002).
[5.] A. Mabin, personal communication, 25.9.01.

Box 11.4   Urban growth, 1985–91, in selected (non-agglomerated) urban settlements in South Africa

Urban area	Population 1985	Population 1991	AAGR 1985–91 (%)	Part of agglomeration	Agglomeration AAGR 1985–91 (%)
Cape Town	776,617	854,616	**1.6**	Cape Town	3.5
Durban	634,301	715,669	2.0	Durban	2.5
Johannesburg	632,369	712,507	2.0	Johannesburg	3.0
Vereeniging	60,584	71,255	**2.7**	Vereeniging	6.2
Pretoria	443,059	525,583	2.9	Pretoria	4.6
Port Elizabeth	272,844	303,353	**1.8**	Port Elizabeth	4.9
Benoni	94,926	113,501	3.0	East Rand	**12.3**
Alexandra	67,155	124,586	**10.8**	Johannesburg	3.0
Bloemfontein	104,381	126,867	3.3	Bloemfontein	7.0
East London	85,699	102,325	3.0	East London	5.7
Guguletu	63,893	54,635	**−2.6**	Cape Town	3.5
Khayelitsha (Lingelethu West)	. . .	189,586	. .	Cape Town	3.5
Kwa Mashu (Inanda)	111,593	156,679	**5.8**	Durban	2.5
Lekoa (Sharpeville)	218,392	217,582	**−0.1**	Vereeniging	6.2
Madadeni	65,832	95,931	6.5	Newcastle	5.7
Mamelodi (Wonderboom)	127,033	154,845	3.4	Pretoria	4.6
Newcastle	34,931	38,767	**1.8**		5.7
Nyanga (Crossroads)	148,882	92,896	**−7.6**	Cape Town	3.5
Orange Farm	. . .	49,838	. .	. . .	
Soweto	521,984	596,632	2.3	Johannesburg	3.0
Umlazi	194,933	299,275	**7.4**	Durban	2.5

The data in the table above illustrate a variety of dynamics which were operating within the South African urban hierarchy towards the end of the apartheid era. Influx controls were lifted in 1986, and given a natural increase rate of around 2.2 per cent at the time, which would have been lower in the urban areas, rapid net in-migration appears to have occurred in a number of urban agglomerations. This is exactly as expected. The table shows that this growth was usually concentrated in the non-white suburbs around the main city, while the already highly developed cities themselves (excluding their surrounding settlements) grew more in line with natural increase. Thus Cape Town, Durban, Johannesburg and Port Elizabeth all grew at 2 per cent or less. Khayelitsha in metropolitan Cape Town, however, mushroomed from virtually nothing to a sizeable settlement between 1985 and 1991; and Umlazi and KwaMashu, Durban's main African townships, were growing about three times faster than natural increase. Alexandra's growth was the most outstanding and this makes good sense given its very good (and anomalous – under apartheid policies) central location in Johannesburg in relation to urban employment opportunities. The intermediate agglomeration of Newcastle also illustrates this pattern well: with the town of Newcastle growing at around the rate of natural increase, but its associate township of Madadeni growing three times as fast, at 6.5 per cent.

This pattern has notable exceptions, however. Overcrowding in some townships may have led to some outmigration from them as the option of informal, more spacious, housing became increasingly possible. Cape Town's old-established township of Guguletu, for example, lost residents over the period, and Soweto was not growing particularly fast. Political violence must also have played a part in making some settlements unattractive at this time, as may have been the case in Sharpeville and Crossroads.

Source: Table data derived from: Brinkhoff (2002); natural increase rate for 1986 from Department of Welfare (1998).

---

urban economic shocks associated with externally imposed structural adjustment (nor have Botswana, Lesotho, Swaziland or Namibia). Zimbabwe, however, adopted structural adjustment from 1991 (see Chapter 3) and by the end of the century its urban situation closely mirrored that in most of the rest of the region at the end of the 1980s: the informal labour market has exploded, urban agriculture is rife, and real incomes have been decimated (Potts and Mutambirwa 1998; Kanji and Kajdowska 1993; Bijlmakers *et al.* 1996; Tevera 1995). The disaffection of the country's urban population was made evident in their strong vote for the opposition MDC in parliamentary elections in 2000 and presidential elections in 2002, although the rural vote ensured ZANU (PF) remained in power. Zimbabwe's economic decline has been exacerbated by its involvement in the DRC war and its exclusion from many sources of external finance because of international resistance to that involvement and its new fast-track land resettlement programme, which defies capitalist property norms. The country's urban sectors are undergoing severe stresses and there is therefore good reason to expect that net in-migration will have fallen during the 1990s.

In summary, the dynamics of urban growth in these countries over the past 40 years cannot be understood without reference to the overarching context of the impact of white minority rule which imposed specific controls over the location and nature of urban growth in South Africa, Namibia and Rhodesia, and helped to create massive political instability which influenced urban patterns in Mozambique, Angola and Namibia. By the beginning of the twenty-first century, while the legacies of apartheid are still very evident, the forces driving current urban growth in these countries are now generally the familiar factors of the rural–urban income gap and the state of the urban economy, and the dynamics of urban fertility and mortality. If the peace established in Angola in 2002 proves sustainable, this should also become true there – and it will be interesting to see how many of Luanda's 'refugee' population leave town, as happened in post-war Zimbabwe and Mozambique.

Urban livelihoods in eastern and southern Africa

The majority of the people who live in the urban areas of the countries in the region are poor and surveys often indicate that most live below the local poverty datum line. For most urban households, the past 20 years or so have seen a fall, in many cases of a dramatic nature, in their real incomes.

There had been real increases in formal sector wages for many urban workers in sub-Saharan African countries in the 1960s. Some felt that this created an 'urban labour aristocracy' – an overprivileged group that was receiving an unwarranted share of national resources. Yet, to some extent, the pay rises were due to the redressing of long-standing colonial policies whereby (particularly in southern and eastern Africa) permanent urban residence by African *families* was rendered difficult (or even illegal): short-term male migration was favoured. Once independence was attained it was natural and inevitable that many male workers would establish urban families, with the corollary that female migration flows and urban natural increase would both surge. Such changes have been reflected in the increasing normalisation of urban sex ratios throughout the region from the strong male bias evident throughout the colonial period and 1960s. Wage rises often only reflected this change in the nature of the labour force, to allow workers to support families rather than just themselves, and disposable incomes may have changed very little. In any case this situation only pertained to the formal sector and, although many new urban jobs were generated in the early post-independence years, formal wage employment progressively lagged behind urban growth. In the absence of a welfare sector, urban dwellers had to find some way of getting by, and the result was the rapid growth of the informal sector where incomes tended to be much lower, and often irregular.

The subsequent decline in urban living standards occurred in two phases. The first occurred against a background of ever-increasing national debts and shortages of foreign exchange, often accompanied by the development of massive parallel market activity in goods which were in short supply in official (legal) markets. Shortages occurred in a vast range of commodities from basic foodstuffs, salt and kerosene to foreign currencies. During this phase the frustrations and problems associated with trying to track down or 'arrange' for vitally needed commodities was part and parcel of everyday urban life. At one scale this would include urban women and children spending hours queuing for food. On another scale it included industrialists desperately trying to find an outlet which had the spare part they needed to keep their factory running, or attempting to secure (either legally or illegally) the foreign exchange or the government licence needed to import it. Controls on prices and imports (via tariffs and the strict regulation of access to foreign exchange) were normally central aspects of government economic policy during this phase. As a result many basic goods (and services) were pegged at prices affordable to poor urban households, but once these goods were in short supply it became very difficult and time-consuming to obtain them at the controlled price. Other sources of supply on parallel markets were often far too expensive for low-income households. An important aspect of this phase of economic decline was that income alone was not always the key determinant of consumption patterns, because social and political relations (whom you 'knew') could often affect access to urban (and rural) goods and services. Thus personal contacts with parallel market traders or, at the other end of the spectrum, civil servants or politicians with control over foreign exchange or import licences could greatly influence what, or how much, one could purchase. This greatly facilitated corruption in the bureaucracy and political life.

These economic difficulties which faced many African countries by the end of the 1970s meant that urban people's livelihoods were already highly constrained at the time that SAPs were implemented in the 1980s. The specifically anti-urban policies of structural adjustment were disastrous for the urban poor and ushered in the second phase of decline in urban living standards. A whole range of negative impacts were experienced as government austerity programmes and trade liberalisation were introduced. Since towns are inevitably centres for service and administrative provision, the reduction of government expenditure on salaries and services was particularly serious for urban employment. Food subsidies were ended (often leading in the short term to urban riots as in Zambia in the 1980s and Zimbabwe in the 1990s) which greatly increased the share of urban incomes spent on food and was far more serious for urban people than it was for rural people who grow food (although it is important to realise that in very many countries, including Zimbabwe, Malawi and South Africa, most rural households are net food purchasers, so this policy, which was theoretically designed to improve rural incomes, has often not had the desired impact) (Potts 1997; Potts and Mutambirwa 1998). Urban wage protection was ended and many price controls were removed. There were major formal job losses in most cities as government expenditure was curtailed, parastatals were overhauled, and formerly protected industries felt the whole weight of international competition. SAPs invariably ushered in long periods of high inflation and the price of everything urban people had to buy (e.g. rent, food, transport) shot up. Any salary increases that waged people managed to obtain (often by collective action) have tended to be less than inflation and extremely belated – the trend is therefore for such incomes to continue to decline in real terms and for the relief offered by increases to be swiftly eroded. Urban people's advantages over their rural peers had not however only been measured in income terms – they had had much better provision of services and infrastructure. The infrastructure was already deteriorating by the time SAPs were introduced, but since then there has been an increasing emphasis on cost recovery in service provision. In services such as health and education this policy has been devastating across the continent for the poor, wherever they are (and in any case, frequently government resources are so constrained that the schools have few books and the clinics hardly any medicines). Similar policies for low-income housing and transport have also been implemented widely. The advantages of urban residence have thus been undermined in many ways.

During the 1980s it became clear in many countries that urban people's monthly salaries could not support their family's food needs for more than a week or two and this left nothing over to cover other basic needs such as rent, fuel, transport, clothing, schooling and health needs. The drastic falls in urban people's incomes have been documented in, for example, Jamal and Weeks (1993) and Potts (1997) and are examined in Box 11.5. The gap between incomes and survival needs was deemed 'the wages puzzle' by Jamal and Weeks (1993).

Self-evidently, this gap had to be closed somehow or people would have starved. There have been three main sets of adaptations by urban residents to the situation. Family members who previously did not earn income, such as women and children, have had to enter the informal sector to add to the family's income, so this sector has expanded significantly. In addition, many of those in formal (and informal) sector jobs have had to take on secondary jobs, again often in the informal sector, to help make ends meet (see e.g. Potts 1997; Rogerson 1997; Tripp 1990, 1997). Participation in the labour force has thus increased significantly. It should be noted that this increase runs counter to a popular view, often encountered in the literature, that structural adjustment has been accompanied by a massive *rise* in urban unemployment. In reality, negative trends in urban *formal* employment have been mirrored by large increases in *informal* sector work. The 'luxury' of open unemployment is rarely an option in African towns for anyone but young, unmarried people (although among that group

---

Box 11.5   Declining urban living standards/incomes

Country	Index of real minimum wage (date)			
Uganda	29 (1957)	100 (1972)	6 (1980)	10 (1990)
Zambia	56 (1960)	128 (1970)	83 (1985)	
Tanzania[1]	100 (1957)	206 (1972)	37 (1989)	130 (1997)

1. Minimum wage for 1997 is for the government sector only.
Note: indices are not comparable across countries.

The rise and fall of *formal sector* urban incomes in three countries in the region during the 1960s, 1970s and 1980s is shown in the table above. In all three cases real minimum wages by the 1980s were drastically lower than they had been in the 1970s, and in Uganda and Tanzania they were lower than the level of the 1950s. In Uganda the minimum wage in 1990 was only a tenth of its level in 1972, and only a third of its 1957 level. It was, however, almost double the level it had fallen to in 1980 after a long period of political instability, which had devastated the country. In Uganda, therefore, the urban economy was actually improving during the 1980s, as the situation normalised, but people's living standards still compared very badly to the levels attained in earlier decades and were extremely constrained and precarious. In Tanzania it can be seen that there was a recovery in minimum wages which began from the mid-1990s. This undoubtedly reflects the improvement in the national economy's macro-economic indices by this time, as inflation fell and GDP growth rates improved. However, unfortunately this has not meant much improvement in Tanzania's urban living standards because, by 1997, the minimum wage had ceased to have much meaning for most urban people's lives since well under a fifth of urban workers were then in the formal sector, compared to about 90 per cent in 1960. Furthermore, the 1997 index only refers to government jobs and it is estimated that unskilled workers in the private *formal* sector would have received less. The vast majority of urban workers were, however, in the informal sector, where most (but not all) monthly incomes would have been even lower.

These falls in real incomes have meant that many urban families have had to make drastic alterations to their lives, which have often had serious implications for their health. The food consumed has changed – 'luxuries' such as bread have often been squeezed out and replaced by more traditional staples such as maize, often in a less refined form than before. While this may not affect people's health too much (although it involves more cooking and fuel must be purchased), many surveys also show that urban families can no longer afford protein elements such as meat or milk – so their diets are becoming much more like those typical of rural areas. Surveys throughout the region's urban areas also show that many people have had to *reduce* the number of meals eaten per day, perhaps to only one. The gap between incomes and food needs (let alone other basic needs) is indicated by the data in the table below.

Urban incomes versus basic needs

Country	Income measure (date)	Monthly urban purchasing power (for family of five)
Tanzania	Minimum wage (1985)	Insufficient for one month's maize meal
Tanzania	Minimum wage (1990)	Just enough for one month's maize meal
Tanzania	Minimum government wage (1997)	1.3 times minimum food basket (maize, beans, oil, sugar)
Kenya	30% of urban wage earners (1988)	Insufficient income to buy minimum calorific needs
Angola	Minimum wage (1994)	5 loaves of bread OR 10 buckets of clean water

Sources: Jamal (2001), Jamal and Weeks (1993), Jespersen (1990), Potts (1997), Tripp (1990).

very high rates of unemployment are often found). The false estimates of African urban unemployment come from an inappropriate focus on the formal job market and job creation (Potts 2000). A more useful focus for any analysis of trends in contemporary urban livelihoods, and policies for poverty alleviation, would be the increasing proportion of jobs that yield low and irregular incomes (in either sector).

There has also been a great increase in urban agriculture (i.e. growing food within the boundaries of the city or just on its fringes) as people try to bring down their expenditure on increasingly costly food (Chimbowu and Gumbo 1993; Drakakis-Smith *et al.* 1995; Egziabher *et al.* 1993; Freeman 1991; Grossman *et al.* 1999; Maxwell 1995; Mbiba 1995; Mlozi 1996). Urban people have also drawn to an increasing extent on their linkages with rural areas to ameliorate their livelihoods' impasse. Such linkages have always been important in sub-Saharan Africa, and they have been newly dynamised by the exigencies of urban economic decline. The new dynamics include sending family members to rural areas to decrease urban household costs (including sending urban children to rural schools), bringing in food grown in rural areas on land belonging to urban residents or the land of their relatives, reducing the amount and frequency of remittances sent to rural relatives and, if all else fails, leaving town altogether and 'returning' to a rural area. In some smaller towns residents may base their livelihoods *mainly* on farming, combining some urban-based activities with agricultural production on a nearby *shamba* (farm) or regular visits to their 'home' *shamba*. These adaptations have been analysed and documented in a growing body of literature (e.g. Holm 1992, 1995; Jamal and Weeks 1994; King 1996; Potts 1997; Rogerson 1997; Tripp 1997).

These adaptations are the solution to the urban 'wages puzzle'. While they may be regarded as a testimony to the initiative and determination of the urban poor, it is important to recognise that such strategies can be associated with heavy costs for the individuals involved – and many are precarious coping strategies which leave people very insecure and exhausted (Rakodi 1991). For example, there is abundant evidence that women have borne a very high share of the burdens involved in taking on extra jobs, growing food and going short of food themselves to try and feed their families (Kanji 1993; Kanji and Kajdowksa 1993; Sparr 1993). Increased competition in the informal sector and their customers' worsened poverty sometimes means that their incomes from vending in particular have fallen (an exception being second-hand clothes sales which have flourished) (e.g. Field 2000; Hansen 2000). Men and youths have often moved into sectors which had previously largely been women's realms, and they in turn have sometimes been squeezed out and into less lucrative and more dangerous occupations. A telling analysis of the gendered shifts in occupations in Mathare Valley, Nairobi, over this period is provided by Nici Nelson (1997). The strategies of the poor can also involve removing children from school, which has exceedingly negative long-term implications for Africa's future 'human capital'. It may also involve reducing or ending remittances to rural family members with evident negative effects for the rural poor, and 'return' migration to rural areas can burden rural livelihoods if the returnees are unable to produce at least their food needs. Box 11.6 illustrates further

Box 11.6   Livelihood adaptations and urban welfare in Zimbabwe and Malawi

In 1994, three and a half years after structural adjustment was introduced in Zimbabwe, research was conducted among a sample of recent in-migrants to the city. They were asked their views on how they had been affected by structural adjustment. Their responses were overwhelmingly negative as they recited a litany of jobs being lost, retrenchment packages not being paid, incomes falling as rents, food prices, school fees and other necessities increased dramatically, and strains being placed on their linkages with their rural relatives. The results are summarised in the table below. As elsewhere in Africa, Zimbabwean urban households were adapting to the new exigencies with recourse to the informal sector, urban agriculture and significant revisions of their expenditure patterns. They were very clear, however, that in nearly all cases the quality of their lives had reduced sharply. Two people (out of 171) did however feel that their new informal sector jobs had benefited: both were involved in the second-hand clothing trade which had exploded as people could no longer afford new clothes.

Perceptions of the impact of structural adjustment in Harare, Zimbabwe, 1994

Type of comment	% of responses[1]	Number of responses
Negative impact	93	159
of which:		
1. *Prices and affordability issues:*	57	97
of which:		
Accommodation problems	14	24
Food (e.g. not enough; consumption reduced)	5	9
Clothing (e.g. cannot afford to clothe children)	5	9
Devalued Zimbabwe dollar	4	6
Taxes too high	4	6
Transport costs	3	5
2. *Job-related* (e.g. pay, retrenchment)	23	39
3. *Education problems* (e.g. children had to leave school; high fees)	16	27
4. *Adverse effect on rural linkages*	9	16
of which:		
'Forced to live separate from family to make ends meet'	4	7
'Had to join family in Harare to cut expenses'	1	2
'Cannot go "home" regularly because of high transport costs'	2	3
'Cannot afford to support rural relatives properly'	1	2

353

Type of comment	% of responses[1]	Number of responses
Neutral impact	4	6
Informal sector/second job-related comments of which:	8	14
1. Positive	2	4
2. Negative	6	10

1. 171 respondents made specific comments about the effect of ESAP on themselves personally. The percentages indicate how frequently specific issues were included in these comments.

In Malawi structural adjustment began at the beginning of the 1980s. Research in the Traditional Housing Areas (low-income residential areas) of the two main cities, Blantyre and Lilongwe, at the end of the decade found that there had been a serious deepening of urban poverty. This was particularly indicated by a huge increase in the proportion of household income spent on food. The adjusted average for the two cities shown in the table below results from further detailed surveys in which careful diaries of expenditure were kept. These indicated that the households tended to underestimate their true expenditure on food, which accounted for nearly two-thirds of total income on average. It is important to note that this was the share of income from *all* members of the household, including informal sector jobs – and not the share of, say, one formal sector income – and that food expenditure would be in addition to any food grown by the household. In other words, even after urban households' adaptations to structural adjustment, they were still desperately poor. Expenditure on non-food items had been squeezed to almost nothing.

Expenditure on food: Traditional Housing Area households in Blantyre and Lilongwe, Malawi, 1988–89

	Blantyre	Lilongwe
Household income spent on food 1987–88 (%):		
(i) Reported	50	45
(ii) Adjusted (average for both cities)	63	63
Household income spent on food 1980 (%):	24	26

Sources: Compiled from data in: Chilowa and Roe (1990), Roe and Chilowa (1990), Potts and Mutambirwa (1998).

some of these problems drawing on research in Harare in Zimbabwe, and Blantyre and Lilongwe in Malawi.

Urban services: the example of housing

Trends in policies towards urban housing in the region have broadly mirrored the trends in development approaches more widely and are outlined in Table 11.3. As can be seen, there has been a shift from strong opposition to informal housing (which is sometimes, but not always, also illegal) to greater, if reluctant, acceptance, mainly because most governments in East and southern Africa had insufficient resources to house their growing urban populations in formal housing. This was true in most cases even when most public housing expenditure went to subsidising civil servants' housing (see Table 11.3). Site-and-service schemes, in which housing plots are laid out with access to some services and residents have the responsibility of building the houses, became a common solution to urban housing problems throughout the region by the late 1970s. The upgrading of unauthorised settlements was also adopted, with Zambia's experience often regarded as something of a showcase (Schlyter 1991). Such schemes nevertheless often suffered problems related to expense, and their occupation by residents with higher incomes than planned for (e.g. Amis and Lloyd 1990; Amis 1996; Potts and Mutambirwa 1991). However, even funding for this sort of housing option is rarely available in the current era of austerity.

The current agency and donor emphasis, in line with neo-liberal trends, is that unsubsidised provision by the large-scale private sector must be the answer. In fact small-scale, but mass, private informal provision of rented rooms has been the 'solution' for most poor people in most large towns in the region for decades, but this is rarely recognised by, let alone incorporated into, housing policies. The current focus on the formal private sector and home ownership faces a central contradiction: the incomes of the urban majority are too low for this sector to be interested in providing housing for the market that they represent. Essentially affordable housing for the poor is insufficiently (or un-) profitable. As a result unauthorised settlements, in which people build their own houses, remain, or have yet again become, the most important supplier of new housing for the poor urban majority.

The rest of this chapter provides an overview of post-apartheid housing policies in South Africa, both to exemplify some of the dilemmas faced in this sector in the region, and to indicate ways in which the housing legacies of the southern white minority regimes affect contemporary urban housing issues. The 'new' South Africa inherited virtually every type of low-income housing found within the region. It had hostels and formal township housing; it had vast unplanned settlements, some within municipal boundaries usually on public land (known in South Africa as free-standing shack areas (FSSAs)), some on 'white' farms, and many within the former homelands on customary land (which are therefore not 'squatter' settlements); it had lodgers within formal township housing; it

Table 11.3 Common aspects in the phases of housing policy in eastern and southern Africa

Phase	Effect
Independence: inherit policy of subsidising heavily housing for middle- and high-income civil servants (expatriates and local)	Most of housing budget tied up in subsidies; high expectations of civil servants
Government unable to meet the demand for low-income housing with very rapid post-independence urbanization	Very widespread unplanned, informal housing – often squatting. Often bulldozed
1970s: many governments in tandem with donors adopt self-help approaches to housing	(i) Large-scale planned site-and-service settlements (ii) Upgrading of existing unplanned settlements
1980s: most countries, swamped by economic crisis, can no longer afford/manage even the most basic plot lay-outs	Once again overwhelmed by informal, unplanned housing 'solutions' created by the homeless poor
IFIs + market-oriented ideology intervene in African economies	Strong emphasis on: (i) private sector (i.e. large-scale financial institutions particularly building societies) (ii) ending subsidies for civil servants (iii) charging for/privatising utilities which are part and parcel of housing such as water and electricity (iv) site-and-service schemes remain a part of housing policy, but on very ad hoc and occasional basis

had a huge number of backyard shacks on formal township plots; and it had many site-and-service schemes developed in the latter years of apartheid in a desperate attempt to 'direct' the low-income housing process.[6]

As had also occurred in Zimbabwe and Namibia in 1980 and 1990 respectively, ambitious early promises were made in South Africa once majority rule was attained. This was within the framework of South Africa's first economic development policy, the Reconstruction and Development Programme (RDP), although this was soon to be replaced with the far more austere and market-oriented Growth, Employment and Redistribution strategy (GEAR). The ANC pledged to deliver 1 million houses in five years to tackle a 1994 backlog of 1.5 million units. The first Housing Minister, Joe Slovo, hoped for 5 per cent of the national budget but in March 1995 (after his death) the first budget allocated only 2.3 per cent. Much was made of very slow progress in the first year after the elections as only 5,000 new houses had been constructed (Tomlinson 1995a), yet by the end of 1999 1,169,354 subsidies had been approved, and 920,891 housing options built or under construction at a cost of R12bn (Tomlinson 2000).

South Africa's housing institutions experienced major changes in 1993. The racial housing departments were replaced with one National Housing Board (NHB) and four regional housing boards, which then became nine provincial housing boards (PHBs). Their role is to allocate housing subsidies and advise provincial Members of Executive Councils (MECs) on housing policy. After an important summit on housing held at Botshabelo in 1994 the NHB was restructured: it comprises representatives of both the government and civil society and formulates housing policy, and then advises the Department of Housing. Previously the National Housing Forum (NHF) – an important negotiating body of political groups, businesses, banks and development organisations launched in 1992 – had been involved in policy formation, but this ended and its role is now more in lobbying and monitoring (Tomlinson 1995a).

The granting of *cash subsidies* to prospective homeowners, the major plank of post-apartheid housing policy, was a radical departure from regional experience, and would be anathema in the current era of structural adjustment (although the focus on homeownership is familiar). However, the approach had strong historical precedents in South Africa: subsidies for first-time buyers were available under the apartheid regime (mainly aimed at white purchasers) and bond subsidies were available for all state employees (Hendler and Parnell 1987). In 2000, the subsidies were between R5,000 and R16,000 on a sliding scale for households with monthly incomes of R2,500–R3,500 to less than R800 (Tomlinson 2000). The subsidies are part of an approach in which households generally purchase freehold serviced plots (and perhaps a 'starter' unit) from private or public developers (project-linked subsidies) – the whole concept being essentially

---

[6.] To avoid confusion, it is worth noting that *planned* site-and-service settlements in South Africa are designated as *informal* settlements, along with *unplanned* and perhaps illegal housing areas. Elsewhere in Africa the term 'informal' usually excludes any schemes which have been planned by the authorities.

similar to projects initiated prior to 1994 by South Africa's private 'think tank', the Urban Foundation, under the auspices of the Independent Development Trust. The emphasis on private developers for plot delivery is another key difference from other countries in the region, although the overall approach rests on the usual site-and-service concept, with households responsible for gradually 'building' their own house (known in South Africa as 'incrementalism').

The subsidies can be seen as at least an attempt to address the affordability problem which seems now to be being abandoned in many other parts of southern and eastern Africa. The government's 1994 White Paper on housing recognised explicitly that '45–55% of households . . . [are] unlikely to be able to afford or access credit and [are] therefore entirely dependent on own (limited) resources and State subsidisation' (Department of Housing 1995). However, the government also expected banks and building societies to be heavily involved in lending money to households to complete their houses, in line with the theoretical stance elsewhere in the region. Slovo was credited with getting a commitment from such institutions to the low-income housing process at the Botshabelo summit when a degree of consensus on housing policy seemed to have been achieved (Tomlinson 1995a). However, their subsequent involvement has been very much less than hoped, despite great efforts by the government to meet their concerns.

A major problem in South Africa has been a legacy from the apartheid era whereby African township residents, angered at the illegitimacy of the government, refused to pay many of their bills including bonds (mortgages). This discouraged banks from lending to this sector – whole areas were 'red-lined (i.e. private lenders refused to lend within these areas). In response the ANC government set up a Mortgage Indemnity Scheme (MIS) indemnifying lenders from bond boycotts if these occur because of 'abnormal political conditions' (Tomlinson 1995a, b), and launched a major programme (twice) – the Masakhane Campaign ('Let us build together') – to try and encourage people to abandon the so-called 'culture of non-payment' although this has not been very successful (Gilfellan 1997, pers. comm.). However, non-payment had also occurred because of 'a lack of quality control in the building industry that resulted in . . . poorly constructed housing' (Tomlinson 1995b: 9). Thus the government provided a 'product defect warranty' to protect lenders if boycotts occur because of poor construction, and a Home Builders' Registration Council to enhance building standards (Tomlinson 1995b). Concerns about political repercussions should repossession become necessary are yet another issue (*Economist* 1996) which, in the light of the Zimbabwean experience where repossessions in the early years of structural adjustment were banned by the government, is quite realistic, although a service organisation was designed to help defaulting households move to more affordable accommodation (Department of Housing 1995; Tomlinson 1995b). However, none of these pragmatic policies to bring private finance into the low-income housing process could overcome the *basic* problem with this option – that the very poor are perceived as very bad *risks* and are *competing* for loans against many other, much more profitable investments. Hence the South

African banks refused to lend to households earning less than R1,500 per month in the mid-1990s (Tomlinson 1995b) which ruled out 70 per cent of South African households (Department of Housing 1995). Since the most basic approved house in, for example, Gauteng was then estimated to cost about R45,000 to complete, the poor clearly faced major problems (*Economist* 1996). Thus, despite the subsidy element, the regional parallels are evident: the current World Bank-inspired focus on providing recoverable loans for housing tends only to reach middle-income people and this appears to be true also of South Africa.

Very similar issues have become evident in relation to service delivery in the 'new' South Africa. A real political commitment to improving both rural and urban households' access to various services, including water and electricity, led to some 7 million additional people being connected to clean water and 3.5 million to electricity by February 2002 (Mbeki 2002 cited in McDonald 2002). However, in line with the government's neo-liberal orientation, cost recovery has been aggressive. A survey in 2001 indicated that 57 per cent of households were earning less than R1,000 per month yet median monthly bills for water, electricity, sewerage and rubbish removal were R224–R400. Monthly electricity bills alone for almost a fifth of these very low-income households were over R200. Not surprisingly, a significant proportion of the households surveyed were struggling to meet these bills. It was estimated that, of the 7 million benefiting from the new water connections, for example, 1.26 million simply could not afford it and 1.2 million would have to choose between water bills and food, rent and clothing, etc. The result has been a massive rate of service cut-offs, significantly undermining, or apparently even negating, the government's achievements. On the basis of average household sizes, the survey estimates that nearly 10 million people have been cut off from water and the same number from electricity. Around 2 million have even been evicted from their homes for non-payment of water and electricity bills. Importantly, in line with the arguments made above with reference to housing, it is argued from the evidence of the qualitative aspects of the survey that *affordability* is the problem, rather than any culture of non-payment. A new approach to service delivery is therefore called for since 'no amount of moralizing or threatening is going to alleviate the payments crisis. . . . You cannot squeeze blood from a stone' (McDonald 2002). In 2001 a new policy to give a basic allocation of *free* water and electricity to all South African households was announced which, if successfully implemented, seems a very positive step (see also Chapter 7).

Another contrasting aspect of South African policy towards housing when compared to elsewhere in the region is the very flexible attitude to building standards on site-and-service schemes (Potts 1994; Department of Housing 1995). Such standards as are required relate mainly to plot size and house layout (Tomlinson 1995b). On the other hand, there has been huge controversy over housing standards in general, with much resistance to the whole concept of 'incrementalism' (i.e. the site-and-service approach). This has often come from the regional political level, but also to some extent from South African people themselves. Very similar sentiments were expressed in Namibia and Zimbabwe in their immediate post-independence years. In mid-1995, for example, the Housing Minister Mthembi-Nkhondo expressed doubts about 'incrementalism' and ordered a review (Tomlinson 1995a). The element of a degree of regional political autonomy in South Africa has exacerbated the issue there, as opposing provincial Members of Executive Councils (MECs) raise expectations beyond what central government funds could deliver (Tomlinson 1995b). In contradiction to stated government policy, 'double' subsidies have some-times been implemented by providing free serviced land, for example, even in the regions which are more accepting of 'incrementalism'. While provinces are free to address their housing problems within the guidelines set down by the central Department of Housing, they are theoretically constrained to promising only what is feasible according to national policy and budgetary considerations (Tomlinson 1995b).

Policies on rental housing were at first similar to those elsewhere in the region: essentially the idea that large-scale public or private sector pro-vision of some rental housing might fulfil a specific set of needs in the housing market was largely ignored. But towards the end of 1996 this was addressed and rented social housing was planned to account for about 10 per cent of dwellings with the government paying R15,000 per dwelling which would then be matched by the developer – a housing corporation or the provincial authority (*Economist* 1996). Progress in production of such rental housing was extremely limited (Tomlinson 2000) and in 1998, under major pressure from organised labour, the government also agreed to policies and mechanisms to produce significantly more rented 'social housing' at a job summit. These included a commitment by the govern-ment to make R750m. available specifically for rental housing and business to provide R1.3bn at a rate significantly below the mortgage rate (Tomlinson 2000: 6) – these monies then to be lent on to social housing associations which would construct the units, using labour-based methods wherever possible. The first pilot rental projects were begun in 1999 and a total of 50,000 units were planned to be completed by mid-2003. While the rental option is vital, the average cost per unit in the pilots was estimated at R41,000 for a monthly rent of R575 which necessitated a subsidy increase of R5,000 per unit. There are serious cost and budget issues involved in the policy, therefore, especially given that the government had just cut the housing budget by R1bn at the time the new policy was agreed.

South Africa's stance on community participation at first went far beyond anything even NGOs implement in other countries in the region. Community participation was originally obligatory before a housing development could be subsidised, in the form of a 'social compact' – an agreement signed by 'the community, the developer and other stakeholders' (Tomlinson 1995b). Not surprisingly, it was unpopular with private devel-opers (Tomlinson 1995b), as community-based development is a difficult thing to achieve as it requires time-consuming explanations and negotia-tions (Tomlinson 1995c; Tomlinson *et al.* 1995; Watson 1999), and can easily fail if it is assumed that 'communities' are homogeneous and likely to have a common goal. Identifying acceptable leaders is often problematic, and the whole approach is very difficult, or perhaps unachievable, in areas where people have no prior common experience, or the development area is very large and heterogeneous and consists of perhaps dozens of 'communities' all with different aspirations (e.g. see Tomlinson 1995b; Hendler 1996). All these issues caused problems with social compacts in South Africa and, from the end of 1996, when provincial administrations were allowed more choice on this issue, such 'compacts' were often bypassed (Watson 1999).

South Africa also faced a dilemma over the hostels which provided mass, theoretically single-sex, accommodation for migrant workers. Conditions within them were rightly condemned, and they have very uncomfortable associations with apartheid influx control and the institutionalised migrant labour system, not to mention political violence in parts of Gauteng in the early 1990s. However, they represent an important element of fixed housing stock in circumstances of acute housing shortage. Some have been converted to accommodate families, although this was already occurring unofficially on a large scale before 1994 (Minnaar 1993; Ramphele 1993). Others still house 'single' workers. Minnaar (1993) argued that they should not be demolished as there may still be a 'demand' for this sort of accommodation from single migrants although, rather contradictorily, it was also argued that the conditions in them are demeaning and dehumanising. Administration by hostel cooperatives was suggested as a means of improving the situation, but this relied on the existence of a community spirit created by 'whole families [having] taken up residence' (i.e. *not* the *single* workers for whom the accommodation was deemed 'suitable' (Minnaar 1993: 239)).

In conclusion, while South Africa learnt some lessons from the experiences of other countries in delivering low-income housing, it still faces major problems. The affordability issue was recognised, but the emphasis on private developers and homeownership cannot but end in the exclusion of the poor majority, although it reflects current norms in the region. Undoubtedly, private developers have a role, but in South Africa too much has been expected of them. In 1997, for example, the Ministry of Housing reported that only about 20 per cent of those receiving housing grants had also managed to obtain a bank loan. There is also increasing evidence that laudable efforts to extend services such as electricity and water connections to people's homes on a mass basis have also run into affordability issues, as many families have been cut off by private suppliers for non-payment with some estimates suggesting that such disconnections are now running at the rate of new connections. Many projects have 'succeeded' only because there has been far more public subsidy than theoretically allowed, which may not be sustainable.

If South Africa follows regional trends it is probable that the opposition voiced to 'incrementalism' will recede. Indeed by the end of the 1990s a new policy allowed grant funding, in addition to the individual housing subsidy, to encourage more self-building (rather than relying on core houses delivered via private developers) by communities (Napier 1998). On the other hand, one of the driving forces behind the recent 'social housing' rental policy described above was very much a perception that the owner-occupied houses delivered by the subsidy and housing projects system were too rudimentary, and less adequate than the apartheid 'matchstick' township houses. Nevertheless, the government cannot afford to provide four-roomed houses (the desired 'norm') for all those needing to be housed (Simkins *et al.* 1992). It does, however, have a financial capacity that gives it some room for manoeuvre and for significant impacts to be made in the housing sector, which many other countries in the region do not.

Conclusion

This chapter has shown that the dynamics of urbanisation, in its demographic sense, in eastern and southern African countries are now different in many key respects from those operating in the first 10–20 years of the post-independence era. This is also true of wider processes of urbanisation – that is, in terms of the economic processes underpinning urbanisation and, in particular, the livelihood strategies that the urban poor have to adopt to sustain themselves. The overarching context of these shifts is globalisation, played out in the lives of the urban poor of eastern and southern Africa in the guise of SAPs. There are also regionally specific factors, particularly the impact of AIDS on urban demography and urban household economies. War has also played a part.

The shifts that have occurred are far-reaching and it is no longer adequate to use the old theoretical or conceptual frameworks within which urbanisation in the region (and sub-Saharan Africa more widely) has long been studied. Urban growth is *not* unabated; there is no unilineal trend in the stabilisation of circular migration; the urban poor are not *a priori* better off than all their rural compatriots. It is widely recognised in the rural livelihoods literature (e.g. Bryceson and Jamal 1997; Bryceson *et al.* 2000; Ellis 1999; Francis 2000; Francis and Murray 2002) that an urban 'string' to the rural 'bow' is an important element of rural multiple livelihood strategies. The rural 'string' to many urban families' urban 'bow' under current conditions of heightened urban poverty is perhaps not as widely recognised, although there is far more conceptual awareness of the blurring of the rural–urban divide (cf. Tacoli 1998). Indeed, logically, the urban straddling strategy of a rural family is, at the same time, the rural straddling strategy of an urban household.

It is now widely recognised that structural adjustment in Africa has had its worst impacts on the urban poor. This does not deny that rural poverty is still generally more severe, particularly because of lack of access to basic services. Downward trends in urban income provide unpromising circumstances for reliance on the profit-oriented large-scale and formal private sector for service delivery, yet this is exactly what has been happening over the past 20 years. As shown in this chapter, this has compounded the problems faced by the urban poor. Thus studying urbanisation in contemporary eastern and southern Africa requires greater sensitivity to poverty issues than in the past (see also Simon 1999). Political patterns across the region indicate also how resistant urban people are to these new circumstances, with opposition to incumbent governments strongly concentrated in the cities. New patterns of urban association and social networking are also developing (Zeleza and Kalipeni 1999; Tostensen *et al.* 2001) which sometimes help to alleviate the services crisis to some extent. Economically, socially and politically, the development problems associated with urban areas in eastern and southern Africa today are thus highly challenging.

References

Africa Confidential (1997) Special issue on Angola, **16**: 22 April.

Amis, P. (1996) Long run trends in Nairobi's informal housing market, *Third World Planning Review* **18**(3): 271–86.

Amis, P. and Lloyd, P. (eds) (1990) *Housing Africa's Urban Poor*. Manchester University Press, Manchester.

Barnett, T. and Blaikie, P. (1992) *AIDS in Africa: Its present and future impact*. Pinter, London.

Beall, J. (2002) Globalization and social exclusion: framing the debate with lessons from Asia and Africa, *Environment and Urbanization* [Globalization and Cities] **14**(1): 41–52.

Bijlmakers, L., Bassett, M. and Sanders, D. (1996) *Health and Structural Adjustment in Rural and Urban Zimbabwe*. Nordiska Afrikainstitutet, Uppsala.

Brinkhoff, T. (2002) City population, South Africa [www.citypopulation. de (accessed on 11.11.01)].

Bryceson, D. and Jamal, V. (eds) (1997) *Farewell to Farms: De-agrarianization and employment in Africa*. Ashgate, Aldershot.

Bryceson, D., Kay, C. and Mooij, J. (eds) (2000) *Disappearing Peasantries: Rural labour in Africa, Asia and Latin America*. IT Publications, London.

Chilowa, W. and Roe, G. (1990) Expenditure patterns and nutritional status of low income urban households in Malawi, in Roe, G. (ed.) *Workshop on the Effects of the Structural Adjustment Programme in Malawi*. Vol. 2: *Papers Presented*. University of Malawi, Centre for Social Research, Zomba.

Chimbowu, A. and Gumbo, D. (1993) *Urban Agricultural Research in East and Southern Africa II: Record, Capacities and Opportunities*. Cities Feeding People, Series Report No. 4, IDRC, Ottawa.

Crankshaw, O. and Parnell, S. (2002) Urban change in South Africa: report prepared for IIED. IIED, London.

CSO (2002) *2000 Census of Population and Housing*, Preliminary Report. Central Statistical Office, Lusaka.

Department of Housing (1995) *White Paper: A new housing policy and strategy for South Africa* (Department of Housing, Pretoria).

Department of Welfare (1998) *Population Policy for South Africa*, Pretoria.

de V Graaff, J.F. (1987) The present state of urbanisation in the homelands: rethinking the concepts and predicting the future, *Development Southern Africa* **4**(1).

Drakakis-Smith, D., Bowyer-Bower, T. and **Tevera, D.** (1995) Urban poverty and urban agriculture: an overview of linkages in Harare, *Habitat International* **19**(2): 183–93.

Economist (1996) South Africa: building slowly for the future, 16 November.

Egziabher, A.G., Lee-Smith, D., Maxwell, D., Memon, P., Mougeot, L. and **Sawio, C.** (eds) (1993) *Cities Feeding People: An examination of urban agriculture in East Africa.* IDRC, Ottawa.

Ellis, F. (1999) Rural livelihood diversity in developing countries: evidence and policy implications, *Natural Resource Perspectives* **40**.

Ferguson, J. (1999) *Expectations of Modernity: Myths and Meanings of urban life on the Zambian Copperbelt.* University of California Press, Berkeley.

Field, S. (2000) The internationalization of the second-hand clothing trade: the Zimbabwe experience. Unpublished PhD thesis, Coventry University.

Francis, E. (2000) *Making a Living: Rural existence in Africa.* Routledge, London.

Francis, E. and **Murray, C.** (eds) (2002) *Journal of Southern African Studies* [special issue on Livelihoods] **28**(3).

Freeman, D.B. (1991) *A City of Farmers: Informal urban agriculture in the open spaces of Nairobi, Kenya.* McGill-Queen's University Press, Montreal and Kingston, London and Buffalo.

Grossman, D., van den Berg, L. and **Ajaegbu, H.** (1999) *Urban and Peri-urban Agriculture in Africa: Proceedings of a workshop,* Netanya, Israel 23–27 June 1996. Ashgate, Aldershot.

Hansen, K.T. (2000) *Salaula: The world of secondhand clothing and Zambia.* University of Chicago Press, Chicago.

Hendler, P. (1996) *Weighting Costs and Benefits: Maximising participation and output in housing the homeless.* Centre for Policy Studies, Social Policy series: Policy issues and actors, **9**(1), Johannesburg.

Hendler, P. and **Parnell, S.** (1987) Land and finance under the new housing dispensation, in Moss, G. and Obery, I. (eds) *South Africa Review* Vol. 4, Ravan Press, Johannesburg.

Holm, M. (1992) Survival strategies of migrants to Makambako – an intermediate town in Tanzania, in Baker, J. and Pedersen, P.O. (eds) *The Rural–urban Interface in Africa: Expansion and adaptation.* Scandinavian Institute of African Studies, Uppsala, in co-operation with Centre for Development Research, Copenhagen, Seminar Proceedings No. 27: 223–38.

Holm, M. (1995) Rural–urban migration and urban living conditions: the experience of a case study of Tanzanian intermediate towns. PhD thesis, Royal Holloway London.

Jamal, V. (2001) Chasing the elusive rural–urban gap in Tanzania, *Journal of Contemporary African Studies* **19**(1): 25–38.

Jamal, V. and **Weeks, J.** (1993) *Africa Misunderstood: Or whatever happened to the rural–urban gap?* Macmillan, London.

Jauch, H. (2002) Export processing zones and the quest for sustainable development: a southern African perspective, *Environment and Urbanization* [Globalization and Cities] **14**(1): 3–12.

Jenkins, P., Robson, P. and **Cain, A.** (2002) Local responses to globalization and peripheralization in Luanda, Angola, *Environment and Urbanization* [Globalization and Cities] **14**(1): 115–28.

Jespersen, E. (1990) Household responses to the impact of the economic crisis on social services, in Roe, G. (ed.) *Workshop on the Effects of the Structural Adjustment Programme in Malawi. Vol. 2: Papers Presented.* University of Malawi, Centre for Social Research, Zomba.

Johnson, L.L. (1970) Luanda, Angola: the development of internal forms and functional patterns. PhD thesis, University of California, Los Angeles.

Kanji, N. (1993) Gender and structural adjustment policies: a case study of Harare, Zimbabwe. PhD thesis, University of London.

Kanji, N. and **Kajdowska, N.** (1993) Structural adjustment and the implications for low-income urban women in Zimbabwe, *Review of African Political Economy* **56**: 11–26.

King, K. (1996) *Jua Kali Kenya: Change and development in an informal economy, 1970–95.* James Currey, Oxford.

Krige, D.S. (1996) *Botshabelo: Former fastest-growing urban area in South Africa approaching zero population growth.* Department of Urban and Regional Planning, University of the Orange Free State, Bloemfontein.

Kulaba, S. (1989) Local government and the management of urban services in Tanzania, in Stren, R. and White, R. (eds) *African Cities in Crisis: Managing Rapid Urban Growth.* Westview Press, Boulder: 203–46.

McDonald, D.A. (2002) The bell tolls for thee: cost recovery, cutoffs, and the affordability of municipal services in South Africa, in McDonald, D. and Pape, J. (eds) *The Crisis of Service Delivery in South Africa.* Zed, London; HSRC, Weaver, Cape Town.

Maxwell, D.G. (1995) Alternative food security strategy: a household analysis of urban agriculture in Kampala, *World Development* **23**(10): 1669–81.

Mbiba, B. (1995) *Urban Agriculture in Zimbabwe: Implications for urban management, urban economy, the environment, poverty and gender.* Avebury, Aldershot.

Minnaar, A. (ed.) (1993) *Communities in Isolation: Perspectives on hostels in South Africa: Goldstone hostels report.* Human Sciences Research Council, Pretoria.

Mlozi, M. (1996) Urban agriculture in Dar: its contribution to solving the economic crisis and the damage it does to the environment, *Development Southern Africa* **13**(1): 47–65.

Mtatifikolo, F. (1992) Population dynamics and socioeconomic development in Tanzania, in Toure, M. and Fadayomi, T. (eds) *Migrations, Development and Urbanization Policies in sub-Saharan Africa.* CODESRIA, Dakar.

Murray, C. (1987) Displaced urbanization: South Africa's rural slums, *African Affairs* **86**(344): 311–30.

Napier, M. (1998) Core housing and residents' impacts: personal experiences of incremental growth in two South African settlements, *Third World Planning Review* **20**(4): 391–418.

Nelson, N. (1997) How men and women got by and still get by, only not so well: the gender division of labour in a Nairobi shanty-town, in Gugler, J. (ed.) *Cities in the Developing World*. Oxford University Press, Oxford.

Palmer, R. (2001) Land tenure insecurity on the Zambian copperbelt, 1998: anyone going back to the land? Oxfam, UK [www.oxfam.org.uk/landrights/CBLand.rtf].

Parnell, S. and **Beall, J.** (2001) Paper presented at Societies of Southern Africa in the nineteenth and twentieth century seminar, ICS, London, 15 June.

Piermay, J.-L. (1997) Kinshasa: a reprieved mega-city? in C. Rakodi (ed.) *The Urban Challenge in Africa: Growth and management of its large cities*. United Nations University Press, Tokyo.

Pottier, J. (1988) *Migrants No More: Settlement and Survival in Mambwe Villages*, Zambia. Manchester University Press, Manchester.

Potts, D. (1985a) Capital relocation in Africa: the case of Lilongwe in Malawi, *Geographical Journal* **151**(2): 182–96.

Potts, D. (1985b) The development of Malawi's new capital at Lilongwe: a comparison with other new African capitals, *Comparative Urban Research* **X**(2): 42–56.

Potts, D. (1994) Urban environmental controls and low-income housing in southern Africa, in Main, H. and Williams, S. (eds) *Environment and Housing in Third World Cities*. John Wiley, Chichester: 207–23.

Potts, D. (1995) Shall we go home? Increasing urban poverty in African cities and migration processes, *Geographical Journal* **161**(3): 245–64.

Potts, D. (1997) Urban lives: adopting new strategies and adapting rural links, in Rakodi, C. (ed.) *The Urban Challenge in Africa: Growth and management of its large cities*. United Nations University Press, Tokyo: 447–9.

Potts, D. (2000) Urban unemployment and migrants in Africa: evidence from Harare, 1985–94, *Development and Change* **31**(4): 879–910.

Potts, D. and **Marks, S.** (2001) Fertility in Southern Africa: the silent revolution, *Journal of Southern African Studies*, Special issue on Fertility in Southern Africa **27**(2): 189–206.

Potts, D. and **Mutambirwa, C.C.** (1990) Changing patterns of African rural–urban migration and urbanisation in Zimbabwe, *Eastern and Southern African Geographical Journal* **1**(1): 26–39.

Potts, D. and **Mutambirwa, C.C.** (199?) High-density housing in Harare: commodification and overcrowding, *Third World Planning Review* **13**(1): 1–26.

Potts, D. and **Mutambirwa, C.C.** (1998) 'Basics are now a luxury': perceptions of ESAP's impact on rural and urban areas in Zimbabwe, *Environment and Urbanization*, special issue on 'Beyond the Rural–Urban Divide', **10**(1): 55–76.

Rakodi, C. (1991) Women's work or household strategies? *Environment and Urbanization* **3**(2): 91–145.

Ramphele, M. (1993) *A Bed Called Home: Life in the migrant labour hostels of Cape Town.* David Philip, Cape Town.

Roe, G. and **Chilowa, W.** (1990) A profile of low income urban households in Malawi: results from a baseline survey, in Roe, G. (ed.) *Workshop on the Effects of the Structural Adjustment Programme in Malawi.* Vol. 2: *Papers Presented.* University of Malawi, Centre for Social Research, Zomba.

Rogerson, C. (1997) Globalization or informalization? African urban economies in the 1990s, in Rakodi, C. (ed.) *The Urban Challenge in Africa: Growth and management of its large cities.* United Nations University Press, Tokyo.

Satterthwaite, D. (1996) *The Scale and Nature of Urban Change in the South.* IIED, London.

Satterthwaite, D. (2002) Globalization and cities, *Environment and Urbanization,* special issue on Globalization and Cities **14**(1): 3–12.

Schlyter, A. (1991) *Twenty Years of Development in George, Zambia.* Swedish Council for Building Research, Stockholm.

Simkins, C. (1981) The distribution of the African population of SA by age, sex and region-type: 1960, 1970 and 1980, Southern Africa Labour and Development Research Unit, SALDRU Working Paper 32, Cape Town.

Simkins, C., Oleofse, M. and **Gardner, D.** (1992) Designing a market-driven state-aided housing programme for South Africa, *Urban Forum* **3**(1): 13–38.

Simon, D. (1992) *Cities, Capital and Development: African cities in the world economy.* Pinter, London.

Simon, D. (1997) Urbanization, globalization, and economic crisis in Africa, in Rakodi, C. (ed.) *The Urban Challenge in Africa: Growth and management of its large cities.* United Nations University Press, Tokyo: 74–110.

Simon, D. (1999) Rethinking cities, sustainability and development in Africa, in Zeleza, P. and Kalipeni, E. (eds) *Sacred Spaces and Public Quarrels: African cultural and economic landscapes.* Asmara, Trenton, NJ; Africa World Press, Eritrea.

Simon, D. and **Närman, A.** (1999) *Development as Theory and Practice: Current perspectives on development and development co-operation.* DARG Regional Development Series No. 1. Longman, Harlow.

Sparr, P. (ed.) (1993) *Mortgaging Women's Lives.* Zed Press, London.

Tacoli, C. (1998) Rural–urban interactions: a guide to the literature. *Environment and Urbanization,* special issue on 'Beyond the Rural–Urban Divide **10**(1): 147–66.

Tevera, D. (1995) The medicine that might kill the patient: structural adjustment and urban poverty in Zimbabwe, in Simon, D., van Spengen, W., Dixon, C. and Närman, A. (eds) *Structurally Adjusted Africa: Poverty, debt and basic needs.* Pluto Press, London and Boulder.

Tomlinson, M.R. (1995a) *The First Million is the Hardest: Will the consensus on housing endure?* Policy Review series: Policy issues and actors, 8(8). Centre for Policy Studies, Johannesburg.

Tomlinson, M.R. (1995b) *From Principle to Practice: Implementers' views on the new housing subsidy scheme.* Social Policy Series, Research Report No. 44. Centre for Policy Studies, Johannesburg.

Tomlinson, M.R. (1995c) *Problems on the Ground?: Developers' perspectives on the government's housing subsidy scheme.* Social Policy Series, Research Report No. 46. Centre for Policy Studies, Johannesburg.

Tomlinson, M.R. (2000) New housing delivery model: the Presidential Job Summit Housing Pilot Project. Paper presented at Urban Futures 2000 Conference, July. University of Witwatersrand, Johannesburg.

Tomlinson, M.R., Sivuyile, B. and **Mathole, T.** (1995) *More than Mealies and Marigolds: From homeseekers to citizens in Ivory Park.* Centre for Policy Studies, Johannesburg.

Tostensen, A., Tvedten, I. and **Vaa, M.** (eds) (2001) *Associational Life in African Cities: Popular responses to the urban crisis.* Nordiska Afrikainstitutet, Uppsala.

Tripp, A.M. (1990) The informal economy, labour and the state in Tanzania, *Comparative Politics* 22(3): 253–64.

Tripp, A.M. (1997) *Changing the Rules: The politics of liberalization and the urban informal economy in Tanzania.* University of California Press, Berkeley.

United Republic of Tanzania (1988) *1988 Provisional Census.* Bureau of Statistics, Dar es Salaam.

Urban Strategy Department, City of Durban (1995) *Settlement Areas and Population Estimate Project, Durban Metropolitan Area 1995.* Urban Strategy Department, Durban.

van Zyl, J. (1988). *Adjustment of the 1985 Census Population of RSA by District.* Research Report No. 149, Bureau of Market Research, University of SA, Pretoria.

Watson, V. (1999) L'accès au logement en Afrique du Sud, in Gervais-Lambony, P., Jaglin, S. and Mabin, A. (eds) *La question urbaine en Afrique australe. Perspectives de recherches.* Karthala, Paris: 227–42.

Zeleza, P. and **Kalipeni, E.** (eds) (1999) *Sacred Spaces and Public Quarrels: African Cultural and Economic Landscapes.* Africa World Press, Asmara, Eritrea; Trenton, NJ.

CHAPTER 12

Conclusion

Deborah Potts and Tanya Bowyer-Bower

A comparison of the analyses of the three developing world regions so far covered in this series of the Developing Areas Research Group of the Institute of British Geographers, including this volume on eastern and southern Africa, indicates both commonalities and differences in their experiences of 'development'. Different physical environments and different social and economic histories help to explain these differences and the concomitant variance in capacity to respond or adapt to changing external circumstances. The theme of globalising economic forces and ideologies is an important one in this volume, and is echoed in McIlwaine and Willis's (2002) *Challenges and Change in Middle America: Perspectives on development in Mexico, Central America and the Caribbean,* and Bradnock and Williams's (2002a) *South Asia in a Globalising World: A reconstructed regional geography.* It is also of central significance to the analysis of changing development policy and practice presented in the introductory volume to the series, Simon and Närman's (1999) *Development as Theory and Practice: Current perspectives on development and development co-operation.* Issues relating to social inequality and social change, the politics of development and environmental challenges are also examined in each volume. In each case this provides not only detailed analysis of the current realities in each region, but also helps to elucidate theoretical debates by providing 'ground-truthing', bringing together conceptualisations of development and development policy and empirical research (Bradnock and Williams 2002b).

On the issue of globalisation, as alluded to in the introductory chapter, both Dixon (1999) in the first volume of this series, and Corbridge and Harris (2002) in the Southeast Asia volume, have pointed out that, in relation to Southeast Asia and India respectively, some room for manoeuvre plus established development impetus have mitigated, to some extent, the potentially negative impacts of trade liberalisation and structural adjustment. In the Middle America volume one chapter focuses on the role of economic diversification, including export-oriented manufacturing, and provides an analysis of the pros and cons of the type and nature of formal employment thus generated (Willis 2002). At a very broad level of generalisation these contributions to the series help to identify an important difference in the development policy context between those major world regions and eastern and southern Africa. For, as Stoneman (Chapter 3, this volume) and Gibb (Chapter 10, this volume) show, the capacity of countries in eastern and southern Africa to balance out the negative impacts of economic liberalisation with some benefits has proved to be much weaker than elsewhere in the less developed world. The term 'capacity' has thus become significant in many contemporary development policy initiatives in the region which have, correctly, identified the fact that without the institutions and educated human resources to 'deliver' or manage positive, developmental programmes or initiatives, major if not insuperable difficulties arise. This, in itself, is one indication of the relatively recent appreciation by neo-liberal economists that 'rolling back the state' is not always and necessarily good for the realisation of economic potential. There is a paradox in this focus on capacity building, however. For, before the era of structural adjustment, many African governments had struggled to build up capacity in many of their institutions – for example, their health services, their schools and universities, their urban planning offices and agricultural extension services. Yet under structural adjustment the lesson has been taught that the level of expenditure on the public services was unaffordable for the national economy of most sub-Saharan African countries. Even if staff were not retrenched, inevitably their pay was reduced very sharply in real terms, often below any feasible level of subsistence (see Potts, Chapter 11, this volume). The results have been inevitable. More 'corruption' by those who choose, sometimes from some sense of commitment, to remain in the public sector, as they sell their 'services' or public resources to keep their families alive (e.g. nurses selling drugs; planners selling land leases). Others leave to work elsewhere, probably in the informal sector, or remain on the payroll but actually spend most of their time on other income-generating activities. And some of the best educated leave the country altogether and probably thereby support a far larger number of their 'local' extended family via hard currency remittances. Even more paradoxically, a few will end up in foreign or local development organisations, or working for consultancy firms, attempting to address the lack of local 'capacity' (see Zeleza (2002) for a discussion of these sorts of paradoxes among the academic community in Malawi).

Returning to the issue of employment generation and economic growth based on participation in competitive world markets, McIlwaine and Willis (2002: 273) point out, with respect to Middle America, that the

> challenge in this field, which is very difficult in the face of the movement of global capital, is to create sustainable employment opportunities that provide decent and non-exploitative wages in companies that are not entirely dependent on the pursuit of cheap labour and the

provision of extensive tax, tariff and regulatory benefits on the part of Middle American governments.

This is surely one of the most fundamental development struggles of the contemporary age and is of key significance to all developing world governments and labour movements. But, if Asia and Middle and South America feel threatened by the exploitative connotations of the desperate competition between developing countries to attract foreign direct investment (FDI), how much more difficult is the situation faced by nearly every country in eastern and southern Africa? A telling analysis of the role of export processing zones (EPZs) in southern Africa has typified the situation as 'a race to the bottom', with governments offering ever lower and more exploitative labour conditions and tax breaks to foreign corporations, having been advised that this is the way 'forward' (Jauch 2002). Yet in some cases, such as mining companies or refineries, EPZ status may be unnecessary to attract FDI. The employment created also tends to fall far below what was anticipated. In Namibia, for example, an EPZ which was expected to provide 25,000 jobs in the first three years, produced only 400 (Jauch 2002).

Another theme emerging from the volumes in this series is the vital role of regional specialists, be they geographers or from other social sciences, in the development of both theory and policies relating to developing world regions (see especially Parnwell 1999). As the preceding paragraphs indicate, there are important variations in the impact and implications of neo-liberal policies at a global as opposed to a regional scale. Within regions these variations are again very marked, and the variability between and within countries in eastern and southern Africa has been stressed throughout this volume. The point that policy packages need to be tailored to take into account national specificities has been well rehearsed over the past 20 years and the IFIs have not been entirely deaf to these arguments. Only deep familiarisation with the area or country in question can allow appropriate and sensitive interpretation of data, indices and situations to suggest policy prescriptions which might not only 'work' but also might actually be developmental. The ability to take a multidisciplinary approach to understanding and analysing development issues in specific locations is also highly desirable, for rarely can one discipline provide all the answers or even the right questions to ask.

Examples of the value of area expertise in understanding the 'actual workings of development' (Parnwell 1999: 84) and interpreting development models and statistics abound in all the volumes in this series. A very simple example from this book is the water stress index which is widely used as an indicator of resource constraints in different countries and which depicts both Botswana and Namibia as not being water-stressed. Yet, as indicated in Taylor's chapter on water resources, this is highly misleading and occurs because the index does not factor in the geography of water resources in relation to the location of human demands. Another obvious issue which pervades the contributions to this volume is the necessity of factoring in the tragic unfolding of the AIDS epidemic into virtually all aspects of research and theorising on development in the

371

region of eastern and southern Africa. Another example is the level of urbanisation in Botswana. The levels reported in many sources indicated that by 1991 it was one of the most urbanised countries in the region, second only to South Africa, with an extraordinary leap occurring in the 1980s. Yet as explained in Potts's chapter on urbanisation, definitional issues need to be taken into account when interpreting these statistics and an alternative, much lower, level is also used by the country's Central Statistical Office. Paradoxically, Botswana's highly favourable government revenues have allowed the government now to extend its investment in infrastructure and employment generation to the large agro-villages which historically had been essentially rural in their economic character and thus, arguably, not regarded as 'urban'. The result has been that in many of these 'villages' the majority of working adults are now involved in non-agricultural activities (Selolwane forthcoming) and thus these settlements might well be regarded as 'urban' today. An interpretation that the data suggest ongoing massive migration to a capital city (although this occurred up until the end of the 1980s) and policies insensitive to rural development would thus be incorrect – in fact it is 'rural development' that has made these settlements more 'urban'.

In the introductory chapter a question was posed about the factors underlying Zimbabwe's tragic transition to the status of a political pariah with an imploding economy and it was stated that this would be returned to in this concluding chapter. It provides further proof of the need for regional expertise to deconstruct complex situations. On one level it is possible to attribute the crisis to the President's and ruling party's lack of democratic principles and willingness to use violence to retain political power when facing a viable and active opposition challenge. This provides a simple prescription – get rid of Mugabe – but evidently to ascribe complex situations affecting millions of people's lives to the simplistic interpretation that 'men make history' is unsatisfactory and unlikely to yield an understanding that might prove useful to other parts of the region. However, an investigation of how political opposition to the nature of economic and political developments in Zimbabwe has developed since independence, an investigation which demands the skills of area specialism, can help to provide clues to the structural issues underlying the economic collapse and the nature of the expectations of Zimbabwe's black majority. One is the extreme inequality in living standards which is still highly racialised. Another is the related land inequality which still existed in 1997, the year when the war veterans successfully forced Mugabe to deliver payments and pension packages which totally wrecked the budget and set the scene for the ensuing economic implosion. Another is the impact of structural adjustment from 1990 to 1997 which had completely undermined the urban-based formal economy and the nationwide social welfare improvements which had been the country's greatest achievement in the 1980s (see Stoneman, Chapter 3 this volume). These made the land issue much more important than it had been as access to that means of production became even more critical. It also set the scene for the veterans' demands which developed from low-level grumbling to sudden, vociferous and threatening insistence that their livelihoods be improved. None of these factors were specifically of Mugabe/ZANU(PF)'s making. The political opposition to these factors was extreme by the end of the 1990s, however, and the real possibility that this would lead to a change in the party in power was the trigger for the government's subsequent descent into violent authoritarianism. This analysis might be countered with the argument that if democracy had prevailed and the opposition MDC had come to power, the depth of the economic implosion could have been lessened and the situation stabilised. There is much merit in this argument. However, the economic decline was already extremely severe, and the power of the war veterans and disaffected urban youth well entrenched, by the time of the presidential elections in 2000 and the MDC's first opportunity to influence the political process. Given global policy constraints it is difficult to see how they could make real inroads into the crisis in formal urban-based employment and social welfare delivery. The MDC also recognised that more rapid land reform had become politically essential, although via a transparent and legal process, and therefore recognised the structural issues which underpinned the crisis. Any assumption that the MDC could bring about a return to the more positive economic and social characteristics of Zimbabwe in the 1980s, and that major restructuring of the agrarian economy could be avoided, is therefore misplaced.

In the context of a sub-region characterised by extreme, often racially based, economic inequalities, such an interpretation of the Zimbabwe crisis suggests that development strategies which do not deliver some easing of inequality and a modicum of economic security or, at the very least, not a diminution in security, are recipes for future *political* disaster. Simon and Närman (1999: 271) make a similar argument in the first volume of this series where they argue that part of the power of modernisation theory lies in the demonstration effect of the material trappings of modernisation, so that 'the aspirations of most poor people are framed . . . by the conspicuous consumption of their wealthier compatriots. Frustration at prolonged inability to meet their aspirations, particularly when recession or structural adjustment policies and their associated hardships bite deep, often give rise to protests and riots'. There are clear issues here for societies such as South Africa, Zimbabwe and Namibia. In South Africa the adoption by the first democratic government of a neo-liberal economic policy, fortunately not enforced by the IFIs and thus allowing some local adaptations, has been accompanied by hundreds of thousands of formal job losses. Evidently this is a dangerous configuration for one of the most unequal societies in the world, with its specific, and extremely recent, history of struggle against white political and economic supremacy.

In Power's chapter in this volume he writes of the putative political 'African renaissance' of which much was made in the 1990s. Reference is also made in several chapters to the still current emphasis in development strategies on participation and empowerment. In the economic sphere, one outcome of this has been the shift from blatantly externally imposed SAPs to new Poverty Reduction Strategy Programmes (PRSPs), whereby country governments write their own programmes for 'development' thereby giving them 'ownership' of their policies and allowing them to prioritise appropriate sectors given local conditions. These are criticised,

however, as appearing to be more empowering than they really are because their implementation remains conditional on their acceptance by the
IFIs, and thus PRSPs have to keep firmly to the neo-liberal agenda still,
although there is potentially more sympathy for pro-poor policies (Pender
2001). All these threads were combined in the New Economic Partnership
for African Development (NEPAD) which was formally presented to the
G8 summit in June 2002. In 1999 President Thabo Mbeki of South Africa
and President Bouteflika of Algeria were mandated by an Extraordinary
Summit of the Organisation of African Unity to try and persuade Africa's
creditors to cancel all outstanding debts. Although this was, unsurprisingly,
unsuccessful, from this and various other initiatives to engage the developed
North with African development issues, NEPAD emerged as Africa's
'principal agenda for development providing a holistic, comprehensive
integrated strategic framework for the socio-economic development of
the continent, within the institutional framework of the African Union'
(Department of Foreign Affairs 2002).

The five initiating states for NEPAD are South Africa, Nigeria, Algeria,
Senegal and Egypt. Thus, within the region of southern and eastern
Africa, Thabo Mbeki is a key player. Regional representation on the implementation committee from the region as defined in this volume includes
Mozambique, Botswana and Rwanda. The significance of African 'ownership' of NEPAD is stressed. The essential components of the initiative
as perceived by the North are that it is a partnership between Africa and
the developed countries of the world that 'commits African countries to
setting and policing standards of good governance across the continent,
respecting human rights and working for peace and poverty reduction
in return for increased aid, private investment and a reduction of trade
barriers by rich countries' (Tillin 2003) and that the developed countries
will only enhance their partnerships with African countries that are producing 'measured results' in terms of these standards (G8 Africa Action
Plan 2002). NEPAD seeks to increase economic growth rates across the
continent to 7 per cent per year via the investment of an extra US$64bn,
most of which it is hoped will come from donors (G8 Africa Action Plan
2002) and also highlights the importance of democratic values. However,
the NEPAD documentation also emphasises that the 'partnership must be
based on *mutual* respect, dignity, shared responsibility and *mutual* accountability' (Department of Foreign Affairs 2002, emphasis added). That the
developed North should also be accountable for its policy mistakes is not
something that the IFIs have yet acknowledged (see Stoneman, Chapter 3
this volume) and the outcomes of this aspect of NEPAD's aims remains to
be seen. African countries have been quick to associate with this theme
however, an example being Tanzania's initiative to institute independent
monitoring of aid programmes that would include evaluation of donor,
as well as Tanzanian, performance (Wangwe 2002). The presentation of
NEPAD to the G8 conference also sought a commitment to ensuring that
mining and oil companies behaved more ethically in African countries
which were experiencing conflict (Katzenellenbogen and Mvoko 2002).

The significance of the 'policing' aspect of NEPAD has been high for
South Africa. As discussed in the introductory chapter, Mbeki was drawn

into negotiations to end the DRC war. Analysts were quick to make the links with NEPAD, arguing that 'African leaders, foremost among them Thabo Mbeki, have had to accept the fact that dreams of modernizing the continent will go nowhere until Congo is stabilized' (*Washington Post* 2002). Mbeki's extremely cautious approach to the crisis in Zimbabwe, which is certainly not inexplicable in the light of the analysis offered above of the dangers inherent in South Africa's and Zimbabwe's inherited structural similarities, has, however, drawn heavy criticism from the international community.

It is too early to say whether NEPAD will provide a really positive new kickstart to development in eastern and southern Africa. African critics point out it is still firmly within the Washington consensus (Katzenellenbogen and Mvoko 2002). The new or renewed emphasis in many development policies from a variety of agencies and governments on pro-poor policies is certainly more important for Africa than any other global region and, if realised, is welcome. On a note of optimism, at the time of writing in early 2003, the eastern and southern African region has less open conflict than has been characteristic of this part of the world for decades. Without peace very little 'development' can be accomplished.

References

Bradnock, R. and **Williams, G.** (eds) (2002a) *South Asia in a Globalising World: A reconstructed regional geography*. DARG Regional Development Series No. 3, Pearson Education, Harlow.

Bradnock, R. and **Williams, G.** (2002b) Conclusion: development in South Asia. Ground-truthing constructions of regional change, in R. Bradnock and G. Williams (eds) *South Asia in a Globalising World: A reconstructed regional geography*. DARG Regional Development Series No. 3, Pearson Education, Harlow: 270–3.

Corbridge, S. and **Harriss, J.** (2002) The shock of reform: the political economy of liberalisation in contemporary India, in R. Bradnock and G. Williams (eds) *South Asia in a Globalising World: A reconstructed regional geography*. DARG Regional Development Series No. 3, Pearson Education, Harlow.

Department of Foreign Affairs, Republic of South Africa (2002) *New Partnership for Africa's Development (NEPAD)*. Department of Foreign Affairs, Pretoria.

Dixon, D. (1999) The Pacific Asian challenge to neoliberalism, in Simon, D. and Närman, A. (eds) *Development as Theory and Practice: Current perspectives on development and development co-operation*. DARG Regional Development Series No. 1, Longman, Harlow: 205–29.

G8 Africa Action Plan (2002) [www.sarpn.org.za./NEPAD/june/2002/g8/action_plan.php accessed 23.2.03]

Jauch, H. (2002) Export processing zones and the quest for sustainable development: a southern African perspective. *Environment and Urbanization* [Globalization and Cities] **14**(1): 3–12.

Katzenellenbogen, J. and **Mvoko, V.** (2002) NEPAD is under fire from African experts, *Business Day*, 27 June [allafrica.com/stories, accessed 23.2.03]

McIlwaine, C. and **Willis, K.** (2002) *Challenges and Change in Middle America: Perspectives on development in Mexico, Central America and the Caribbean.* DARG Regional Development Series No. 2, Pearson Education, Harlow.

McIlwaine, C. and **Willis, K.** (2002) Conclusion, in McIlwaine, C. and K. Willis (eds) *Challenges and Change in Middle America: Perspectives on development in Mexico, Central America and the Caribbean*, DARG Regional Development Series No. 2, Pearson Education, Harlow: 270–5.

Parnwell, M. (1999) Between theory and reality: the area specialist and the study of development, in Simon, D. and Närman, A. (eds) *Development as Theory and Practice: Current perspectives on development and development co-operation.* DARG Regional Development Series No. 1, Longman, Harlow: 76–94.

Pender, J. (2001) From 'structural adjustment' to 'comprehensive development framework': conditionality transformed? *Third World Quarterly* **22**(3): 397–411.

Selolwane, O. (forthcoming) From infrastructural development to privatization: the challenges of employment creation and poverty reduction in Gaborone, in Potts, D. and D. Bryceson, *African Urban Economies: Viability, vitality or vitiation of major cities in East and Southern Africa?*

Simon, D. and **Närman, A.** (1999) Conclusions and prospects, in Simon, D. and Närman, A. (eds) *Development as Theory and Practice: Current perspectives on development and development co-operation.* DARG Regional Development Series No. 1, Longman, Harlow: 267–75.

Tillin, M. (2003) Africa's development plan: analysis [news.bbc.co.uk/1/hi/business 12.2.03, accessed 23.2.03]

Wangwe, S. (2002) *NEPAD at Country Level – Changing aid relationships in Tanzania.* Mkuki na Nyota Publishers, Dar es Salaam.

Washington Post (2002) A chance for Congo, 29 October [www.iht.com/articles/75123 accessed on 23.2.03]

Willis, K. (2002) Open for business: strategies for economic diversification, in McIlwaine, C. and K. Willis, *Challenges and Change in Middle America: Perspectives on development in Mexico, Central America and the Caribbean.* DARG Regional Development Series No. 2, Pearson Education, Harlow: 136–58.

Zeleza, P. (2002) The politics of historical and social science research in Africa, *Journal of Southern African Studies* **28**(1): 9–24.

Index

Acquired Immunity Deficiency Syndrome (AIDS) *see* HIV/AIDS
abortion, 35
African, Caribbean and Pacific (ACP) group, 298–306, 308–10, 312, 323 *see also* trade: Cotonou Agreement
Africa's great war, 263
African Development Bank, 70, 234, 313
African renaissance, 285, 373
African socialism, 8, 138
agriculture, 123, 167–93, 233, 240, 245: and population growth, 177–9; agroforestry, 234; capital, 187–8; commercial farms, 82, 179, 184, 191, 209, 245; credit, 15, 184, 188; export crops, 12, 60–1, 72–3, 76, 179, 181–2, 184–6, 300, 305; floodplain, 203, 209, 220; HIV/AIDS, 11, 191–2; irrigation, 15, 28, 169, 175, 177, 187, 199, 206, 208–11, 219–20, 230, 240; labour force, 173, 190–1; marketing boards, 65, 68, 76, 181–3, 187–8; policy, 63, 138, 179–85; production, 15, 19, 104, 167–9, 174–5, 178, 187, 190–2, 206, 221; rainfall, 174, 176, 200; rainfed, 125, 206, 208, 235; smallholder communal farms, 15–16, 135, 137, 139, 170, 175, 184, 188, 209, 235, 243–4, 247; smuggling, 168; socialist policies, 179 *see also* food
aid, 13, 60, 67, 278, 374: agencies, 106, 276, 299; conditionality, 8, 278; food, 67, 104
Albedo Effect, 247
Algeria, 374: NEPAD, 374; white settlers, 59
Allbright, M., 285
Amin, Idi, 76
Anglo-American Corporation: Zambian copper mines, 71
Angola, 2–3, 262, 319; agriculture, 169, 173; Cabinda exclave, 265; child nutrition, 106; civil war, 71; Cold War, 71, 261, 265; corruption, 72, 95, 277; debt, 72; demographic data, 29; diamonds, 265, 272–3; DRC, 262; economy, 67, 71, 276;

education, 71, 111, 277; elections, 8, 265; environmental conservation, 143; fertility, 30; food consumption, 104–5; foreign investment, 101; health, 107, 277; HIV/AIDS, 42; incomes, 6–7; IMF programme, 72; IFIs, 277; independence, 3; mortality, 41, 98; internally displaced people (IDPs), 72, 92, 280; land, 15; land-mines, 281; liberation war, 3–4; life expectancy, 104; livestock, 172; Lusaka Protocol, 265; Marxism-Leninism, 3, 62; military expenditure, 277; minerals, 265–6; oil, 72, 265; peace, 348; population, 6, 335: density, 7, 28, growth, 91, 335; poverty, 72, 95, 266, 280, 351; refugees, 280–1; rural-urban migration, 280; SADC, 217, 275, 301; SADCC, 271; South African destabilisation, 4, 71; transit routes, 265; urbanisation, 72, 331, 344, 347; USA, 71; war, 7, 89, 91, 95, 104, 167, 264–5, 274, 277, 281–2, 344; and DRC, 4, 263; and minerals, 280
apartheid, 3–4, 77, 138, 185, 209, 218, 235, 256–7, 259, 271, 274–5, 340, 342, 344–5, 347, 361; child mortality, 39; removals, 343; rent boycotts, 358; urbanisation, 342, 346, 357; victims, 281
Arusha, 132, 187; growth, 338

bananas, 192, 300, 303
Banda, Hastings Kamuzu, 8–9, 258, 270, 278
Bangladesh, 300; child mortality, 98
Bangweulu swamps, 203
banks, 245; domestic, 61; western, 59, 61, 112
Bantustans *see* homelands
Batoko Gorge: hydro-electricity, 211
Bayart, 267–9, 271, 282, 322
beef, 126, 300, 304–5, 308, 312, 319
Beira Corridor, 79n, 80
Benguela Current, 175
biodiversity, 120, 140–1, 147, 233